MARITIME HYDRAULICS

Flow Structure Interaction

ADVANCED SERIES ON OCEAN ENGINEERING

ISSN 1793-074X

Published

Vol. 47 Coastal Engineering: Theory and Practice
by Vallam Sundar and Sannasi A. Sannasiraj

Vol. 48 Introduction to Coastal Engineering and Management (3rd Edition)
by J. William Kamphuis

Vol. 49 Design of Coastal Hazard Mitigation Alternatives for Rising Seas
by David Basco

Vol. 50 Tsunami: Engineering Perspective for Mitigation, Protection and Modelling
by V. Sundar, S. A. Sannasiraj, K. Murali and V. Sriram

Vol. 51 Turbulence in Coastal and Civil Engineering
by B. Mutlu Sumer and David R. Fuhrman

Vol. 52 Ocean Wave Dynamics for Coastal and Marine Structures
by V. Sundar

Vol. 53 Headland-Bay Beaches: Static Equilibrium Concept for Shoreline Management
by John Hsu, Jung L. Lee, Antonio H. F. Klein, Mauricio González and Raúl Medina

Vol. 54 Numerical Modeling of Tsunami Waves
by Juan Horrillo, William Knight and Zygmunt Kowalik

Vol. 55 Fine Sediment in Open Water: From Fundamentals to Modeling
by Johan C. Winterwerp, Thijs van Kessel, Dirk S. van Maren and Bram C. van Prooijen

Vol. 56 An Introduction to Hydraulics of Fine Sediment Transport (2nd Edition)
by Ashish J. Mehta

Vol. 57 Coasts and Estuaries: Management and Engineering
by V. Sundar, S. A. Sannasiraj and K. Murali

Vol. 58 Maritime Hydraulics: Flow Structure Interaction
by V. Sundar, H. Schüttrumpf, V. Sriram and S. A. Sannasiraj

*For the complete list of titles in this series, go to https://www.worldscientific.com/series/asoe

Advanced Series on Ocean Engineering — Volume 58

MARITIME HYDRAULICS

Flow Structure Interaction

V Sundar
Indian Institute of Technology Madras, India

H Schüttrumpf
RWTH Aachen University, Germany

V Sriram
Indian Institute of Technology Madras, India

S A Sannasiraj
Indian Institute of Technology Madras, India

World Scientific

NEW JERSEY • LONDON • SINGAPORE • BEIJING • SHANGHAI • HONG KONG • TAIPEI • CHENNAI

Published by

World Scientific Publishing Co. Pte. Ltd.

5 Toh Tuck Link, Singapore 596224

USA office: 27 Warren Street, Suite 401-402, Hackensack, NJ 07601

UK office: 57 Shelton Street, Covent Garden, London WC2H 9HE

Library of Congress Cataloging-in-Publication Data
Names: Sundar, V. (Vallam), author. | Schüttrumpf, Holger, author. |
Sriram, V. (Venkatachalam), author. | Sannasiraj, S. A., 1968– author.
Title: Maritime hydraulics : flow structure interaction / V. Sundar, Indian Institute of
Technology Madras, India, H. Schüttrumpf, RWTH Aachen University, Germany,
V. Sriram, Indian Institute of Technology Madras, India,
S.A. Sannasiraj, Indian Institute of Technology Madras, India.
Description: Hackensack, New Jersey : World Scientific, [2025] | Series: Advanced series on
ocean engineering, 1793-074X ; volume 58 | Includes bibliographical references and index.
Identifiers: LCCN 2024032853 | ISBN 9789811294167 (hardcover) |
ISBN 9789811294174 (ebook for institutions) | ISBN 9789811294181 (ebook for individuals)
Subjects: LCSH: Coastal engineering. | Fluid-structure interaction. |
Hydraulic structures. | Ocean waves. | Wave resistance (Hydrodynamics)
Classification: LCC TC205 .S847 2025 | DDC 627/.58--dc23/eng/20240907
LC record available at https://lccn.loc.gov/2024032853

British Library Cataloguing-in-Publication Data
A catalogue record for this book is available from the British Library.

For any available supplementary material, please visit
https://www.worldscientific.com/worldscibooks/10.1142/13867#t=suppl

Desk Editors: Balasubramanian Shanmugam/Steven Patt

Typeset by Stallion Press
Email: enquiries@stallionpress.com

Preface

The book aims to provide a thorough exploration of the fundamental physical processes involved in the interaction between waves and coastal structures. It delves into the complexities of wave–seawall interaction, including stability issues and failure mechanisms, along with the effects of waves on rubble mound structures. The following topics are covered in this book:

Wave–Structure Interaction: This book an in-depth exploration of the physical processes governing how waves interact with coastal structures. It discusses phenomena such as wave–seawall interactions and the stability and failure mechanisms of coastal structures under wave action. Special attention is given to understanding how different types of coastal structures, like seawalls and rubble mound structures, respond to wave forces.

Scale Effects and Model Considerations: One of the key challenges in studying wave–structure interactions is the scale at which experiments and models are conducted. This book explores the intricacies of scale effects, particularly within porous breakwaters, and the limitations and complexities of scale modelling, especially when dealing with breaking waves. It highlights the critical importance of understanding and mitigating scale effects in order to accurately predict extreme wave behaviour.

Probabilistic Estimation of Extreme Events: Coastal areas are vulnerable to a range of extreme events, including floods. This book focuses on probabilistic methods for estimating the likelihood and impact of such events, taking into account site-specific conditions. It discusses strategies for mitigating flood risks, drawing on current practices in flood control and management.

Multifunctional Coastal Structures: Traditional coastal structures are often designed for a single purpose, such as flood protection. This book advocates for the integration of additional functionalities, such as energy extraction systems, into coastal infrastructure design. It explores the potential benefits and challenges of multifunctional coastal structures and their role in sustainable coastal development.

Numerical Modelling in Coastal Engineering: Numerical modelling is a crucial tool in coastal engineering for simulating wave behaviour and predicting coastal dynamics. This book examines different numerical modelling approaches, focusing on their applicability to large spatial and temporal scales. It discusses the strengths and limitations of linear and weakly nonlinear models in both frequency and time domains.

Artificial Reefs and Hydrodynamic Performance: Artificial reefs play a significant role in coastal protection and habitat restoration. This book traces the evolution of artificial reefs and evaluates their hydrodynamic performance using insights from physical modelling experiments and real-world experiences. It discusses the various types of artificial reefs and their effectiveness in mitigating coastal hazards.

Field Monitoring and Health Assessment: The book explores field monitoring techniques for assessing the effectiveness of coastal hazard mitigation measures. It discusses methods for monitoring the health and performance of coastal structures over time, providing valuable insights into optimizing maintenance and management strategies.

Overall, this book offers a comprehensive exploration of various aspects of coastal engineering, ranging from theoretical principles to practical applications. It bridges the gap between academic research and real-world engineering practice, making it an invaluable resource for students, researchers and practitioners in the field.

About the Authors

Dr. V. Sundar is former Professor Emeritus in the Department of Ocean Engineering at IIT Madras and currently serves as an Advisory Consultant. He has guided 33 Ph.D. and 14 M.S. students and has been honoured with ten distinguished awards from the scientific community, including an Honorary Doctorate from the University of Wuppertal, Germany. From 2007 to 2011, he chaired the Asia-Pacific division of the International Association of Hydro-Environment Research (IAHR). Dr. Sundar has an extensive publication record, with approximately 600 papers in conferences and journals. He has participated in over 100 conferences worldwide and organized five such major events. He has also served on the International Scientific Committee for various global conferences and is a member of the editorial boards of 10 international journals. His research portfolio includes around 30 completed projects, of which 11 are international. His research expertise covers a wide range of topics, including breakwaters (semi-circular, skirt-type and rubble mound), wave loads on structures, sediment transport under waves and currents, coastal protection and scour around structures. Notably, the instrumentation he developed for measuring scour has been widely cited by other researchers.

Dr. Sundar has successfully completed 300 plus consultancy projects and has also delivered lectures to media outlets. He has played a pivotal role in and prepared the master plan for tsunami mitigation measures for the tsunami-affected maritime states of India. For more detailed information on his contributions, please visit the following: https://sites.google.com/alumni. iitm.ac.in/vallam-sundar/home

Dr.-Ing. H. Schüttrumpf has been Director of the Institute of Hydraulic Engineering and Water Resources Management at RWTH Aachen University, Germany, and Full Professor at RWTH Aachen University since 2007. He had previously worked at the Technical University of Braunschweig, Germany, as a research assistant until 2001 and at the Federal Waterways Engineering and Research Institute (BAW) in Hamburg, Germany, as a senior researcher until 2007. His research interests are related to flood risk management, hydraulic structures, coastal engineering, waterways engineering and water resources management with a special interest in interdisciplinary research and international cooperation. He has successfully supervised more than 50 Ph.D. students.

For complete details, please visit the following: https://www.iww.rwth-aachen.de/cms/iww/Das-Institut/Mitarbeitende/Institutsleitung/~mivv/Holger-Schuettrumpf/?lidx=1

Dr. V. Sriram is Professor at the Department of Ocean Engineering, Indian Institute of Technology Madras. After experience in the industrial domain at HR Wallingford, UK, he re-entered the research sphere through the Newton International Fellowship from Royal Society, London. After completing the fellowship, he received the prestigious Alexander Von Humboldt Fellowship from Humboldt Foundation, Germany. He is also a recipient of the RJ Garde Research Award from the Indian Society of Hydraulics, Pune; the Young Faculty Recognition Award, 2018 from IIT Madras; and the ISOPE Conference Best Session Organizer Award, 2020, and DFG Mercator Professorship. His research interests include numerical model developments (such as the particle method, FVM and FEM for wave–structure interactions) applied to a wide variety of problems such as ship simulations, offshore structures and coastal engineering applications. He has also been involved in the development of a state-of-the-art experimental facility. He is currently carrying out research to understand extreme wave interactions with marine structures. He has more than 100 international peer-reviewed journal and conference publications. He is Associate Editor of *International Journal of Offshore and Polar Engineering (IJOPE)* and is an editorial board member of Ocean Engineering. For complete details, please visit the following: https://doe.iitm.ac.in/vsriram1/

Dr. S. A. Sannasiraj is Chair Professor at the Department of Ocean Engineering, Indian Institute of Technology Madras, and also served as its Head from 2016 to 2019. He is the lead Indian Coordinator of the DAAD-sponsored Global Water and Climate Adaptation Centre in India. At present, he is the Vice-Chair of the Executive Council of the Asia-Pacific division of the IAHR. His area of specialization includes climate change impacts on coastal infrastructures and structural adaptation, port and harbour structures, wind–wave modelling, and coastal erosion and protection. He has supervised 28 Ph.D. and 11 M.S. students. Since 2003, he has completed 22 sponsored research projects, including nine international collaborative projects. He has 140 peer-reviewed journal publications and has participated in over 150 technical conferences at both national and international levels. He has co-authored five books in his area of expertise. He was awarded the following: *DAAD fellowship in Germany in 2006; Endeavour India Executive Award from the Australian Government in 2007; Fulbright-Nehru Senior Research Fellowship in 2011; and the Melbourne School of Engineering Guest Faculty Award in 2023.* He has successfully completed more than 300 industrial consultancy projects related to ports and harbours, intake/outfall systems, the design of coastal protection structures, health audit of berthing structures and wind–wave prediction. Further, he has organized 25 short courses and workshops for researchers, field engineers, faculties and senior managers. For complete details of his contributions, please visit the following: https://doe.iitm.ac.in/sasraj/

Acknowledgements

The authors extend their heartfelt gratitude to their families and friends whose unwavering support and encouragement have been invaluable throughout the journey of completing this book. Special recognition is also due to the dedicated efforts of the former and current research scholars from the Department of Ocean Engineering, Indian Institute of Technology Madras, including Dr. Ravindar, Dr. Shagun, Dr. Shaswat, R. Divya, C. Sruthi, Sukanya and Golda Percy, whose technical contributions have greatly enriched this work. Moreover, we express our appreciation to individuals such as Anne Mozer, Lisa Burghardt and Philipp Meier from the Institute of Hydraulic Engineering and Water Resources Management, RWTH Aachen University, Germany, whose input has significantly enhanced the content of the chapters.

The authors express their gratitude to the Indian Institute of Technology Madras, India, and RWTH Aachen, Germany, for their constant encouragement and the facilities provided for conducting the studies reported in this book.

The authors express their gratitude to the ABCD "Global Water and Climate Adaptation" center funded by the German Academic Exchange Service (DAAD) for supporting the exchange of faculty and student visits between India and Germany, which facilitated in-depth discussions on the contents of this book.

The significant findings from physical and numerical modeling presented here were made possible through the support and generous funding for equipment provided by the German Research Foundation (Deutsche Forschungsgemeinschaft, DFG).

The field and case studies discussed in this book are the result of collaborative efforts, with substantial support from the Kerala Infrastructure Investment Fund Board (KIIFB) and Kerala State Coastal Area Development Corporation Limited (KSCADC). Lastly, the authors humbly apologize to all those whose names may have inadvertently been omitted but whose presence and assistance have been deeply appreciated throughout the years.

Contents

Chapter 1

Introduction

1.1 Background

In this chapter, we delve into the foundational physical processes governing the interaction between waves and coastal structures. The forces and pressures induced on structures by different types of waves are briefly discussed. The progress made in understanding the physical processes through tools such as experiments and numerical aspects will also be discussed.

Climate change will have a direct impact on the wind patterns worldwide, which in turn will affect global ocean wave characteristics. The alteration in wave conditions can lead to heightened wave impact along the coastline and a reduction in the protective efficacy of nearshore bank systems. This could expose the coast to larger waves, posing threats to critical infrastructure such as roads, bridges, subways, water sources, oil and gas wells, power plants, sewage treatment plants, landfills and various other structures. The potential repercussions extend to endangering people's livelihoods.

As an example, considering the extensive coastline of India, spanning over 7,500 kilometres across nine states, two Union territories and two island territories, approximately 171 million people — equivalent to a seventh of the country's population — reside in 70 coastal districts within these regions. Furthermore, 14.5 million people depend on fishing activities. Notably, four of India's ten most populous cities, including Greater Mumbai, Kolkata, Chennai, and Surat, are located on or in close proximity to the coast. A report from Climate Central, featured in the journal Nature Communications, predicts that by 2050, the risk of persistent flooding will extend to 36 million Indians, a stark increase from the initial projection of five million. This global trend could affect over 300 million people, nearly quadrupling prior estimates (UNHCR, 2020).

With the escalating wind and wave activity attributed to climate change, the threat of wave overtopping intensifies. This heightened risk renders currently secure areas increasingly susceptible as protective coastal

Fig. 1.1. Erosion near north Chennai, Ennore, India, leading to compromise of coastal buildings.
Source: Sundar and Sannasiraj (2019).

structures such as dikes, seawalls, and similar installations contend with elevated floods and waves. Consequently, instances of wave overtopping and the resulting flooding have become more prevalent. Beyond causing inundation, extreme overtopping events propel water over crests at high velocities, presenting substantial hazards to both human life and infrastructure. Such occurrences have, in some cases, led to the tragic outcomes of people, vehicles, and even trains being swept into the sea. This is an exceptionally perilous situation. An example is the long-term coastal erosion due to longshore sediment transport that swept the national highway in Tamil Nadu, India, as seen in Fig. 1.1.

1.2 Coastal Structures

1.2.1 *General*

The Federal Emergency Management Agency (FEMA) defines coastal structures as structures constructed along or parallel to the shoreline to maintain, fill and protect against flooding and waves. A coastal structure can alter flood levels and topography on both the landward and seaward sides of the structure and adjacent to it. Coastal protection structures can be classified into natural and human-made structures along the coasts. Several coastal defence schemes, such as hard or soft measures, are shown in Fig. 1.2. The seawall, breakwater and groyne fall under the categorization of hard structures.

Seawalls serve as onshore structures strategically designed to prevent land and structures from succumbing to flooding caused by storm surges

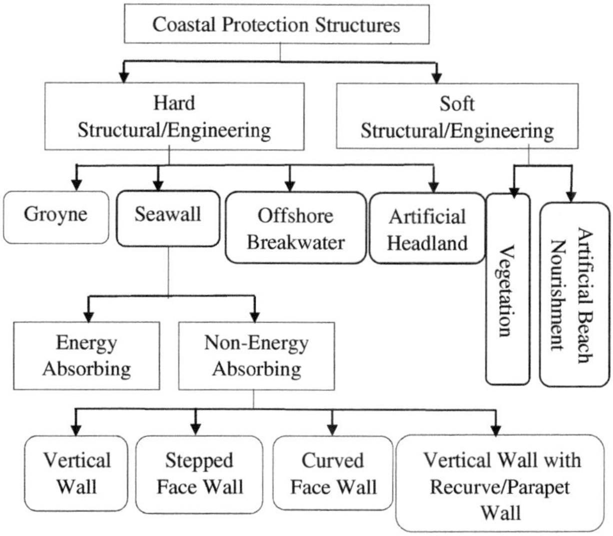

Fig. 1.2. A flowchart capturing various measures for coastal protection.

and waves. Positioned parallel to the shoreline, these structures function as a reinforcement to specific sections of the coastal profile. They are typically deployed to shield promenades, sidewalks and residences situated seaward of the natural beach profile. In scenarios where protection is needed beyond the natural beach profile, seawalls are meticulously constructed.

The design of seawalls encompasses a spectrum of options, ranging from substantial gravity concrete walls, bound structures featuring steel or concrete pilings and stone-filled cribwork to more sloping configurations incorporating reinforced concrete slabs, concrete armour units or standard stone rubble surfaces. Consequently, the threat of erosion to the beach profile landward of a seawall can be effectively halted or, at the very least, significantly reduced. Breakwaters, jetties and groynes use similar materials to seawalls, with the exception that the breakwater is designed to minimize the strength of waves so that harbours and marinas can be covered, jetties are used to reduce the current generated by waves and groynes are built to protect the eroding shore by increasing sediment deposition. The construction of seawalls offers various methods, including non-energy-absorbing options like vertical walls, wave return walls and vertical walls with parapets, as well as energy-absorbing approaches like rubble mound stones. Parapets may come in recurve or plain designs. While Chapter 2 delves into the specifics of these structures, this chapter provides a brief overview.

The use of the nearly vertical seawall in the past had the unintended consequence of reflecting wave energy, aggravating erosion and hastening beach disappearance. To minimize or to avoid overtopping, the crest elevation must be elevated as the sea level increases. Sloped walls or berms are used to dissipate the incident wave energy by shoaling progressively. Sloped walls, though more stable, are said to experience higher pressures and run-up, necessitating a higher crest elevation. Furthermore, compared to a vertical seawall, it takes up more room. A curved front structure is adapted with smooth curvature, which guides the wave over its curvature and returns it to the sea to overcome those aforementioned disadvantages. During this process, a significant amount of energy gets dissipated. These structures are called curved seawalls.

1.2.2 *Breakwaters*

Breakwaters are marine structures constructed to break the incoming sea waves, thereby creating a sheltered area for the safe manoeuvring of marine vessels. They are adopted for the formation of artificial harbours as they dissipate the incident wave energy that impinges on them and create tranquillity on their leeward side. The degree of tranquillity required depends on the types of crafts and vessels used in the harbour, and maximum tranquillity is usually the aim. However, rubble mound breakwaters are obsolete for deeper waters wherein the extent of tranquillity required is more for an emerged vertical breakwater. The dissipation of incident wave energy, which is predominantly through reflection causes excessive pressures due to waves on the seaside. In order to reduce the consequences of high pressures that dictate the stability of vertical breakwaters, several types of breakwaters such as sloping breakwater, cylindrical caisson breakwater, pile breakwater, semi-circular caisson breakwater, quarter circular caisson breakwater and skirt breakwater are adopted to reduce wave forces, reflection and overtopping.

The stability of these breakwaters under rough sea conditions plays a vital role in their design. The choice of breakwater type is contingent upon factors such as wave height, wave period, water depth, underlying soil conditions and the accessibility of personnel, materials and equipment in close proximity to the site.

In general, breakwaters are classified as (1) conventional breakwaters, (2) non-conventional breakwaters and (3) composite breakwaters. Takahashi (1996) classified them into four types according to their structural features: (a) sloping (mound) breakwaters, (b) vertical breakwaters, (c) composite breakwaters and (d) special (non-gravity) types of breakwaters. Rubble mound breakwaters are the most commonly used conventional type of breakwaters worldwide. They are constructed in the nearshore areas in relatively shallow water. They are permanent structures constructed by arranging different layers of natural stones. The inner layer known as a core is mainly made up of porous sand while the outermost layer known as an armour is made up of large stones or man-made artificial blocks.

More importantly, the size of the rubble mound has to be designed in such a way that it should not be displaced during rough weather conditions. Rubble mound breakwaters experience high wave reflection, dissipation and lower transmission through the pores. During high tides and rough weather conditions, waves may overtop the structure. In order to maintain the stability of the breakwater, the seaward side of the breakwaters is sloped. The sloping side of the breakwater reduces the height of the waves as they travel from the toe to the berm and thereby into the highly permeable core, thus dissipating the incident wave energy by forcing the waves to break on the slope. This causes lower reflection and higher dissipation, thereby increasing the efficiency. Rubble mound breakwaters are economical in areas where the availability of stones and skilled labourers is high.

Vertical breakwaters are gravity structures constructed as box-type structures enclosed with vertical walls and a central cavity filled with stone or gravel. This type of breakwater is also known as vertical wall-type breakwater. This type of breakwater can be constructed in intermediate to deep waters, where the availability of stones is high and in areas of rough sea conditions. These types of breakwaters are good at resisting overturning by virtue of their weight. Vertical breakwaters are impermeable and have no porosity. Therefore, waves interacting with this specific breakwater type demonstrate both reflection and dissipation on the seaward side. The wave transmission on the leeward side is almost absent in this type of breakwater, provided no overtopping takes place. The main drawback of this type of breakwater is the high wave reflection experienced on the seaward side of the breakwater. The waves in front of the breakwater are very high because of superposition of incident and reflected waves, thus creating very high wave forces on the vertical walls, causing sudden failure of the breakwater. The increased wave height in front of the breakwater causes difficulty in the movement of marine vehicles.

A composite breakwater is a combination breakwater, which uses the advantages both of rubble mound breakwaters and vertical breakwaters.

These breakwaters are constructed in two or three types depending upon their performance: (a) vertical composite, (b) horizontal composite and (c) special type.

1.2.3 *Vertical composite breakwater*

This type of breakwater variant comprises a vertical caisson positioned atop a rubble mound (low, medium or high) as illustrated in Fig. 1.3. This type of breakwater is preferred in areas of water depth greater than 20 m. This breakwater acts as a rubble mound breakwater during low tides and as a vertical wall breakwater during high tides. The leeward side of this type of breakwater is used for berthing vessels.

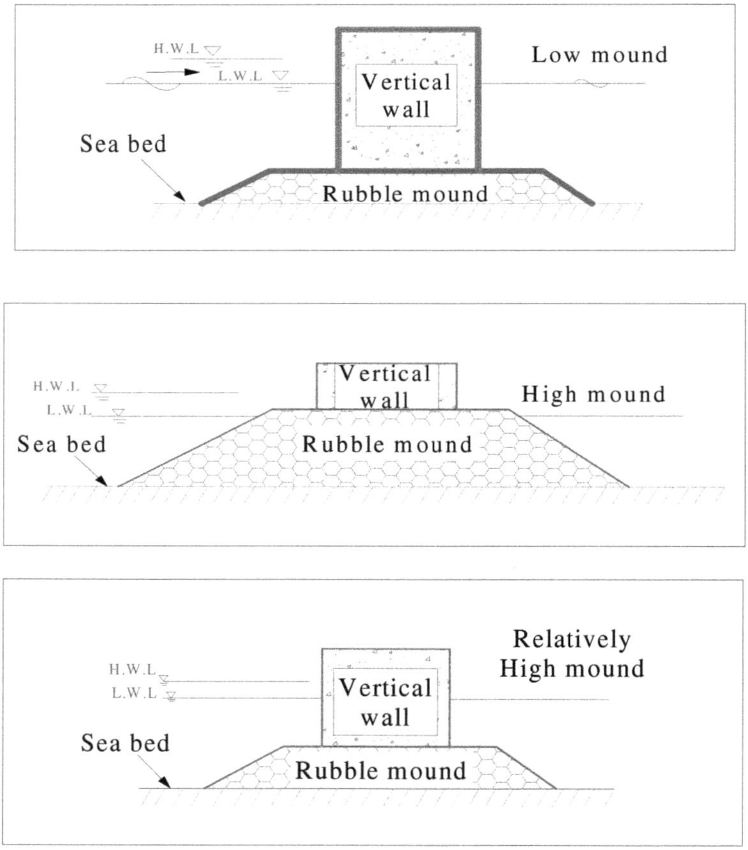

Fig. 1.3. Broad classification of vertical composite caisson breakwaters.

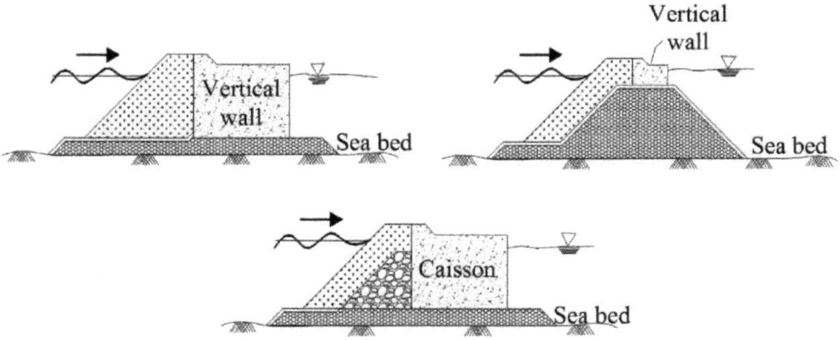

Fig. 1.4. Types of horizontal composite caisson breakwaters.

1.2.4 *Horizontal composite breakwater*

This breakwater design features a vertical caisson situated on a rubble mound accompanied by a protective armour layer facing the sea, as depicted in Fig. 1.4. This armour layer consists of a rubble mound on the seaward side, which dissipates the incident wave energy and helps in the reduction of wave reflection and forces acting on the caisson. These types of breakwaters are generally constructed as an upgradation and modernization of vertical breakwaters to accommodate new modern vessels. However, these breakwaters are very expensive.

1.2.5 *Special type of breakwater*

In areas designated for specific purposes where the underlying conditions on the site are unique and uncommon in practice, special types of breakwaters are constructed. These breakwaters mostly combine the advantages of two or more types of breakwaters as discussed by Tanimoto and Takahashi (1994).

Krishnakumar *et al.* (2010a) conducted research into the effects of pressure and wave forces on a partially submerged solid wall shielded by one or two slotted barriers. Their findings indicated a notable reduction in wave-induced pressures and forces on the solid wall as the incident wave angle increased. In a complementary study, Krishnakumar *et al.* (2010b) performed comprehensive experiments on single- and double-wave screens composed of horizontally arranged circular bars. The double-wave screens comprised combinations of bottom-sitting and suspended slotted barriers with varying degrees of submergence. Through their experiments, Krishnakumar *et al.* derived design formulae for reflection and transmission coefficients, as well as the pressures acting on the slotted barriers of the single- and double-slotted breakwaters.

A partially perforated breakwater is characterized by a perforated front wall positioned in front of the caisson, situated at approximately half the height above the caisson breakwater. A classic example is the breakwater at the Wolnae fishing port at Busan, South Korea, the details of which are discussed by Lee *et al.* (2017).

1.2.6 *Gravity type*

These types of breakwaters are classified into two types:

(a) Conventional gravity-type breakwaters

These are conventional gravity-type rubble mound breakwaters that are widely adopted for harbour formations and coastal protection as mentioned in Section 1.2.4.

(b) Non-conventional gravity-type breakwaters

These are conventional non-gravity-type breakwaters with site-specific improvement to reduce the massive weight of the structural stability measurements. Examples include Jarlan-type breakwaters.

The Jarlan-type breakwater was the first perforated front wall caisson breakwater proposed by Jarlan (1961).

Huang *et al.* (2011) reviewed studies on perforated/slotted breakwaters with and without a rear wall. For surface-piercing slotted breakwaters without a back wall, small transmission coefficients can be achieved over a wide range of wave frequencies by using multi-slotted walls, with the porosity of each wall carefully chosen. Impermeable curtains/crowns can also effectively reduce the transmission coefficients. Coastal currents may also help reduce the transmission coefficients.

The following are categorizations of perforated breakwaters according to Huang *et al.* (2011):

(i) Fully perforated breakwater,
(ii) Partially perforated breakwater,
(iii) Breakwater with multiple perforated walls,
(iv) Perforated breakwater with a top cover,
(v) Perforated breakwater with rock core,
(vi) Perforated breakwater with a horizontal porous plate.

A fully perforated breakwater is composed of a perforated front wall positioned in front of the caisson breakwater, spanning the entire height of the caisson. The perforations, which can take the form of horizontal slits, vertical slits, circular holes or curved slits in caisson breakwaters, serve to augment the dissipation of incident wave energy. The first perforated wall caisson breakwater also known as Jarlan-type breakwater was constructed in Naples, Italy, with a perforated front wall and impermeable back wall (Jarlan, 1961). Vertical slit breakwaters were constructed in the Port of Yobuko and curved slit breakwaters are constructed in the Port of Funakawa in Japan. A partially perforated breakwater is characterized by a perforated front wall positioned in front of the caisson, situated at approximately half the height above the caisson breakwater.

Several other creative configurations of breakwaters are reported in the literature and few have also been established on site. As stated earlier, apart from the concept of introducing perforated screens or slits to reduce the force on the caisson breakwaters, there have been efforts made to change the seaside facing design (with or without perforations or slits) in order to dissipate the incident wave energy through an increased run-up. Cylindrical breakwaters, hollow cylinder breakwaters, semi-circular breakwaters (SCBs) and quarter-circular breakwaters (QCBs) are some of the modified forms of caisson breakwaters that have been adopted to reduce wave forces, reflection and overtopping.

Dual-cylinder breakwaters (DCBs) consist of an inner impermeable cylinder covered by a porous outer cylinder as shown in Fig. 1.5. The empty space between the outer and inner cylinders is known as a wave chamber,

Fig. 1.5. A view of a porous cylinder breakwater.
Photo: Schüttrumpf.

wherein the process of energy dissipation occurs when the waves impinge the outer porous cylinder and enter the wave chamber, thereby weakening. The cylindrical cross-section of the breakwater resist the forces to a large extent. DCBs have been designed and effectively constructed to form harbours at the ports of Nagashima, Sakara and Shibayama in Japan. Field experiments on this type of breakwater are discussed in detail by Tanimoto and Takahashi (1994).

1.2.7 *Non-Gravity type*

In the development of new ports and the upgradation of existing ports, the need for more marine vessels has led to the development of non-gravity-type breakwaters. The need for a cost-effective, deep-water breakwater led to the development of non-gravity-type breakwaters. Pile breakwaters are the most popular among the non-gravity-type breakwaters, the details of which are discussed by Karthik Ramnarayanan *et al.* (2020a). Photographs of pile breakwaters are shown Fig. 1.6.

1.3 Tubular Members Supporting Marine Structures

1.3.1 *Vertical and inclined members*

Cylindrical tubular members often constitute a vital element in marine structures, including bridges, wharves, piers, drilling and production platforms, floating breakwaters and articulated mooring towers. The extensive use of cylindrical members in the marine environment has spurred ongoing research over several decades, particularly focusing on understanding the loads imposed on these structures by ocean waves. The structures in the marine environment are subjected to a variety of waves. The structures in the marine environment are subjected to a variety of waves. In laboratory modeling of wave-structure interaction problems, regular, random, and multi-directional waves are used to simulate operational conditions. For survival conditions, extreme waves such as design waves or freak waves are considered.

The force exerted by waves on slender tubular members is characterized by a combination of drag force, dependent on orbital velocity, and inertial force, contingent upon orbital acceleration. Morison *et al.* (1950) initially introduced this approach for a vertical circular cylinder with a diameter (D) resting on the seafloor and extending through the sea surface. Although this method is commonly employed for estimating wave forces on tubular

Fig. 1.6. Photographs of pile breakwater.
Photos: Schüttrumpf.

members, challenges persist in accurately determining the hydrodynamic coefficients of drag (C_D) and inertia (C_M).

Several researchers have depicted the fluctuation of hydrodynamic coefficients as a function of non-dimensional parameters, specifically the Keulegan–Carpenter number, $\mathrm{KC} = u_{\max}T/D$ ($u_{\max}$: maximum horizontal velocity and T: wave period), and Reynolds number, $\mathrm{Re} = u_{\max}D/\nu$ (ν: kinematic viscosity). Several studies showed a wide variation in the relationships between the drag and inertia coefficients with the KC and Re.

Sarpkaya and Isaacson (1981) compiled their experimental results and reported the strong dependence of C_D and C_M on KC and Re. Anand Kumar *et al.* (1995) and Sundar *et al.* (1997) carried out comprehensive experimental investigations in a wave flume and basin to measure the dynamic pressures around a 20-cm-diameter cylinder exposed to the action of regular waves. The cylinder inclination up to $\pm 45°$ in $15°$ increments was considered where, the positive and negative values indicate inclination in the direction of the wave and against the direction of the wave, respectively. The pressures exerted on the inclined cylinder were integrated to determine the sleeve force. From this, the drag coefficient (C_D) and inertia coefficient (C_M) were derived. The variations of drag coefficient C_D with KC number is observed to be increasing with an increase in angle of orientation of the cylinder along and against the direction of waves for KC less than 4. The ratio of in-line to transverse force was reported to be four to five, while the ratio of in-line to vertical force was found to be around two to three.

Sundar *et al.* (2002) investigated the impact of wave directionality on pressures and sleeve forces, in the form of root mean square pressure and force ratios. These ratios were defined as the normalized root mean square pressures/forces in short-crested waves relative to those in long-crested waves. Notably, the pressure ratio was reported to be at a minimum for a cylinder inclined at $45°$ along the wave direction.

In addition to experiments, numerical modelling gained popularity, especially for simple geometric shapes like cylinders. Qu *et al.* (2020) employed unsteady Reynolds-Averaged Navier–Stokes (URANS) equations with the $k-\omega$ SST turbulence model to simulate the interaction of breaking waves with a stationary cylinder at various transverse inclined angles. Their findings revealed that the normalized high-order wave force attains its minimum value when the cylinder is inclined transversely at $45°$. The results were validated by comparing them with the numerical outcomes of Choi *et al.* (2015) and the laboratory results of Irschik *et al.* (2005), showing a satisfactory level of agreement. Mayilvahanan *et al.* (2017), through numerical simulations, showed that for the maximum force on a cylinder induced by breaking solitary waves, the relative distance and the breaking point play an important role.

Freak waves and rogue waves, first reported by Draper (1965), are extreme events that can sustain only for a brief duration without causing huge accidents in the ocean. In practice, freak waves occur when the ratio of the maximum wave height and the significant height is greater than 2, whereas rogue waves have ratios of 2.1 to 3.0. Fei Deng *et al.* (2019) carried out tests in a wave flume to measure the forces on a cylinder induced by nonlinear freak waves and compared the measured results with predictions made using a formulation by Morison *et al.* (1950). Wavelet-based analysis methods were adopted to identify the local energy structures and quadratic phase coupling in freak wave forces.

In recent years, the design challenge of offshore wind turbine foundations has been compounded by the variability of the wave-current climate, which must be taken into consideration. Specifically, the wave loads may be regular, random or even slamming (owing to wave breaking), coexisting with a following or opposing current. Wave–current–structure interactions may be classified as being *current-dominant* ($\mathrm{KC} > 10$) or *structure-dominant* ($\mathrm{KC} < 2$) where $\mathrm{KC} = \frac{|u_0 \pm U|T}{D}$ is the Keulegan–Carpenter number where, u_o is the amplitude of orbital velocity, U is the current speed (+ve for following and −ve for opposing), T is the wave period and D is the characteristic dimension of the structure. Through the judicious use of appropriate assumptions, the analytical treatment of the wave–current–structure interaction problem is indeed possible. There are two ways of estimating the forces on the slender cylindrical members due to wave and current interactions based on whether the flow field is dependent or independent. This is estimated based on the KC number and Vorley–Moe Number as shown in Fig. 1.7.

In the dependent flow field, flow fields acting on the structure *cannot* be physically separated and the forces at an elevation are estimated as follows:

$$f = C_M \left\{ \rho \frac{\pi}{4} D^2 \right\} \dot{u} + C_D \left\{ \frac{1}{2} \rho D \right\} |u \pm U|(u \pm U). \tag{1.1a}$$

This is a linear superposition of the wave kinematics. For instance, a coexisting wave–current incident on a fixed cylinder is a typical example as shown in Fig. 1.8. In the case of independent flow field formulations, the

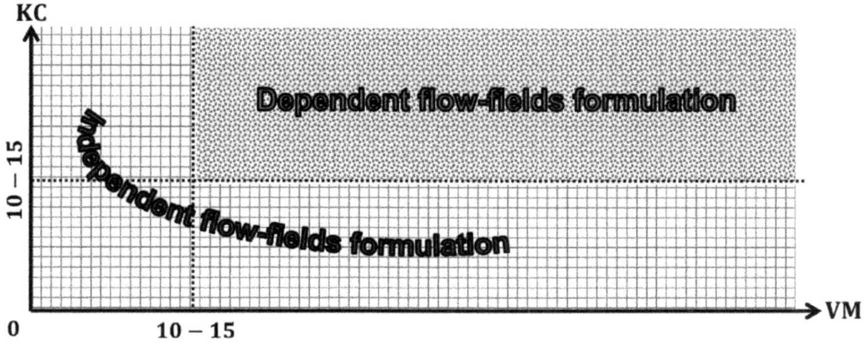

Fig. 1.7. Vorley–Moe vs. Keulegan–Carpenter number for flow field formulation.
Source: Saincher *et al.* (2022).

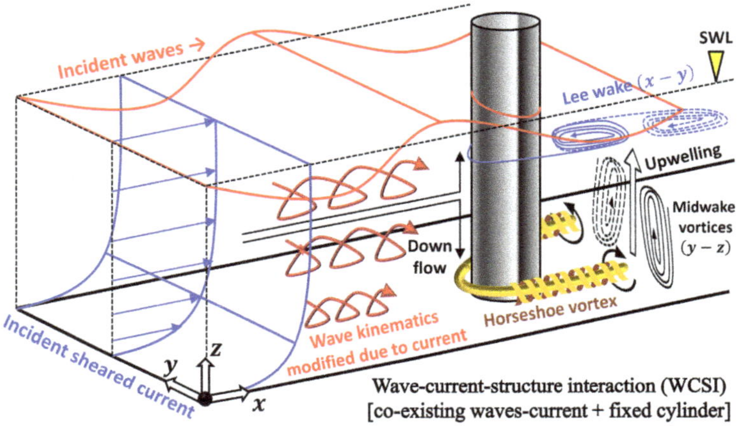

Fig. 1.8(a). Linear superposition of wave kinematics for wave–current interaction with a fixed cylinder (dependent flow field formulation).
Source: Sainchet *et al.* (2022).

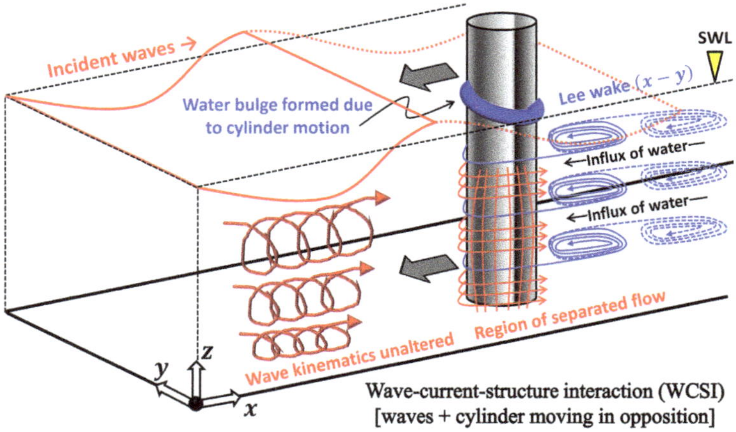

Fig. 1.8(b). Linear superposition of wave kinematics for wave–current interaction with a fixed cylinder representing wave interaction with a moving cylinder (independent flow field formulation).
Source: Sainchet *et al.* (2022).

forces are estimated using

$$f = C_M \left\{\rho \frac{\pi}{4} D^2\right\} \dot{u} + C_D \left\{\frac{1}{2}\rho D\right\} |u|u + \bar{C_D} \left\{\frac{1}{2}\rho D\right\} U^2, \tag{1.1b}$$

wherein $\overline{C_D}$ is the steady-drag (or current-drag) coefficient. This condition can be a representation of wave interaction with a moving cylinder. The pioneering works by Iwagaki and Asano (1984) and Koterayama (1984) primarily focused on estimating hydrodynamic inertia (C_M) and drag (C_D) coefficients with a linear superposition of wave and current loads being considered within the Morison framework. These works complement each other in that an *opposing current* ($-U$) was generated using a power pump in a recirculating flume by Iwagaki and Asano (1984), while a cylinder was towed opposite to the wave direction by Koterayama (1984) resulting in a *following current* ($+U$) in the cylinder frame of reference.

Nonetheless, the range of Keulegan–Carpenter numbers considered by Iwagaki and Asano ($0 < \mathrm{KC} < 50$) was much wider in comparison to that considered by Koterayama ($0 < \mathrm{KC} < 20$). It is also worth mentioning that both studies report significant scatter (~50%) in the experimentally determined values of C_D and C_M, which means that any Morison theory-based computation of the total wave–current force using these data is also deemed to exhibit scatter at constant KC.

Shaswat *et al.* (2022) reported a comprehensive experimental and numerical work on moving cylinder interaction with regular, steep breaking waves. They highlighted that the steep nonlinear focused wave and locally-imposed current loads on a moving structure can be linearly superimposed; despite focused wave formation itself being an inherently non-linear process.

1.3.2 *Horizontal pipelines over a sloping bottom*

As there is sufficient information on the forces on pipelines available in the literature, such information is avoided here, and the readers if interested, may refer to Sundar and Sannasiraj (2019). Sundar and Roopsekhar (2003) measured the dynamic pressure on a pipeline of diameter, D, 0.2 m, positioned normal and parallel to the wave direction near a sloping (1:10) rigid boundary induced by regular waves. Inline and transverse forces by integrating the pressures in terms of maximum and root mean square (rms) coefficients were reported as a function of the KC number. The instrumented pipeline was rigidly fixed over the sloping bed in a local water depth at the location of the pipeline, d', which is the depth from free surface to the top of the pipeline at 0.72 m. The water depth, on the seaside of the slope was 1.0 m. The studies covered pipelines normal and parallel to the wave direction at $x/X = 0.72$, for which three e/D, gap ratios of 0.25, 0.50 and 0.75 were considered, where e is the clearance of the pipe from the flume bed, x is the horizontal distance from the beach end to the test section and X is the total horizontal distance of the sloping bed. The definition sketches are shown in Fig. 1.9.

It was reported that the maximum dimensionless pressures on the pipeline for both the configurations at $\theta = 90°$ were found to be close,

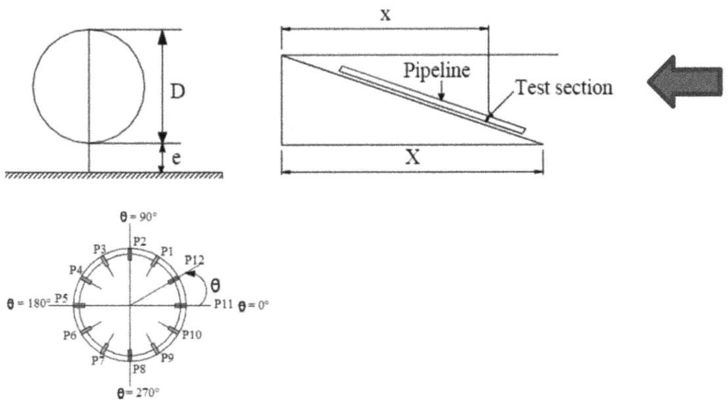

Fig. 1.9. Definition sketches of the pipeline, pipeline over sloping bed and orientation of pressure sensors.
Source: Sundar and Roopsekhar (2003).

whereas, at other locations, the pressures around the pipeline normal to wave direction, with the variations in the pressures being significant, were found to be slightly higher. The pressure gradient is more for the pipeline normal to the wave direction as it experiences a larger force.

Typical comparisons of the root mean square force coefficients, $C_{H_{\mathrm{rms}}} = F_{H_{\mathrm{rms}}}/0.5\rho D u^2_{\max}$ and $C_{V_{\mathrm{rms}}} = F_{V_{\mathrm{rms}}}/0.5\rho D u^2_{\max}$ for the pipeline placed normal and parallel to wave direction for the different e/D, are presented in Fig. 1.10. It is observed that $C_{H_{\mathrm{rms}}}$ for the pipeline parallel to wave direction is almost insignificant compared to that for the pipeline normal to wave direction, whereas the $C_{V_{\mathrm{rms}}}$ for the latter configuration of the pipeline is almost of the same magnitude as that for the former, the difference being only at lower KC values. Herein, $u_{\max}$: maximum orbital velocity at the centre of the pipeline, and $F_{H_{\mathrm{rma}}}$ and $F_{V_{\mathrm{rms}}}$ are the maximum inline and vertical force, respectively.

The probability distribution of random wave forces on a cylinder near a sloping bed was discussed by Sundar *et al.* (2004). It was concluded that the effect of e/D on the probability distribution of peak horizontal and vertical force is not significant for higher e/D. Further, it was reported that the probability distribution of the peak horizontal and vertical forces is higher for larger wave steepness.

1.3.3 *Twin pipelines over a sloping bottom*

Wilson *et al.* (1970) carried out experiments under steady currents to understand the effect of horizontal spacing and vertical spacing of the pipes from

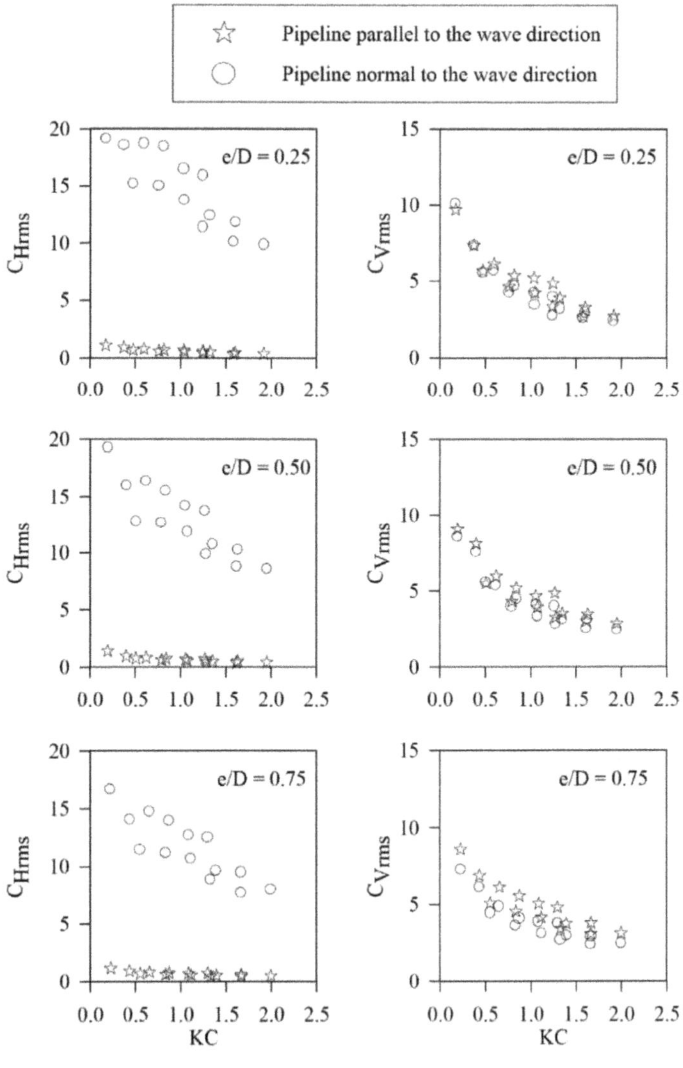

Fig. 1.10. RMS of horizontal and vertical force coefficients for a pipeline near a sloping bottom normal and inline to the wave direction.
Source: Sundar and Roopsekhar (2003).

the bed on forces. Two identical parallel pipelines of diameter, D, separated by horizontal spacing 2 D, 4 D, 7 D and 9 D, and vertical spacing between the pipeline and the boundary in the range of 0.08–1.75 D, were adopted for the tests. It was concluded that the measured lift and drag coefficients, which include the proximity effects, could be used to estimate the pipeline stresses only if the vibrations due to eddy shedding are avoided. Zdravkovich (1985) measured drag and lift forces on circular cylinders by varying the gap between the cylinder and the wall and suggested that the lift coefficient is governed by the gap-to-diameter ratio.

Yüksel and Narayanan (1994a) conducted experiments to measure forces on pipelines positioned on a sloping bottom with a slope of 1:10. Their findings indicated that the maximum inline force was 1.8 times the root mean square inline force, while the maximum vertical force was 1.2 times the root mean square value. Yüksel and Narayanan (1994b) also investigated forces resulting from plunging breakers over a 1:10 slope using the Cnoidal theory and concluded that the maximum forces on the cylinder in the breaker zone occur at the plunging point and can be about four times those experienced by the cylinders at the breaking point.

Chevalier *et al.* (2000) reported on forces acting on a 0.04-m-diameter pipeline laid parallel to wave crests over a bottom with a slope of 1:20, considering different values of e/D. Their observations revealed that both the inline and lift forces decrease with an increase in water depth and the gap-to-diameter ratio. For KC numbers greater than 10, C_D was found to be independent of e/D, whereas C_M was reported to decrease with an increase in e/D.

Cokgor and Avci (2001) studied the forces and pressures on a cylinder in partly buried and exposed conditions in the presence and absence of a neighbouring dummy cylinder. The experiments were replicated with variations in the placement of the dummy cylinder, positioned both upstream and downstream of the test cylinder, across different e/D values. It was reported that the lift force coefficient significantly decreases for $e/D = 0.7$. The force coefficients for the seaside pipe were reported to be higher.

Vijayakumari *et al.* (2006) discussed the variation of force coefficients through the measurement of the circumferential variation of pressure at 12 ports around a pipeline of diameter D of 0.315 m in the presence and absence of another pipeline of the same size, laid side by side (separated by distance, w), over a bed slope of 1:10. The interaction between these waves can be modified by adjusting the gap between them and their distance from the sloping bottom. This manipulation reveals variations in the force coefficients, as evidenced by the integration of pressure around the test pipe. It was reported that the effect of w/D is more predominant in the variation of the horizontal force, which explains the less pronounced effect of w/D on the vertical forces. The horizontal force on the pipe in the presence of an additional pipe on its leeward side was reported to increase by a maximum of about 30% compared to that exerted on a single pipe. On the other hand, the horizontal force on the pipe in the presence of an additional pipe on its seaward side ("leeward side" configuration) decreases by a maximum of also about 30% compared to that exerted on a single pipe. The vertical force on the pipe in the presence of an additional pipe either on its leeward side or on its seaward side does not get significantly altered.

1.4 Physical Processes in Wave–Structure Interaction Problems

1.4.1 *Interaction of waves with vertical structures*

In the previous section, different types of coastal structures were presented. For the physical process, they can be broadly classified into rigid impermeable and permeable/porous structures. In general, coastal structures will undergo interactions with waves under normal and extreme conditions. The physical process during these two scenarios can be different and the coastal structure needs to withstand these environmental conditions. Under normal conditions, during the wave interaction process, these coastal structures will dissipate the wave energy, introduce or reduce the wave run-up and remain intact to the incoming waves. During extreme scenarios, the wave impact leads to the overtopping process. The difficulty is mostly due to the air–water mixture caused by air entrapment and turbulence. According to Oumeraci *et al.* (1993) and Kisacik *et al.* (2012), the breaking scenario is divided into non-breaking, breaking and broken cases based on visual observation of the profile near the structure and pressure measurements. However, only breaking cases lead to maximum impact forces, and non-breaking and broken waves do not produce enough impact. The breaking classification is described with a series of images showing the development of wave surface profile till the point of contact with the structure from Figs. 1.11(a) to 1.11(g).

In the non-breaking case (Fig. 1.11(a)), the incoming wave does not break on the structure; instead, it touches the structure slightly, slides over and reflects without causing any impact on the structure. The pressure profile does not have any impact, and only slowly varying quasi-static pressure is observed. In the slightly breaking wave (SBW) case (Fig. 1.11(b)), the wave transforms itself near the structure with increasing wave height, and when it reaches the structure, it slightly breaks on the structure. The impact pressure is very low for this case, but the impact is observed in all

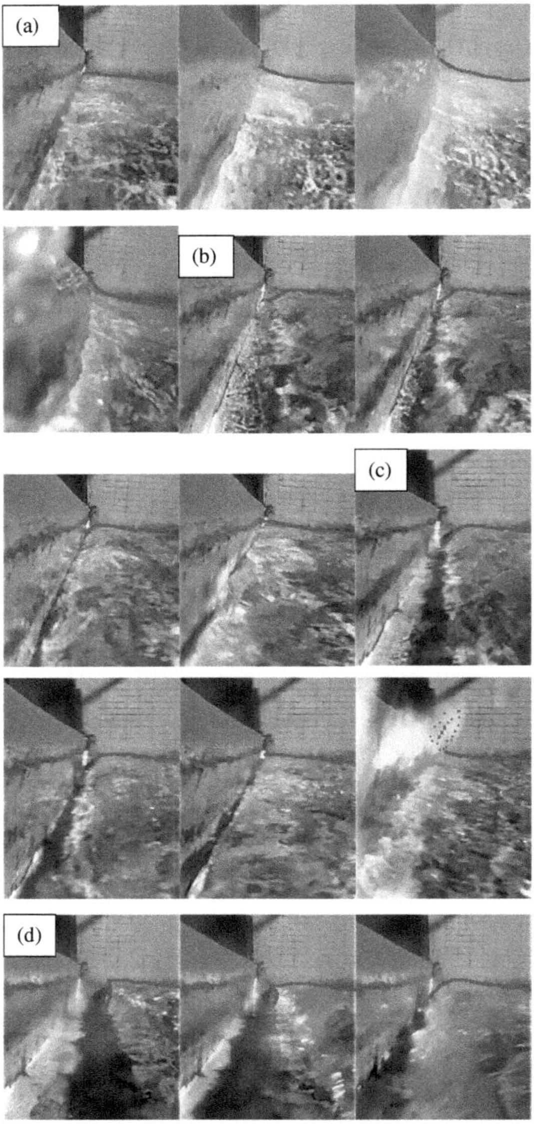

Fig. 1.11. (a) Wave transformation profile for a non-breaking case; (b) wave transformation profile of a slightly breaking wave case; (c) wave transformation profile of a breaking wave with small air trap (BWSAT) case; (d) wave transformation profile of a breaking wave with large air trap (BWLAT) case; (e) wave transformation profile of a broken wave case with wave height; (f) wave transformation profile of a collapsing breaking wave with small air trap (CBWSAT) case; and (g) general characterization of the wave transformation profile near the structure (Ravindar and Sriram, 2021). In the above figures, the waves are propagating from right to left.

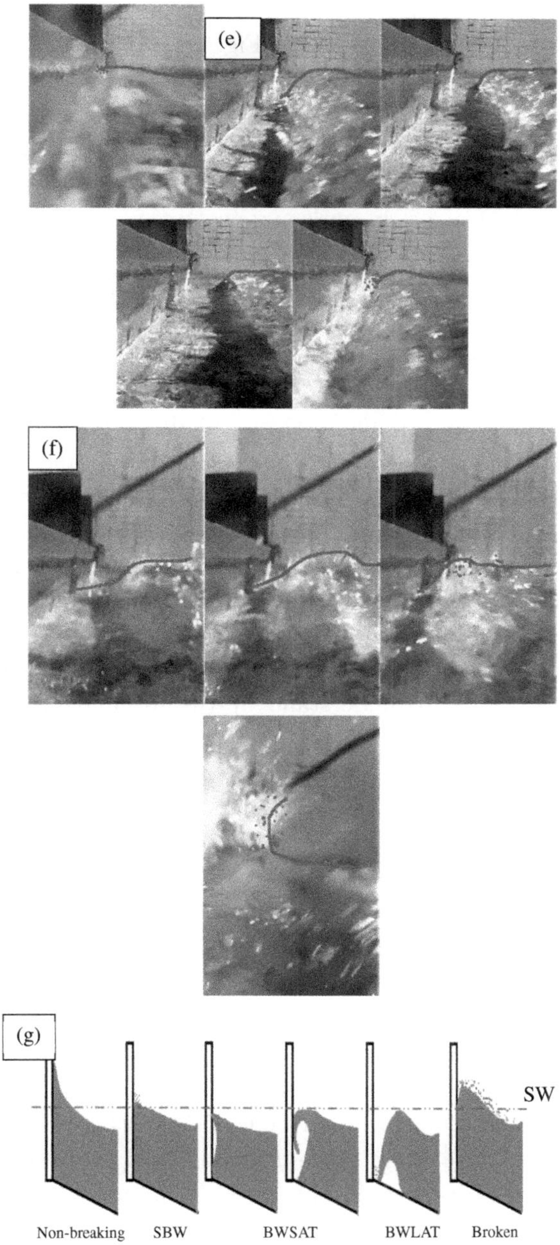

Fig. 1.11. (*Continued*)

the transducers with significantly less magnitude. In the case of breaking waves with a slight air trap (BWSAT) (Fig. 1.11(c)), a smaller quantity of air is entrapped between the structure and incoming wave from development until breaking. Due to small air entrapment, the impact pressure is found to be high among all the cases. The case of breaking wave with a large air trap (BWLAT) (Fig. 1.11(d)) is similar to the BWSAT; however, before breaking, the wave front curls and encloses a large air packet within itself and breaks on the structure. The impact mostly occurs below the still water level (SWL) due to water receding before breaking, as shown in the abstract view in Fig. 1.11(g). In Fig. 1.11(e), it is visible that the waves are broken prematurely before hitting the structure, which is considered to be the case of broken waves.

Finally, the last type is similar to the small air-entrapped case. However, it collapses in front of the structure due to incoming and reflected wave interactions. As it collapses, a considerable mass of water with small air pockets hits the structure. This type of breaking occurs normally for a shorter wave period with higher wave amplitude. Hence, this is classified as a collapsed breaking wave with a small air trap (CBWSAT), as shown in Fig. 1.11(f).

To examine the difference in pressure and force on parapets and vertical walls, it is critical to understand up-rushing and descending water patterns. Figure 1.12 provides a visual representation of the same. The up-rushing and descending pattern for a vertical wall with plain parapet is shown in the top left figure in Fig. 1.12. After wave impact near the SWL, the water jet moves upwards against gravity with overtopping and spillage on the leeward side, and the remaining water separates into droplets and falls freely near the structure. Thus, the interaction between descending water and an incoming wave is minimal in vertical wall models. The toe of the wall, on the other hand, is affected as the wave steepness increases due to the free fall of a large mass of water. In the small parapet (68°), the mechanism is similar to the vertical wall, but up-rushing occurs in an angular trend. The up-rushing water moves at an angle close to the parapet angle (68°) for a longer vertical distance than other models. The water displaced at an angle descends back as droplets and showers over the flume. However, in the case of the small parapet, there is no contact between the deflected up-rush and the incoming waves. The deflection pattern is qualitatively similar for medium (55°) and large (45°) parapets; however, due to the change in exit angle, the interaction between

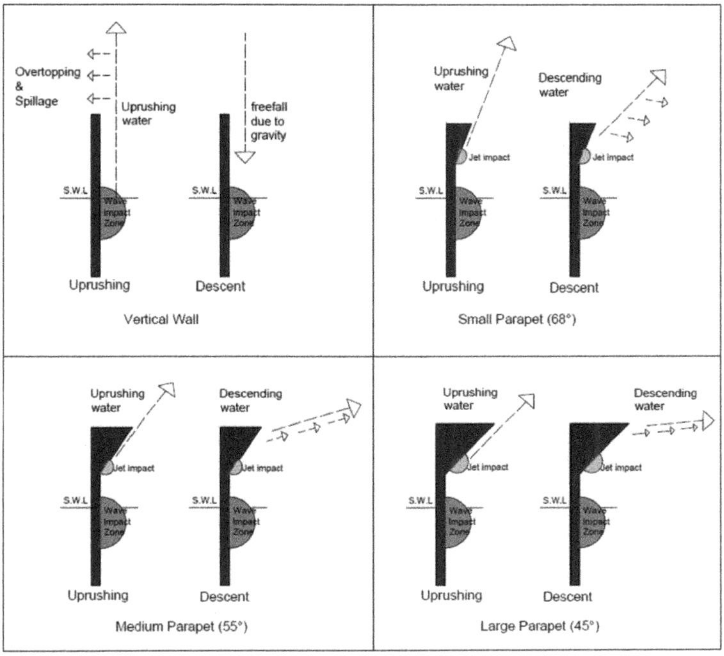

Fig. 1.12. Up-rushing and descending water pattern of the vertical wall with plain parapets.
Source: Ravindar and Sriram (2021).

the descending water and incoming wave increases in relation to the small parapet.

In general, the natural flow of the water jet is hindered by the parapet, and intuitively, the highest pressure is exerted, where a packet of water first hits the parapet, as shown in the top figure of Fig. 1.12. All parapets show this pattern, similar to the results of Castellino *et al.* (2018) who proposed a confined crest impact. Following impact at the parapet, the up-rushing jet of water segregates into droplets and descends back on the seaward side based on the exit angle.

1.4.2 *Overtopping of waves*

The second important aspect during extreme wave interactions, in particular, is wave overtopping. Nevertheless, one can also design a coastal structure that allows overtopping in normal circumstances. Wave overtopping is characterized by the volume of water that surpasses the crest of a coastal structure, such as a seawall, dike or breakwater, due to the influence of wave action (Geeraerts *et al.*, 2007). Direct wave and overtopping effects, along with their potential consequences, are classified into four groups as outlined in the EurOtop Manual (2016): (i) damage to defence structures, either short term or long term, with the risk of breaching and flooding; (ii) immediate hazards of injury or fatality to individuals located directly behind the defence, including pedestrians, cyclists or occupants of vehicles; (iii) impacts on property, operations and infrastructure in the defended area, leading to the loss of economic, environmental or other resources, as well as disruptions to economic activities or processes; and (iv) low-depth flooding, which may be inconvenient but not perilous.

Wave overtopping can be measured and assessed using one of two methods. The individual volume (V) of overtopping per wave is the first approach. The second and more popular method considers the average discharge (q) over a given time and per metre run. An average discharge per metre run is usually expressed in l/s per m. One thousand waves is a common time for averaging the overtopping. However, wave overtopping is a phenomenon that is unevenly distributed in both time and space due to irregular wave motion. As a result, predicting overtopping quantities and assessing or designing the crest level are not easy tasks.

According to EurOtop (2018), there are three forms of overtopping. The "green water overtopping case" is the first mode of overtopping, which happens when wave run-up levels are high enough for water to flow over the crest of the coastal structure.

The crest level and front face configuration of any structure designed to minimize wave overtopping will be dimensioned to provide appropriate levels of wave overtopping under defined extreme conditions or combined conditions (e.g. water level and waves). Setting the appropriate levels of overtopping can be determined by (i) how the defence system is used, including public and personnel access; (ii) how the land (or water) behind it is used; (iii) national and local norms and administrative practice; and (iv) the social and economic basis for funding the defence system. However, studies have investigated the allowable overtopping rates for certain safety conditions. Table 1.1 displays the permissible mean overtopping rates (q) for different conditions according to EurOtop (2018).

1.4.3 *Impact pressures and forces on structures*

The basics of non-breaking and breaking wave-induced pressures and forces on wall-type structures have been covered by Sundar and Sannasiraj (2019)

Table 1.1. Permissible mean overtopping rate.

Mean overtopping rate q (l/s per m) for different conditions	
Pedestrians	
Unsafe for unwary pedestrians since there is no clear view of the sea, they are easily disturbed or scared, and the walkway is narrow or close to the edge.	$q > 0.03$
Unsafe for people who are aware of their surroundings, have a good view of the sea, are not easily disturbed or scared, can handle being wet, and use a wider walkway.	$q > 0.1$
Unsafe for qualified personnel who are well-shod and covered and plan to get wet, overtopping flows are just at lower levels, there is no falling jet, and there is little risk of falling from the walkway.	$q > 1\text{–}10$
Automobiles	
Impulsive overtopping triggers are falling or high-velocity jets, rendering it dangerous to drive at moderate to high speeds.	$q > 0.01\text{–}0.05$
Low-speed driving is unsafe since pulsating flows overtop at low speeds, and there are no falling jets.	$q > 10\text{–}50$
Infrastructure and Buildings	
There is no harm done.	$q > 0.001$
Fitting has minor damage.	$0.001 < q < 0.03$
Harm to the structure.	$q > 0.03$
Behind the seawall, there is damage to the grassed or lightly covered promenade.	$q > 50$
Behind the seawall, there is damage to the paved or armoured promenade.	$q > 200$

Source: EurOtop (2018).

and are not repeated here. However, a few of the methods that are important are presented in this section.

1.4.3.1 *PROVERBS method*

PROVERBS (Probabilistic design tools for vertical breakwaters) was a major European research project funded by the European Union's 4th Framework Programme. The project's overarching goal was to create and incorporate a reliability-based system for vertical breakwater design and associated probabilistic tools. Oumeraci *et al.* (2001) guided the assessment of wave forces on seawalls under breaking and non-breaking conditions within the PROVERBS context.

Nonetheless, the PROVERBS approach is difficult to apply. Even under relatively similar design conditions, there can be substantial scatter in wave

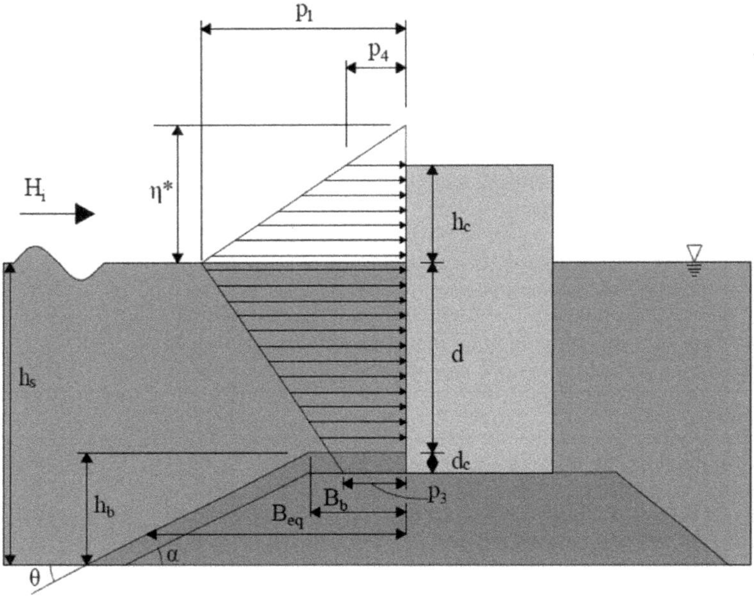

Fig. 1.13. Distribution of pressure based on PROVERBS method.
Source: Based on Goda (2000) and Oumeraci *et al.* (2001).

load projections, with projected design loads often varying significantly from estimates by the most commonly accepted and proven design methods (Cuomo *et al.*, 2010).

Figure 1.13 shows the pressure distribution and main geometric features according to the PROVERBS approach based on the method by Goda (2000). The main geometric and wave parameters in the PROVERBS method are described in Oumeraci *et al.* (2001). The determinations of different quantities are given in the following.

The determination of the breaking wave height is critical and the following method is provided by the PROVERBS approach:

$$H_b = L_{pi}\left(0.1025 + 0.0217\left(\frac{1-k_r}{1+k_r}\right)\right)\left(\tanh\left(\frac{2\pi k_b h_s}{L_{pi}}\right)\right). \qquad (1.2)$$

The linear wave theory defines the local inshore wavelength L_{pi} related to peak period T_p (at h_s). K_r is the overall reflection coefficient from the vertical wall:

$$L_{pi} = \left(\frac{gT_p^2}{2\pi}\right)\left(\tanh\left(\frac{2\pi h_s}{L_{pi}}\right)\right). \qquad (1.3)$$

k_b is the empirical correction factor which can be determined from

$$k_b = 0.0076\left(\frac{B_{\text{eq}}}{d}\right)^2 - 0.1402\left(\frac{B_{\text{eq}}}{d}\right) + 1, \tag{1.4}$$

in which the equivalent berm distance, specified halfway up the berm, is called B_{eq}:

$$B_{\text{eq}} = B_b + \left(\frac{h_b}{2\tan\alpha}\right), \tag{1.5}$$

where $\tan\alpha$ is the steepness of the seaward berm slope, B_b and h_b are the berm width and height, respectively, and d is the water depth at the caisson wall. The percentage of breaking and broken waves is determined according to the results of Equation (1.5). Therefore, P_b should be viewed as an upper limit for the percentage of impact loading. For complete details, refer to Oumeraci *et al.* (2001).

1.4.3.2 *Extended Goda–Takahashi method*

The initial formula of Goda (2000) has a range of benefits, including the ability to withstand standing and slightly breaking waves. This formula was later expanded to include the incident wave path and adjustment factors applicable to other types of vertical walls; the impulsive pressure coefficient to use for impact waves has been detailed by Takahashi and Shimosako (1994) and Takahashi *et al.* (1994).

1.4.3.3 *Method of Allsop et al. (1996)*

Allsop *et al.* (1996) proposed a prediction formula for horizontal wave impact force on vertical walls based on the PROVERBS project's model tests at HR Wallingford. The adopted predicted range was $0.3 < H_{\text{si}}/d \leq 2$ and data were produced on a slope of 1/50 at the 1/250 stage. However, for simple vertical walls, it is recommended to use a value of $0.35 < H_{\text{si}}/d \leq 0.6$. Oumeraci *et al.* (2001) proposed this approach for preliminary design:

$$F_h = 15\rho g d^2\left(\frac{H_{\text{si}}}{d}\right)^{3.134}, \tag{1.6}$$

where the water depth at the model's toe is h_s. The significant wave height at the model's toe is H_{si} and the water depth above the berm is denoted by the letter "d."

1.4.3.4 *Cuomo method (2010)*

Cuomo *et al.* (2010) introduced prediction equations for impact and quasi-static forces acting on vertical-faced coastal structures, such as caisson breakwaters and seawalls. These equations were derived from experimental investigations conducted in the CIEM/LIM broad flume in Barcelona, as part of the VOWS (Violent Overtopping by Waves at Seawalls) project. The total horizontal impact force for a vertical structure at the exceeding 1/250 limit in the seaward direction is given by the following:

$$F_{h,\mathrm{imp},\frac{1}{250}} = C_r^{1.65} \rho g H_{mo} L_{hs} \left(1 - \frac{|h_b - d|}{d}\right), \tag{1.7}$$

where C_r is the reflection coefficient. The significant spectral wave height is H_{mo}. For the mean wave period $T = T_m$, L_{hs} is the wavelength at the structure's toe. "d" represents the water depth in front of the structure. "h_s" represents the water depth at the structure's toe. The expression $\left(1 - \frac{|h_b - d|}{d}\right)$ denotes the magnitude of the structure's breaking to a certain extent. In this case, "h_b" represents the water depth at the time of breaking and is determined by inverting Miche's breaking criteria and assuming $H_b = H_{mo}$:

$$h_b = \frac{1}{k} \mathrm{arctanh}\left(\frac{H_{mo}}{0.14\, L_{hs}}\right), \tag{1.8}$$

where $k = 2\pi/L_{hs}$.

Based on past and new observations, a prediction formula for quasi-static forces was proposed:

$$F_{h,qs,\frac{1}{250}} = 4.8 \rho g H_{m0}^2. \tag{1.9}$$

The prediction formula is valid for $0.5\,\mathrm{m} < h_s < 1.3\,\mathrm{m}$, $0.2\,\mathrm{m} < H_{mo} < 0.7\,\mathrm{m}$ and $0.2\mathrm{s} < T_m < 3.7\,\mathrm{s}$.

1.4.3.5 *Method of Ravindar et al. (2017)*

The pressure–time history of a single wave effect on a vertical wall is depicted in Fig. 1.14. Ravindar *et al.* (2017) and Ravindar and Sriram (2021) analyzed a large amount of small-scale and large-scale experimental data and reported that, overall, the impacts comprise a large-amplitude short-duration peak, an oscillatory pressure and a steadily varying lesser pressure. The impact pressure is now generally referred to as the initial pressure peak portion of this trace, which is due to the initial impact on the structure. Then, because of entrapped air oscillations, the oscillatory

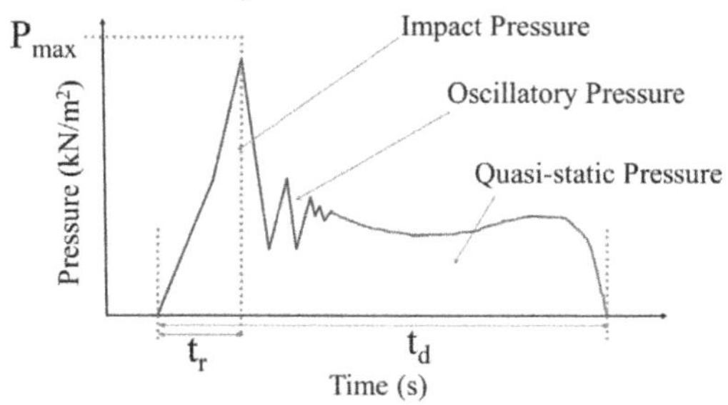

Fig. 1.14. A typical pressure–time history for a single wave impact on a vertical wall.
Source: Ravindar *et al.* (2017).

pressure decays over time. Finally, the quasi-static pressure is the pressure that lasts the longest and continues until a wave crest is reflected off the wall.

In their study, the existing analyses pressures were compared to estimate impact and the PROVERBS method Oumeraci *et al.* (2001) was reported to be in close agreement with the pattern but underpredicted the value.

The proposed equation to estimate the impact pressure for a vertical wall using nonlinear regression analysis is as follows as per the method of Ravindar and Sriram (2021):

$$\frac{P_{\max}}{P_0} = -2.030 + 5.4026\,\frac{H_b}{d_b} - 0.1503\,\frac{H_0}{L_0} + 0.0349\,\frac{(gH_b)^{\frac{1}{2}}T}{d_b} - 0.0101\,\frac{\sigma T^2}{\rho H_b d_b^2}. \tag{1.10}$$

This equation holds good for the following set of parameters: $\frac{H_b}{d_b} = 0.4$ to 1.2, $\frac{H_0}{L_0} = 0$ to 0.05.

However, the proposed equation to estimate the impact pressure for a vertical wall with a recurve parapet using nonlinear regression analysis is as follows:

$$\frac{P_{\max}}{P_0} = -0.1955 + 11.4\,\frac{H_b}{d_b} - 0.0164\,\frac{H_0}{L_0} + 0.0242\,\frac{(gH_b)^{\frac{1}{2}}T}{d_b} - 0.863\,\frac{\sigma T^2}{\rho H_b d_b^2} - 81.7\,\frac{Hl}{L^2}. \tag{1.11}$$

This equation holds good for the following set of parameters: $\frac{H_b}{d_b} = 0.25-1.1$, $\frac{H_0}{L_0} = 0-0.04$, where the overhang length is connected with the radius of the arc and recurve exit angle using the expression $l = r(1-\cos\alpha)$. The following equation is tested for three recurve models with overhang length (l) ranging from 0.2 m to 0.61 m.

The proposed formula for calculating horizontal impact force is as follows:

$$F_{h,I} = \lambda\rho C_b^2 T H_b \left(\alpha - \frac{|h_b - h_s|}{h_s}\right). \tag{1.12}$$

This formula holds well in the range of $0.5 < H_1/h_s < 3.5$, $1.4\,\mathrm{s} < T < 3.5\,\mathrm{s}$.

Further, a new method Cuomo–Froude scaling was proposed to upscale the small-scale experiments to the full scale. Based on the large amount of test data, it was proposed to use the Cuomo scale in the impact and oscillatory phase, the Froude scale in the quasi-static phases and a different scale factor in the impact and non-impact regions.

1.4.3.6 *Maximum impact pressure*

Many researchers have provided information about maximum impact pressure in an experimental investigation. The difference, however, is the height of the maximum impact pressure that is exerted by the waves. It should be noted that different definitions of impact pressure exist.

According to the literature, the highest wave impact pressure has the following characteristics:

- Even with a small amount of air in the water mass's air–water ratio, the overall effect pressure decreases dramatically.
- Experiments demonstrate that the actual pressure effect is equal to the slope angle's tangent.
- The mean impact pressure $P_{\max}$ is inversely proportional to the rise time t_r.
- The time it takes to rise to the highest peak pressure must be much less than the wavelength (roughly 1/10 of the wave period).
- The maximum impact pressure is affected by the slope as well. Steeper slopes generate higher impact pressure than milder slopes.

1.5 Modelling the Coastal Structures

One of the major challenges in a coastal structure is that it is mostly designed as a porous structure to dissipate the wave energy, and one needs to investigate its stability for the extreme sea state conditions. Thus, one needs to do the modelling, either using physical or numerical models, before the installation. In particular, the challenge is due to scaling the prototype scenario to the model scale in the laboratory. One universal scaling law is not applicable for all these scenarios. Further, there are other limitations such as the duration of the tests (due to reflection in the finite tank length), quality of the generated waves similar to field conditions, neglecting the 3D effects and sediment transport phenomenon. These limitations persist irrespective of the model scale effects. On the other hand, the error or limitation in the physical model decreases by increasing the scale, i.e. full-scale modelling. The other advantages of the physical model are that (a) with respect to the numerical/analytical aspects, the physical model incorporates the appropriate equations governing the physical process without simplifying the assumptions and (b) with respect to the field, the small size permits easier data collection throughout the regime at a reduced cost; further, one has control over the scenarios that can be reproduced to understand the physical process along with the visual feedback from the model study. Sometimes, physical models are expensive, so in the recent decade, modelling has been often carried out using numerical models.

In the 21st century, numerical models have been developed to model coastal processes or wave–structure interactions to a greater degree of confidence. Various levels of approximations have been developed from the modelling perspective. Depending on the physics of the modelling and computational aspects, the level of approximation in the governing equation varies. These aspects will be discussed in later chapters of this book.

Any developments or breakthroughs in other allied fields should be utilized to understand the physical process of wave–structure interactions. Particularly, physical modelling should reflect the advancements in the instrumentation field. Traditionally, wave gauges based on resistance or capacitance types have been used to measure wave elevation in the flume. Even though it exists in practice for most of the studies due to its robustness in measurements. In the past decade, ultrasonic probes have been quite popular when non-contact measurements are required. Similarly, for measuring the acceleration or inclination, accelerometers and inclinometers have been replaced to some extent by inertial measurement unit (IMU) sensors to measure the motion of the body. To measure the velocities, traditionally, current meters such as the Price type AA on a round wading rod, the Pygmy current meter, the electromagnetic current meter and the original Woltmann current meter have been used. These current meters are based on propellers and analyze the flow speed based on the rotation of the propellers. Later, Acoustic Doppler velocimetry (ADV) and Laser Doppler Velocimetry (LDV) became quite popular to measure the velocity at a single point. Furthermore, in order to obtain the spatial measurements of the velocity, 2D and 3D Particle Image Velocimetry (PIV) has been used by researchers. PIV consists of a high-speed camera with laser or LED lights with seeding particles such as Titanium dioxide (TiO2). The former is expensive, whereas the latter is a cheap alternative, which is sufficient for most engineering applications. In order to understand the physical processes and explain their characteristics, nowadays, it is very common to use video or camera images, especially high-speed cameras (with underwater capabilities) or normal mobile phones (with underwater shields). Compared to the olden days of the 24-Hz camera, these days, high-speed mobile devices can record up even to 200 Hz. This advancement provides researchers with greater opportunities to understand the physical process and use image processing methods to analyze the run-up and motion of small objects in the non-contact manner. Contact-type measurement devices would otherwise disrupt flow physics.

In the past century, several investigations on the measurement of impact pressure have been carried out by prominent researchers using new technology, and subsequently, new instrumentations evolved in due course. Table 1.2 summarizes the results of the instrumentation that has been used over a period of time to study impact pressure. Stevenson collected and reported field measurements of wave loading as early as 1874. He took measurements of wave loading at Dunbar Harbour and several lighthouse locations. The results of the field measurements were used to create and refine design equations. A spring dynamometer was used to test the wave pressures.

The significant mass and slow mechanical response of the dynamometer rendered it incapable of capturing short-duration, high-magnitude transient pressure, a phenomenon now commonly measured in laboratory settings. Additionally, the design of the dynamometer allowed only one measurement per record at each transducer in the series. Consequently, it remains unclear whether all maxima occur simultaneously or within the same breaking wave. The measured spatial distributions of maximum pressure are also unlikely to have occurred instantaneously at the structure. Consequently, calculations of temporal and spatial pressure variations, as well as the structure's response to wave loading, could not be accurately determined.

A novel spring diaphragm dynamometer that employs fluid-filled pipes for transmitting pressure to manometer tubes was created by

Table 1.2. Various instruments used for measuring impact pressure.

Author	Equipment		Structure	Scale	Spatial Information
Stevenson (1874)	Point-based Pressure transducers	Spring Dynamometer	Dunbar Harbour wall	1:1	Vertical array (6 Nos)
Galliard (1904)		Spring & Diaphragm Dynamometer	Lake Superior Sea Wall		Vertical array (5 Nos)
Rouville *et al.* (1938)		Piezoelectric transducer	Dieppe Vertical breakwater		Vertical array (3 Nos)
Cot (1954)		Pressure cells	Vertical Masonry Breakwater		Cruciform (4 Nos)
Kuribayashi *et al.* (1959)		Strain gauge pressure transducer	Vertical caisson Breakwater		Vertical array (4 Nos)
Bird *et al.* (1998)		Pressure Aeration Unit	Granite Block Wall		Vertical Array (7 Nos)
Stagonas *et al.* (2012)	Pressure Pad	Tactile Sensor	Sea Dike	1:3	Area: 42 cm × 48 cm

Gaillard (1904). These tubes could be visually read to gauge variations in pressure. Despite this advancement, field measurement outcomes using the diaphragm dynamometer were not documented or published.

Bird *et al.* (1998) introduced a novel instrument designed for field measurements of wave impact pressures and seawater aeration, known as the Pressure Aeration Unit (PAU). This instrument combines two transducers, enabling the simultaneous measurement of wave impact pressures and the degree of aeration in the seawater upon striking the structure. Seven PAUs were strategically positioned in a vertical array along the breakwater at Alderney in the Channel Islands. Despite their broad coverage, these PAUs caused interference with the dynamics of the waves. It is worth noting that these studies, however, lack spatial details regarding the variation of impact pressure. To overcome this disadvantage of point-based pressure transducers, Stagonas *et al.* (2012) proposed using tactile sensors to measure wave-induced impact pressure. A tactile sensor was used in 2D large-scale experiments on 1:3 sea dikes, and pressure maps were obtained for different test conditions. This was later adopted to measure the spatial

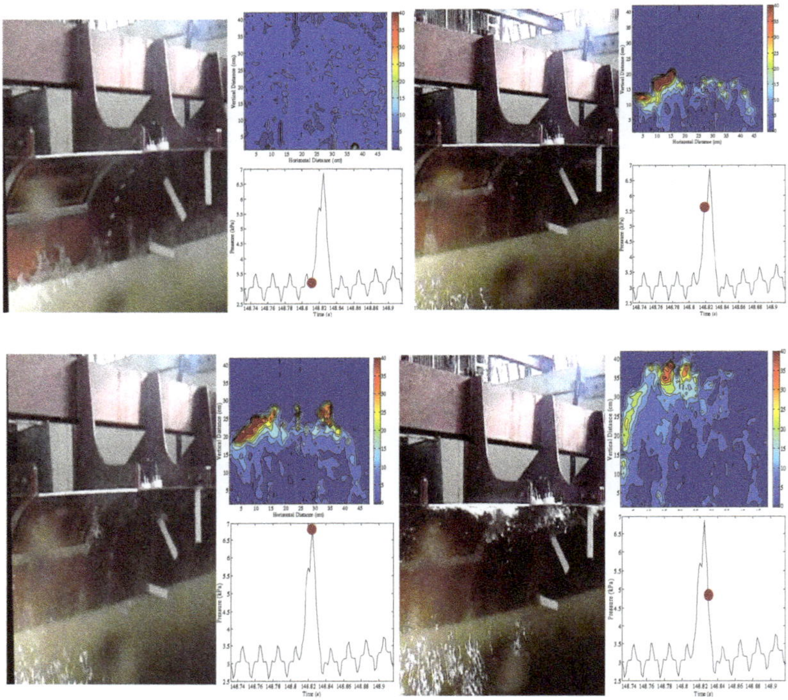

Fig. 1.15. Tactile sensor used for pressure measurement in a recurve wall. *Source*: Ravindar and Sriram (2021).

pressure on a recurve parapet wall as shown in Fig. 1.15. Pressure maps provide information on the variations of pressure over a complete rectangular or square area with spacing as low as 1 millimetre. The vertical and horizontal locations of maximum breaking wave-induced pressures can be obtained based on these pressure maps.

Quantifying suspended sediment concentration (SSC) is a crucial aspect of studying sediment characteristics in rivers or oceans. SSC measurement methods include conventional approaches such as laboratory analysis as well as optical and acoustic methods. Laboratory gravimetric techniques for SSC analysis require multiple water samples from various points and depths within the monitoring station. Water sampling, particularly through suction or pumping, is a traditional, reliable and widely utilized method for SSC measurement. The SSC values obtained from water samples serve primarily as references for sensor calibration (Kineke and Sternberg, 1992; Fugate and Friedrichs, 2002; Gray and Gartner, 2009; Wang *et al.*, 2012;

Druine *et al.*, 2018). However, this method has limitations, failing to provide continuous time series data and spatial characteristics of SSC due to the labour-intensive and cost-ineffective nature. To address the limitations of traditional sample collection methods, several researchers have explored the use of optical backscattering sensors (OBSs) to assess sediment resuspension in confined waterways. OBS instruments measure suspended sediment concentration (SSC) by analyzing backscattered light, providing measurements at a specific point with fixed deployment (Lin *et al.*, 2020). While the spatial distribution of SSC can be improved by employing multiple or movable OBS units, it is important to note that these instruments may disrupt the natural flow field and alter the distribution of SSC within the stream. Moreover, OBSs are highly sensitive to particle size distribution of sediment particles. The reading of the same sensor may increase by as much as ten times for the same SSC with a smaller particle size (Ludwig and Hanes, 1990). Therefore, it is necessary to calibrate the OBS against in-situ SSC obtained from the water samples (Maa *et al.*, 1992; Nauw *et al.*, 2014). Gelinas *et al.* (2013) could not obtain the SSC due to deep-draft vessels as OBSs had reached the maximum limit due to signal attenuation (Haught *et al.*, 2017). Despite its drawbacks, the OBS is a popular instrument for the measurement of SSC and has been utilized by numerous researchers (Rapaglia *et al.*, 2015; Scarpa *et al.*, 2019). Besides OBSs, acoustic tools have been gaining popularity among the coastal engineering community (Thorne and Hanes, 2002).

In riverine environments, backscattering intensity information obtained from instruments like the Acoustic Velocity Profiler (ADV) or Acoustic Doppler Current Profiler (ADCP) is widely utilized to calculate suspended sediment concentration (SSC). Unlike Optical Backscattering Sensors (OBSs), the ADCP can concurrently capture velocity and SSC profiles over several metres, offering synchronized high-frequency measurements of velocity at a specific location (Guo *et al.*, 2018). However, the conversion of acoustic output into SSC introduces ambiguity and substantial uncertainties. Fleit and Baranya (2021) introduced a concept for estimating ship-induced SSC using backscattering information from an ADV, comparing it with the LISST-Acoustic Backscattering Sensor (ABS) in a case study on the Hungarian section of the Danube River. Acoustic backscattering is also applied in large-scale laboratory flumes, such as those at Deltares and GWK. In movable bed experiments, traditionally, a probe or pin mounted on the bed was used to measure and monitor evolution over the period of study. Nowadays, advanced instruments such as acoustic bed profilers are being used. Bed ripple profilers were developed using acoustic pencil beams (Bell and Thorne, 1997) to measure the bottom sand ripples.

An increasing popular approach is the measurement of bed evolution and scour development using several sensors. A multibeam echosounder was used for measuring spatial development. Normally, it is mounted on a rail that transverses along the flume to measure spatial variations. For a point-based depth measurement, acoustic backscatter probes are used. Although this was originally designed as a backscattering sensor, the probes can also be used as single-beam echosounders. In locations where there are obstructions due to multibeam and single-beam echosounders/acoustic backscattering probes, underwater cameras can be utilized by the researchers as they are cost efficient; however, the locations have to be chosen carefully. Using multiple sensors or instruments for measuring the complex bed evolution and scour development will be helpful to validate and verify instruments against each other and provide confidence in the analysis. Moreover, one can use information from multiple sensors to provide overall evolution. Such an analysis was carried out by Stahlmann (2013) using a multibeam, acoustic backscatter probe and camera, and utilizing a spatial interpolation technique to observe scour development under a tripod leg.

Another important instrument that has gained popularity over the last decade is a laser scanner. There are a wide range of commercial scanners, and the coastal community has used their potential to obtain high-quality data in a non-intrusive manner. Terrestrial 3D laser scanning has been adopted for coastal research and field surveys of beaches (van Gaalen *et al.*, 2011). Similarly, the LiDAR (Light Detection And Ranging) instrument has been successfully used in various sectors like surveying, transportation, safety and robotics, and has been employed by the coastal community as well. Apart from using terrestrial LiDAR for topographic measurements, it has also been explored for water surface measurements in the lab (Allis *et al.*, 2011; Blenkinsopp *et al.*, 2012) or in the field (Roshandel *et al.*, 2021) and for monitoring of 2D wave fields. However, there are some limitations to using LiDAR in the lab involving fresh waters (wherein the light beam may penetrate and cannot detect the surface); therefore, some alternate approaches using seeding particles or turbid water were attempted (Allis *et al.*, 2011).

Alternatively, a combination of a 3D scanner, camera images and a 2D LiDAR scanner has been used by Vousdoukas *et al.* (2014) in a large-scale experimental facility to capture and monitor the wave-by-wave swash zone process. The hybrid monitoring system with a combination of different sensors was better and more reliable than the single-sensor measurement system.

Thus, over the past few decades, many new instruments have been used to understand the physical process better. However, it should be noted that these advanced instruments are costly, time-consuming and involve more post-processing effort than traditional measurement techniques. Due to this, it is not quite as popular as the traditional measurement techniques. In the future, one can believe that less post-processing effort will be required to use these advanced instruments and they will be common tools of the coastal community and researchers.

1.6 Summary

In this chapter, a brief description of different types of coastal structures and their functions is provided. Particular emphasis is given to the physical processes involved in the wave–structure interaction. Apart from discussing the physical process, physical modelling efforts over the past few decades and improvements in instrumentation capabilities are also reported.

References

Allis, M.J., Peirson, W.L. and Banner, M.L. (2011). Application of LiDAR as a measurement tool for waves. In *Proceedings of 21st International Offshore and Polar Engineering Conference*, Maui, Hawaii, USA, June. Paper Number: ISOPE-I-11-402.

Allsop, N.W.H., Vicinanza, D. and McKenna, J.E. (1996). Wave Forces on Vertical and Composite Breakwaters. Strategic Research Report SR 443, HR Wallingford, Wallingford, pp. 1–94.

Anand Kumar, G., Sundar, V., Graw, K.U. and Kaldenhoff, H. (1995). Pressure and forces on inclined cylinders due to regular waves. *Ocean Engineering Journal*, 22, 747–759.

Bell, P.S. and Thorne, P.D. (1997). Measurements of sea bed ripple evolution in an estuarine environment using a high-resolution acoustic sand ripple profiling system. In *Oceans '97 MTS/IEEE Conference*, Halifax, Nova Scotia.

Bird, P.A.D., Crawford, A.R., Hewson, P.J. and Bullock, G.N. (1998). An instrument for field measurement of wave impact pressures and seawater aeration. *Coastal Engineering*, 35(1–2), 103–122. https://doi.org/10.1016/S0378-3839(98)00020-9.

Blenkinsopp, C.E., Ian, L.T., Allis, M.J., Peirson, W.L. and Garden, L.E. (2012). Application of LiDAR technology for measurement of time-varying free-surface profiles in a laboratory wave flume. *Coastal Engineering*, 68, 1–5. https://doi.org/10.1016/j.coastaleng.2012.04.006.

Castellino, M., Sammarco, P., Romano, A., Martinelli, L., Ruol, P., Franco, L. and De irolamo, P. (2018). Large impulsive forces on recurved parapets under non-breaking waves. A numerical study. *Coastal Engineering*, 136, 1–15. https://doi.org/10.1016/j.coastaleng.2018.01.012.

Chakrabarti, S.K. (1987). *Hydrodynamics of Offshore Structures*. Computational Mechanics Publications.

Chevalier, C., Lambert, E. and Belorgey, M. (2000). Wave forces on submarine pipelines in the coastal zone. In *Proceedings of the 10th (2000) International Offshore and Polar Engineering Conference*, Seattle, USA, 28 May–2 June, Vol. 3, pp. 267–281.

Choi, S.-J., Lee, K.-H. and Gudmestad, O.T. (2015). The effect of dynamic amplification due to a structure's vibration on breaking wave impact. *Ocean Engineering*, 96, 8–20.

Cokgor, S. and Avci, I. (2001). Hydrodynamic forces on partly buried tandem, twin pipelines in current. *Ocean Engineering*, 28(2001), 1349–1360.

Cot, P.D. (1954). Le laboratoire du havre pour la mesure des efforts dus aux lames. In *Coastal Engineering Proceedings* (in French).

Cuomo, G., Allsop, W., Bruce, T. and Pearson, J. (2010). Breaking wave loads at vertical seawalls and breakwaters. *Coastal Engineering*, 57(4), 424–439.

Draper, L. (1965). 'Freak' ocean waves. *Marine Observation*, 35, 193–195.

Druine, F., Verney, R., Deloffre, J., Lemoine, J.P., Chapalain, M., Landemaine, N. and Lafite, R. (2018). In situ high frequency long term measurements of suspended sediment concentration in turbid estuarine system (Seine Estuary, France): Optical turbidity sensors response to suspended sediment characteristics. *Marine Geology*, 400, 24–37. https://doi.org/10.1016/j.margeo.2018.03.003.

EurOtop. (2007). Pullen, T., N.W.H. Allsop, T. Bruce, A. Kortenhaus, H. Schüttrumpf and J.W. Van der Meer, EurOtop Wave Overtopping of Sea Defences and Related Structures: Assessment Manual.

EurOtop Mannual. (2016). Van der Meer, J.W., Allsop, N.W.H., Bruce, T., De Rouck, J., Kortenhaus, A., Pullen, T., Schüttrumpf, H., Troch, P. and Zanuttigh, B., EurOtop: Manual on Wave Overtopping of Sea Defences and Related Structures: An Overtopping Manual Largely Based on European Research, But for Worldwide Application. www.overtopping-manual.com.

EurOtop Manual. (2018). EurOtop-Wave Overtopping of Sea Defenses and Related Structures. An Overtopping Manual Largely Based on European Research, But for Worldwide Application, p. 320.

Fei Deng, Y., Liand Tian, X. and Li, X. (2019). Study on nonlinear characteristics of freak-wave forces with different wave steepness. *China Ocean Engineering*, 33, 608–617.

Fleit, G. and Baranya, S. (2021). Acoustic measurement of ship wave–induced sediment resuspension in a large river. *Journal of Waterway, Port, Coastal, and Ocean Engineering*, 147(2). https://doi.org/10.1061/(ASCE)WW.1943-5460.0000627.

Fugate, D.C. and Friedrichs, C.T. (2002). Determining concentration and fall velocity of estuarine particle populations using ADV, OBS and LISST. *Continental Shelf Research*, 22(11–13), 1867–1886. https://doi.org/10.1016/S0278-4343(02)00043-2.

Gaillard, D.D. (1904). Wave Action in Relation to Engineering Structures, US Army Corps of Engineers, Reprinted 1935.

Gelinas, M., Bokuniewicz, H., Rapaglia, J., Kamazima, M.M. and Lwiza, K.M.M. (2013). Sediment resuspension by ship wakes in the Venice Lagoon. *Journal of Coastal Research*, 29(1), 8–17. https://doi.org/10.2112/JCOASTRES-D-11-00213.1.

Geeraerts, J., Troch, P., De Rouck, J., Verhaeghe, H. and Bouma, J.J. (2007). Wave overtopping at coastal structures: Prediction tools and related hazard analysis. *Journal of Cleaner Production*, 15(16), 1514–1521. https://doi.org/10.1016/j.jclepro.2006.07.050.

Goda, Y. (2000). *Random Seas and Design of Maritime Structures*. World Scientific, Singapore.

Gray, J.R. and Gartner, J.W. (2009). Technological advances in suspended-sediment surrogate monitoring. *Water Resources Research*, 45, 1–20. http://dx.doi.lorg/10.1029/2008WR007063.

Guo, X.P., Liu, X., Niu, Z.S., Lu, D.P., Zhao, S., Sun, X.L., Wu, J.Y., Chen, R., Tou, F.Y., Hou, L., Liu, M. and Yang, Y. (2018). Seasonal and spatial distribution of antibiotic resistance genes in the sediments along the Yangtze Estuary, China. *Environmental Pollution*, 242(Part A), 576–584. https://doi.org/10.1016/j.envpol.2018.06.099.

Haught, D., Venditti, J.G. and Wright, S.A. (2017). Calculation of in situ acoustic sediment attenuation using off-the-shelf horizontal ADCPs in low concentration settings. *Water Resources Research*, 53, 5017–5037. https://doi.org/10.1002/2016WR019695.

Huang, Z.H., Li, Y.C. and Liu, Y. (2011). Hydraulic performance and wave loadings of perforated/slotted coastal structures: A review. *Ocean Engineering*, 38(10), 1031–1053. https://doi.org/10.1016/j.oceaneng.2011.03.002.

Irschik, K., Sparboom, U. and Oumeraci, H. (2005). Breaking wave loads on a slender pile in shallow water. *Coastal Engineering*, 4, 568–580.

Iwagaki, Y. and Asano, T. (1984). Hydrodynamic forces on a circular cylinder due to combined wave and current loading. *Coastal Engineering*, https://doi.org/10.1061/9780872624382.192.

Jarlan, G.E. (1961). A perforated vertical breakwater. The Dock and Harbour Authority, Vol. 41, n° 486, pp. 394–398.

Karthik Ramnarayan, S., Sundar, V. and Sannasiraj, S.A. (2020a). Hydrodynamic characteristics of concave front pile-supported breakwaters with a tubular wave screen. *Journal of Waterway, Port, Coastal, and Ocean Engineering*, 148, 1. https://ascelibrary.org/doi/10.1061/%28ASCE%29WW.1943-5460.0000682.

Kineke, G.C. and Sternberg, R.W. (1992). Measurements of high concentration suspended sediments using the optical back scatterance sensor. *Marine Geology*, 108, 253–258.

Kisacik, D., Troch, P. and Van Bogaert, P. (2012). Experimental study of pressure distributions due to the breaking wave impacts. In *ICCE 2012*, p. 1–1.

Koterayama, W. (1984). Wave forces acting on a vertical circular cylinder with a constant forward velocity. *Ocean Engineering*, 11(4), 363–379. https://doi.org/10.1016/0029-8018(84)90011-8.

Krishnakumar, C., Sundar, V. and Sannasiraj, S.A. (2010a). Pressures and forces due to directional waves on a vertical wall fronted by wave screens. *Applied Ocean Research*, 32, 1–10.

Krishnakumar, C., Sundar, V. and Sannasiraj, S.A. (2010b). Hydrodynamic performance of single- and double-Wave screens. *Journal of Waterway, Port, Coastal, and Ocean Engineering*, 36(1), 59–65.

Kuribayashi, T., Muraki, Y. and Udai, G. (1959). Field investigation of wave forces on breakwaters. *Coastal Engineering in Japan*, 2, 17–27.

Lee, K.H., Ju-Hyun Bae, J.H., Kim, S.G. and Do-Sam Kim, D.S. (2017). Three-dimensional simulation of wave reflection and pressure acting on circular perforated caisson Breakwater by OLAFOAM. *Journal of Korean Society of Coastal and Ocean Engineers*, 29(6), 286–304. https://doi.org/10.9765/KSCOE.2017.29.6.286.

Lin, J., He, Q., Guo, L., Van Prooijen, B.C. and Wang, Z.B. (2020). An integrated optic and acoustic (IOA) approach for measuring suspended sediment concentration in highly turbid environments. *Marine Geology*, 421. https://doi.org/10.1016/j.margeo.2019.106062.

Ludwig, K.A. and Hanes, D.M. (1990). A laboratory evaluation of optical back scatterance suspended solids sensors exposed to sand-mud mixtures. *Marine Geology*, 94, 173–179.

Maa, P.Y., Xu, J. and Victor, M. (1992). Notes on the performance of an optical backscatter sensor for cohesive sediments. *Marine Geology*, 104(1–4), 215–218. https://doi.org/10.1016/0025-3227(92)90096-Z.

Mayilvahanan, A.C., Bihs, H., Myrhaug, D. and Muskulus, M. (2017). Breaking solitary waves and breaking wave forces on a vertically mounted, slender cylinder over an impermeable sloping seabed. *Journal of Ocean Engineering and Marine Energy*, 3, 1–19.

Morison, J.R., O'Brien, M.P., Johnson, J.W. and Schaaf, S.A. (1950). The forces exerted by surface waves on piles. *Petroleum Transactions AMIE*, 189, 149–154.

Myrhaug, D. and Kjeldsen, S.P. (1987). Prediction of occurrence of steep and high waves in deep water. *Journal of Waterway, Ports, Coastal and Ocean Engineering*, 113(2), 122–138.

Nauw, J.J., Merckelbach, L.M., Ridderinkhof, H. and van Aken, H.M. (2014). Long-term ferry-based observations of the suspended sediment fluxes through the Marsdiep inlet using acoustic Doppler current profilers. *Journal of Sea Research*, 87, 17–29. DOI: 10.1016/j.seares.2013.11.013.

Oumeraci, H., Klammer, P. and Partenscky, H.W. (1993). Classification of breaking wave loads on vertical structures. *Journal of Waterway, Port, Coastal Ocean Engineering*, 119(4), 381–397.

Oumeraci, H., Kortenhaus, A., Allsop, W., de Groot, M., Crouch, R.O.G.E.R., Vrijling, H. and Voortman, H.E.S.S.E.L. (2001). *Probabilistic Design Tools for Vertical Breakwaters*. CRC Press.

Qu, S., Liu, S., Ong, M.C., Sun, S. and Ren, H. (2020). Numerical simulation of breaking wave loading on standing circular cylinders with different transverse inclined angles. *Applied Sciences*, 10(1347), 1–19. DOI: 10.3390/app10041347.

Rapaglia, J., Zaggia, L., Kevin Parnell, K., Lorenzetti, G. and Vafeidis, A.T. (2015). Ship-wake induced sediment remobilization: Effects and proposed management strategies for the Venice Lagoon. *Ocean & Coastal Management*, 110, 1–11. https://doi.org/10.1016/j.ocecoaman.2015.03.002.

Ravindar, R., Sriram, V., Schimmels, S. and Stagonas, D. (2017). Characterization of breaking wave impact on vertical wall with recurve. *ISH Journal of Hydraulic Engineering*, 25, 153–161.

Ravindar, R. and Sriram, V. (2021). Impact pressure and forces on a vertical wall with different types of parapet. *Journal of Waterway, Port, Coastal, and Ocean Engineering, ASCE*, 147(3). https://doi.org/10.1061/(ASCE)WW.1943-5460.0000635.

Roshandel, S., Liu, W., Wang, C. and Li, J. (2021). 3D ocean water wave surface analysis on airborne LiDAR bathymetric point clouds. *Remote Sensing*, 3(19), 3918. https://doi.org/10.3390/rs13193918.

Rouville, M., Besson, P. and Petry, P. (1938). Etat actual des etudes internationales sur les efforts dus aux lambs. *Annates de Ponts et Chaussees*, 108(2), 5–113 (in French).

Saincher, S., Sriram, V., Agarwal, S. and Schlurmann, T. (2022). Experimental investigation of hydrodynamic loading induced by regular, steep non-breaking and breaking focused waves on a fixed and moving cylinder. *European Journal of Mechanics-B/Fluids*, 93, 42–64. https://doi.org/10.1016/j.euromechflu.2021.12.009.

Sarpkaya, T. and Isaacson, M. (1981). *Mechanics of Offshore Structures*. Van Nostrand Reinhold Co.

Scarpa, G.M., Zaggia, L., Manfè, G., Lorenzetti, G., Parnell, K., Soomere, T., Rapaglia, J. and Molinaroli, E. (2019). The effects of ship wakes in the Venice Lagoon and implications for the sustainability of shipping in coastal waters. *Scientific Reports*, 9. Article number: 19014.

Shaswat, S., Sriram, V., Shagun, A. and Schlurmann, T. (2022). Experimental investigation of hydrodynamic loading induced by regular, steep non-breaking and breaking focused waves on a fixed and moving cylinder. *European Journal of Mechanics - B/Fluids*, 93, 42–64. https://doi.org/10.1016/j.euromechflu.2021.12.009.

Stagonas, D., Muller, G., Ramachandran, K., Schimmels, S. and Dane, A. (2012). Distribution of impact induced pressures at the face of uniformly sloped sea dikes: Preliminary 2D experimental results. In *Proceedings of 33rd Conference of Coastal Engineering*, Santander - Spain.

Stevenson, T. (1874). *The Design and Construction of Harbours: A Treatise on Maritime Engineering.* A. and C. Black.

Stahlmann, A. (2013). Experimental and Numerical Modelling of Scour at Offshore Wind Turbines. Dissertation, Leibniz Universität Hannover, Franzius-Institut für Wasserbau, Ästuar- und Küsteningenieurwesen.

Sundar, V. and Roopsekhar, K.A. (2003). Comparison of pressures and forces on pipelines normal and parallel to wave direction near a sloping bed. *Oceanic International Journal*, 7(2), 65–89.

Sundar, V. and Sannasiraj, S.A. (2019). *Coastal Engineering- Theory and Practice.* Advanced Series on Ocean Engineering, Vol. 47. https://www.worldscientific.com/worldscibooks/10.1142/11148, worldscientific.com.

Sundar, V., Vengatesan, V., Anandkumar, G. and Schlenkhoff, A. (1997). Hydrodynamic coefficients for inclined cylinders. *Ocean Engineering*, 25(4–5), 277–294, 1998. EID: 2-s2.0-0032054433.

Sundar, V., Vengatesan, V. and Graw, K.U. (2002). Sleeve forces on inclined cylinders due to long and short crested waves. *Journal of Hydraulic Research, IAHR*, 40(3), 275–287. EID: 2-s2.0-0036430037.

Sundar, V., Roopsekhar, K.A., Schlurmann, T. and Mahmoud, R.A. (2004). Probability distribution of wave forces on pipelines near a sloping boundary parallel and normal to wave direction. *Coastal Engineering Journal (JSCE)*, 46(1), 93–117. https://doi.org/10.1142/s0578563404000902.

Takahashi, S. (1996). Design of Breakwaters. Port and Harbour Research Institute, Japan International Cooperation Agency.

Takahashi, S. and Shimosako, K. (1994). Wave pressure on a perforated wall. In *Proceedings of the International Conference on Hydro-Technical Engineering for Port and Harbor Construction*, Yokosuka, Japan, 19–21 October.

Takahashi, S., Tanimoto, K. and Shimosako, K.A. (1994). Proposal of impulsive pressure coefficient for the design of composite breakwaters. In *Hydro-Port'94, Proceedings of the International Conference on Hydro-Technical Engineering for Port and Harbor Construction*, Yokosuka, Japan, 19–21 October 1994. Port and Harbour Research Institute, Yokosuka, Japan, pp. 489–504.

Tanimoto, K. and Takahashi, S. (1994). Japanese experiences on composite breakwaters. In *Proceedings of International Workshop on Wave Barriers in Deepwater.* Port and Harbour Research Institute, Yokosuka, Japan, pp. 1–22.

Thorne, P.D. and Hanes, D.M. (2002). A review of acoustic measurement of small-scale sediment processes. *Continental Shelf Research*, 22(4), 603–632. https://doi.org/10.1016/S0278-4343(01)00101-7.

UNHCR. (2020). Global Trends 2019. The UN Refugee Agency.

Van Gaalen, J.F., Kruse, S.E., Coco, G., Collins, L. and Doering, T. (2011). Observations of beach cusp evolution at Melbourne Beach, Florida, USA. *Geomorphology*, 129(1–2), 131–140. https://doi.org/10.1016/j.geomorph.2011.01.019.

Vijayakumari, P.L., Murali, K. and Sundar, V. (2006). Wave forces on tandem submarine pipelines in the proximity of a sloping boundary. *The Journal of Ocean Technology*, 1(II), 67–89.

Vousdoukas, M.I., Kirupakaramoorthy, T., Oumeraci, H., Torre, M.D.L., Wübbold, F., Wagner, B. and Schimmels, S. (2014). The role of combined laser scanning and video techniques in monitoring wave-by-wave swash zone processes. *Coastal Engineering*, 83, 150–165. https://doi.org/10.1016/j.coastaleng.2013.10.013.

Wang, J.J., Lu, X.X., Zhou, Y. and Liew, S.C. (2012). Suspended sediment concentrations estimate in highly turbid rivers: A field spectral survey. *Remote Sensing Letters*, 409–417. https://doi.org/10.1080/2150704X.2012.743689.

Wilson, F.J., Duke, U., Caldwell, H.M. and Usaf. (1970). Force and stability measurements on models of submerged pipelines. In *Offshore Technology Conference*, 6200 North Central Expressway Dallas, Texas 75206, Paper No. OTC 1224.

Yüksel, Y. and Narayanan, R. (1994a). Wave forces on horizontal cylinders resting on sloping bottom. In *Proceedings of 4th International Offshore and Polar Engineering Conference*, Vol. 2, pp. 85–92.

Yüksel, Y. and Narayanan, R. (1994b). Breaking wave forces horizontal cylinders close to the sea bed. *Coastal Engineering Journal*, 23, 115–133.

Zdravkovich, M.M. (1985). Forces on circular cylinder near a plane wall. *Applied Ocean Research*, 7(4), 197.

Chapter 2

Coastal Structures: Seawalls, Groynes and Breakwaters

2.1 Introduction

Most stretches of coasts, particularly those that are urbanized or semi-urbanized, are vulnerable to natural coastal hazards. Added to this is the sea level rise due to the melting of ice over the past century, which is anticipated to continue to increase, thereby becoming a big threat to the stability of coasts, especially low-lying zones of coastal stretches around the globe. The increased frequency of coastal flooding due to sea level rise (SLR) has been discussed by several researchers (Brody *et al.*, 2011; Bilskie *et al.*, 2014; Sukop *et al.*, 2018).

The increase in development along the coast due to the migration of people towards the coast for their livelihood requires immediate adaptation due to the sea level rise. The factors that dictate the adaptation strategy are technical, financial, planning and decision-making as well as socio-economic aspects. To combat this problem, the conventional method involves armouring the affected stretch of the coast with hard structures like seawalls, groyne fields, submerged reefs or detached offshore breakwaters. The main purpose of these structures is to mitigate against perennial problem erosion or flooding. Such structures are made of natural rocks or cast as artificial concrete blocks, the weight of which would be dictated by the design wave height.

In addition, a pair of breakwaters are usually preferred for artificial harbours as they provide the desired tranquillity to facilitate smooth loading and unloading operations.

Coastal development, in particular involving coastal structures, henceforth, has to incorporate new methods of planning, designing and monitoring in order to handle the current and future coastal hazards dictated by the changing climate. In order to satisfy requirements of avoiding ecological deterioration, adapting to climate change and achieving multi-functional benefits, the solutions at our disposal are nature-based hard measures that consist of the construction of structures and soft measures like vegetation and sand bypassing. This means that the multi-functional structures should increase the resilience against extreme events, absorb other disturbances that may arise and serve added benefits like relaxation along the beaches.

This chapter briefly introduces conventional structures and discusses the methods of converting them to multi-functional structures.

2.2 Seawalls

2.2.1 *General*

Seawalls classified as rigid, flexible and composite are constructed along the shoreline to protect the coast against the action of waves and coastal flooding. The oldest seawall was uncovered by archaeologists at the Neolithic settlement of Tel Hreiz, near Haifa, Israel, and is assumed to have been constructed about 7500–7000 years ago (Galili *et al.*, 2019). At locations of tidal inlets, seawalls are combined with jetties to stabilize the sides of navigation channels and thereby minimize siltation. Once the seawall is constructed along a particular coast, this would mean surrendering to forcing functions like the wave characteristics and accepting that a beach is lost forever. Hence, great care is needed prior to the decision-making. A variety of seawalls have been adopted in practice. The most commonly prevalent ones are formed by natural rocks. At locations where large rocks are scarce, concrete blocks are employed. The slope of the structure significantly influences the gradual dissipation of the incident wave energy. A flatter slope leads to a more gradual dissipation. Additionally, whether the seaward side of the structure is permeable or impermeable also dictates the extent of dissipation. At locations of frequent flooding or those exposed to a high tidal range, a crown wall is constructed. A wide range of rubble mound seawall cross-sections have been constructed worldwide. In addition, there are other special types of seawalls like those with steps for effective dissipation which at the same time serve as a place for relaxation during the calm season. Where natural rocks of large sizes are scarce, gabions (galvanized wired or rope type) are adopted. A photograph of a permeable seawall is shown in Fig. 2.1. Conversely, with impermeable walls, dissipation occurs through wave propagation over slopes, steps or the shape of the seaside wall.

Seawalls should resist forces due to waves and should occupy less beach width and with as little lower crest elevation as possible so that there is

Fig. 2.1. Photograph of a permeable seawall.
Photo: Schüttrumpf.

very little or no wave overtopping. One of the possibilities is to reshape the seaward side of the structure to enhance the dissipation of wave energy and maintain minimum wave run-up.

The dissipation of energy in waves incident over the seawall is mostly due to the reduction in the kinetic energy while propagating over the front slope of the structure due to partial reflection. Though the vertical seawall requires a lower crest elevation, its high reflection characteristics promote scour near its toe, due to which its stability is in question. As the frequency of natural coastal hazards such as storm surges continues to rise, there is a need to elevate the crest height to mitigate overtopping and coastal flooding risks, especially in light of the sea level rise. The energy in the waves while propagating over sloping walls experiences the phenomenon of shoaling that promotes dissipation. However, Kirkgoz (1991) and Muller and Whittaker (1993) reported that impact pressures and forces on such walls are greater than those exerted on vertical walls. In addition, sloping walls occupy more space compared to vertical seawalls. Hence, seawall should offer multiple benefits and enhance the scenic beauty apart from the minimum beach width, with minimum or no wave overtopping and a lower crest elevation as discussed previously. A curved seaward surface could probably satisfy the above-stated criteria.

2.2.2 *Curved seafront seawalls*

Weber (1934) proposed a conceptual design of a curved seawall with a combination of a parabolic and a circular arch that leads to a smooth change in the direction of propagation from horizontal to vertical and vice versa to reduce wave-induced pressures. A new type of circular arc non-overtopping seawall was proposed by Murakami *et al.* (1996) who concluded that the critical crest elevation is much lower compared to that of a vertical seawall. The hydrodynamic characteristics of the curved seawall through numerical simulation using the finite volume method were discussed by Kamikubo *et al.* (2000), whose results were in good agreement with the results of Murakami *et al.* (1996). An example of a curved seawall from the island of Borkum in Germany is shown in Fig. 2.2. The structure has a wide radius that begins on the promenade and grows in size as the structure's height increases. Due to its deep arch, the Flare shaped seawall (FSS) outperformed the vertical upright seawall. Even today, the FSS is commonly used in Japan. However, if the offshore wave height is significant, this design creates a significant wall height, making it unsuitable for situations where a minimum wall height is needed. In addition, the FSS is more expensive than a traditional composite vertical seawall because of its curved form, which makes it more difficult to build. Furthermore, the FSS is difficult to

Fig. 2.2. Curved Seawall on the island of Borkum, Germany.
Photo: Schüttrumpf.

form in reinforced concrete because the slender sections at the crest and base do not have enough reinforcement cover, which is a flaw (Korenhau *et al.* (2003)).

A comprehensive experimental investigation was carried out by Anand *et al.* (2011) on four different shapes of seawall models (Fig. 2.3): VW: vertical wall, GS: Galveston seawall (USACE, 2006), FSS: Flare-shaped seawall (Kamikubo *et al.*, 2000) and CPS: circular cum parabolic seawall (Weber, 1934). Among the two re-entrant curved walls (FSS and CPS), it was found that since the FSS starts its curvature at an elevation much closer to the sloping bed, it exhibits more reflection than the CPS model.

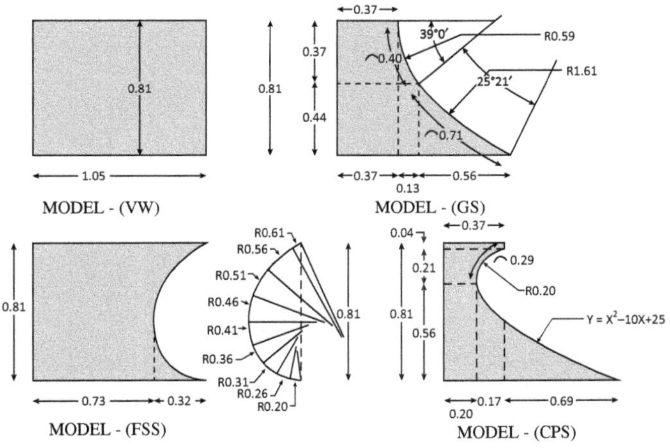

Fig. 2.3(a). The geometry of the curved front seawalls.
Note: VW: vertical wall, GS: Galveston seawall, FSS: Flare-shaped seawall and CPS: circular cum parabolic seawall.
Source: Anand *et al.* (2011).

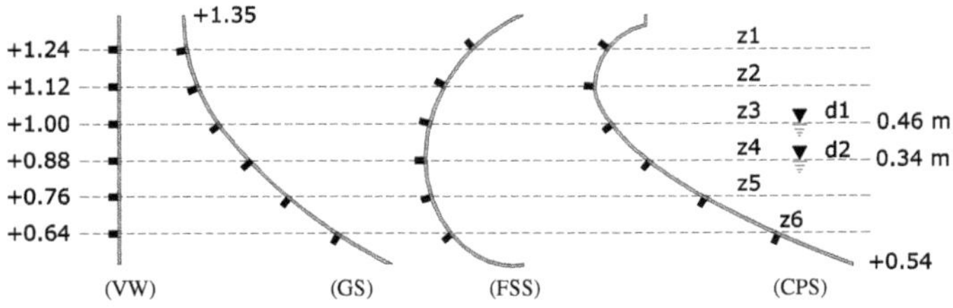

Fig. 2.3(b). Locations of the pressure ports.
Note: *d*1 and *d*2 are the water depths at the toe of the models for the four models.
Source: Anand *et al.* (2011).

The pressures and run-up on curved seawalls exposed to Cnoidal waves have been discussed in detail by Sundar and Anand (2010). In addition, Anand *et al.* (2011) proposed the following regression equations for the dynamic pressures exerted on the walls (Equation (2.1)) as well as for the wave run-up (Equation (2.2)):

Regression equations for dynamic pressures exerted on the four tested models:
p_{sh}: shoreward peak pressure, d: water depth, L: wavelength, H: wave height, z: pressure port location measured negative below the still water level. R_u: wave run-up

Model VW

$$\frac{p_{\text{sh}}}{H} = -1.327\frac{d}{L} - 0.077\frac{H}{d} + 0.091\frac{z}{d} + 0.792. \tag{2.1a}$$

Model GS

$$\frac{p_{\text{sh}}}{H} = -1.576\frac{d}{L} - 0.254\frac{H}{d} + 0.279\frac{z}{d} + 0.922. \tag{2.1b}$$

Model FSS

$$\frac{p_{\text{sh}}}{H} = -1.444\frac{d}{L} - 0.116\frac{H}{d} + 0.216\frac{z}{d} + 0.969. \tag{2.1c}$$

Model CPS

$$\frac{p_{\text{sh}}}{H} = -1.517\frac{d}{L} - 0.207\frac{H}{d} + 0.636\frac{z}{d} + 1.075. \tag{2.1d}$$

where, p_{sh}, H, d, L corresponds to shoreward pressure, wave height, water depth and wave length, respectively.
Regression equations for run-up on the four tested models:

Model VW

$$\frac{R_u}{H} = -2274\frac{d}{L} + 1084\frac{H}{d} + 1433. \tag{2.2a}$$

Model GS

$$\frac{R_u}{H} = -2577\frac{d}{L} + 1556\frac{H}{d} + 1591. \tag{2.2b}$$

Model FSS

$$\frac{R_u}{H} = -1.321\frac{d}{L} + 1.1074\frac{H}{d} + 1.268. \tag{2.2c}$$

Model CPS

$$\frac{R_u}{H} = -0519\frac{d}{L} + 1074\frac{H}{d} + 1422. \tag{2.2d}$$

where R_u is the run-up.

Furthermore, the results of the study of Anand *et al.* (2011) were that both the FSS and CPS yielded favourable results in reducing the crest elevation due to the nature of the recurve. While both the FSS and CPS models can be utilized effectively, the latter is generally less favoured due to its tendency to induce high scour near its toe because of higher velocities.

2.2.3 *Geo-synthetic seawalls*

Although a number of researchers have investigated the behaviour of geo-synthetic material in the marine environment, most of the studies are based on laboratory or field investigations. Applications of such products like geo-tubes, geo-containers and geo-bags for the construction of coastal structures are discussed in detail by Kim *et al.* (2018). Replacement of the core rock material in conventional structures like breakwaters with geo-synthetic products has been studied and reported by Saathoff *et al.* (2007), Jackson *et al.* (2012), Kudale *et al.* (2014), Sundar *et al.* (2009), Sundar (2010) and Oumeraci and Kortenhaus (2011). Dassanayake and Oumeraci (2012) conducted experimental investigations and claimed that, as the geo-synthetic materials have high flexibility and lesser specific gravity, their hydrodynamic response to loads in the marine environment would differ. A case study on the application of geo-synthetic products on a seawall on the coast of Upada, Andhra Pradesh, India (17.05°N and 82.31°E), has been detailed by Sundar and Sannasiraj (2019).

The coast of Shankarpur near Dhiga, West Bengal, India (21° 38′12″N and 87° 34′54″E), had been experiencing erosion and flooding due to waves and tides. The tides, with a range of 5.5 m, also contributed to sediment transport, estimated to be approximately 0.3 million cubic metres per year. Since large-sized natural rocks were not available near the study area, alternate coastal protection measures had to be considered, which resulted in a cross-section with three geo-tubes of 3 m diameter and length 30 m that were adopted as shown in Fig. 2.4(a).

In order to increase their resistance against ultraviolet rays, the geo-tubes were covered with polypropylene rope tubular gabion. In the design of a geo-synthetic seawall, the exposed toe protection stretched over a distance of about 40 m after its installation and sank due to local scour, after which the design was modified with an excavated toe that was constructed in mid-2005 as shown in Fig. 2.4(b), which resulted in greater stability. The installation was successful in withstanding the highest water level in

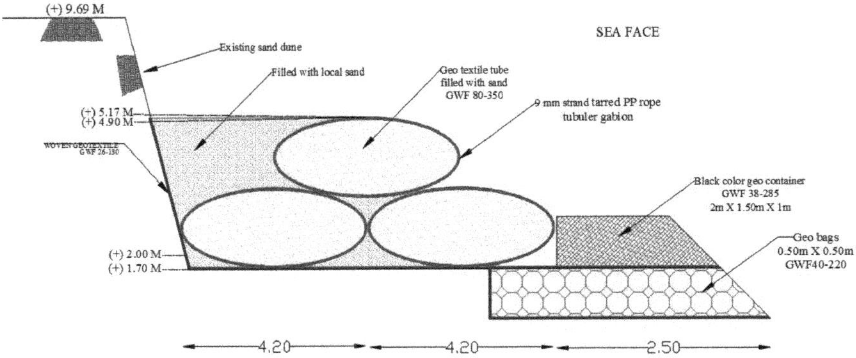

Fig. 2.4(a). Geo-tubes for coastal seawalls with exposed toe.
Source: Sundar *et al.* (2009).

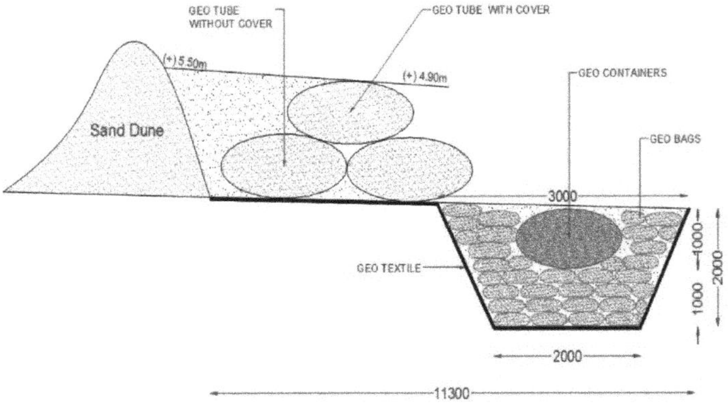

Fig. 2.4(b). Geo-tubes for coastal seawalls with excavated toe.
Source: Sundar *et al.* (2009).

100 years that occurred in September 2008. The protection measure was found to be successful as can be seen in Fig. 2.4(c).

The highest recorded tide level was 5.89 m, whereas the highest high-water level (HHWL) considered in the design of the geo-tube structure was only 3.84 m. The harsh climate eroded the sand surrounding the elevated natural ground and the geo-tubes, exacerbating the issue. Despite the inclusion of a robust UV stabilizer, acts of human vandalism exacerbated the situation, leading to damage to several geo-tubes,

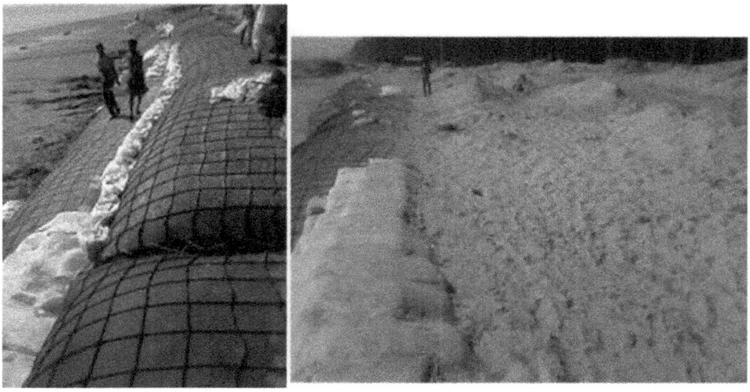

Fig. 2.4(c). Effect of the geo-tubes as a coastal protection measure.
Source: Sundar *et al.* (2009).

Fig. 2.4(d). Human vandalism — a cause for the failure of geo-tubes.
Source: Sundar *et al.* (2009).

as shown in Fig. 2.4(d). For comprehensive insights into this study, readers are encouraged to consult Sundar *et al.* (2009). Sukanya *et al.* (2022) designed, tested and implemented a geo-synthetic seawall along the coast of Pallana, Kerala (19°17′55.19″ N and 76°23′18.55″ E). The seawall was formed with geo-synthetic reinforced sand-filled retaining walls (GRWs). Additional components, such as non-woven geo-textile filter layers, geo-cells, geo-grids, geo-mattresses and geo-bags, were used to design the seawall cross-section that was completely made up of geo-synthetic materials and local beach sediments as shown in Figs. 2.5(a) and 2.5(b). Prior to

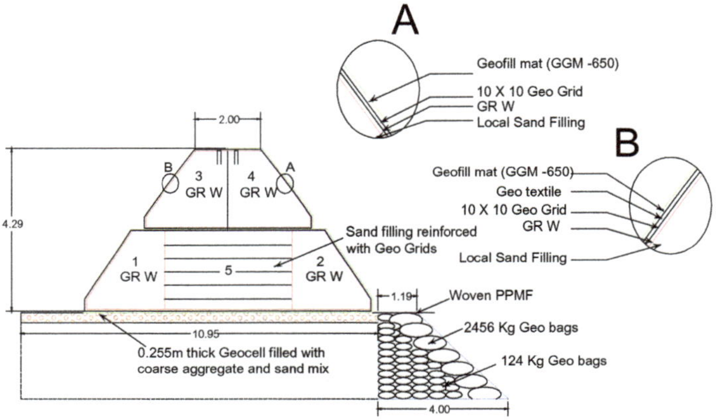

Fig. 2.5(a). Geo-synthetic seawall cross-section for Pallana, Kerala.

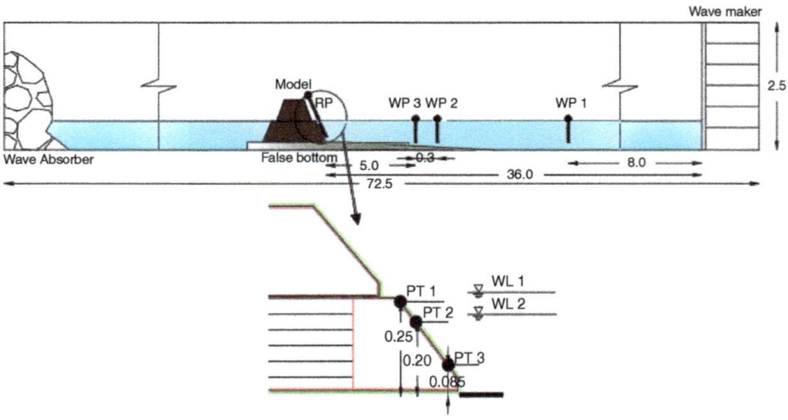

Fig. 2.5(b). The test setup in the wave flume.

implementation in the field, Sukanya *et al.* (2016) analyzed the hydrodynamic stability of the structure through physical model studies at a scale of 1:8.5 and the geotechnical stability through analytical model studies.

The hydrodynamic investigations of Sukanya *et al.* (2022) included regression equations with reflection coefficient, K_r, wave run-up, R_U, and dynamic pressures on the wall, p_S.

The regression equation to compute K_r for random waves is given in the following:

$$K_r = 0.135 + (0.06 * \xi_o) - (0.056 * h_b/d),$$
$$(\text{For } |h_b| < H_s \text{ and } 3 < \xi_o < 12). \tag{2.3}$$

The regression equation to compute the normalized relative run-up, R_s/H_s (R_s is the significant wave run-up elevation and H_s is the significant wave height) is given in the following:

$$R_s/H_s = 0.46 + (0.056 * \xi_o) + (0.11 * h_b/d),$$
$$(\text{For } |h_b| < H_s \text{ and } 3 < \xi_o < 14). \tag{2.4}$$

The regression equation to compute the peak shoreward pressures, p_s, was obtained as follows:

$$p_s/H_s = 1.94 - 3.76 * (d/L_o) - 5.66 * (H_s/L_o)$$
$$+0.43 * (z/d) - 1.91 * (h_b/d),$$
$$(\text{For } |h_b| < H_s \text{ and } 3 < \xi_o < 14). \tag{2.5}$$

Equation 2.5 depicts that p_s/H_s are observed to be a function of h_b/d, d/L_o, H_s/L_o and z/d.

In the above-mentioned equations, d is the water depth at the toe of the structure, h_b is the relative berm height (i.e. the distance between the still water level and the berm of the structure; h_b is negative if the berm is submerged within the water depth), L_o is the deep-water wavelength corresponding to the spectral peak period T_p, z is the vertical coordinate directed negative downwards with origin at the still water level, ξ_o is the surf similarity parameter computed with deep-water wavelength $= [\tan\alpha/(H_s/L_o)^{1/2}]$ and α is the structure slope. The structure constructed in 2015, as depicted in Fig. 2.5(c), has remained intact to date.

Fig. 2.5(c). Views of the geo-synthetic seawall. Photo: Sundar.

2.3 Groynes

2.3.1 *General*

A groyne field consists of a number of groynes, each of which is usually a trapezoidal section formed with natural rocks that serve as the coastal protection measure along a coast dominated by wave-induced longshore sediment transport. Transitional groyne fields (Fig. 2.6) are usually preferred in which the length of the groynes from almost the centre of the affected stretch to be protected stretch reduces on either side. This is to avoid the shifting of the erosion problem to the adjoining stretch of the coast. The longest groyne within a field will have to penetrate at least the surf width, a zone concentrated with sediments. This will promote trapping of the sediments and retain them in between the individual groynes. The general practice is to space the groynes at a distance of 1.5 times to three times their length. At locations where the longshore sediments move as both bed and suspended loads, the crest elevation of the groynes will have to pierce the high-water level, whereas at locations where the bed load is dominant, the crest elevation can be lowered or even submerged. Groynes built with these "technologies" were located on the seaward side of fortified cities in the 16–18th centuries, as shown in several ancient maps (Blaeu, 1652).

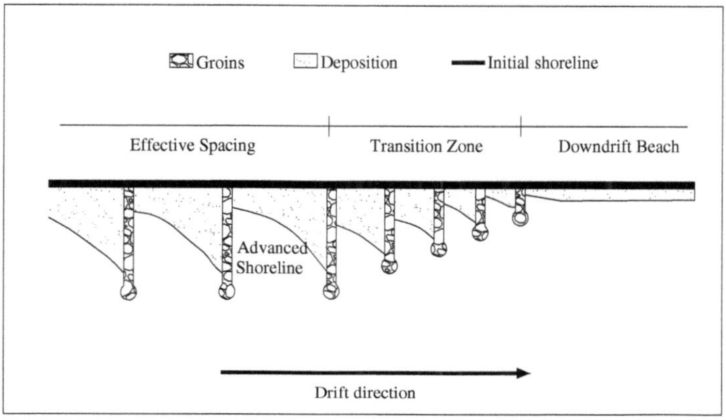

Fig. 2.6. Shoreline evolution due to transition groyne field.

Although there are different plane shapes of groynes, T-groynes are popular and are preferred along coasts that are exposed to both the cross-shore and longshore modes of sediment transport. Groynes not only protect the coast against erosion but also help in getting back lost sediment, which is an additional benefit.

The T-groynes are considered more efficient than straight groynes in holding the local shoreline position (Bruun, 1953) due to the following reasons:

- T-groynes reduce or laterally redirect the rip current that forms at the updrift side of the groyne, thereby reducing offshore losses and sand bypassing.
- The shore-parallel segments shelter the leeward local beach, promoting the accumulation of sediment.
- Accretion behind the structure also reduces the wave height that, in turn, will decrease the wave's steepness.
- As a consequence, with an approach to the stem, the waves will tend to transform from more erosional to more accretionary.

2.3.2 *Touch piles as groins*

A view of the shoreline evolution due to a groin is shown in Fig. 2.7(a). As conventional mound structures require a wider footprint on the seabed, slender structures with a comparable stability are preferred, and thus the composite groynes design was envisaged. Touch piles are made of continuously

Fig. 2.7(a). A view of an impermeable groyne.
Photo: Schüttrumpf.

fixed timber logs driven into tubular steel piles filled with concrete. The groynes are impermeable due to the closely driven steel tubular piles. The penetration depths for the steel piles are calculated considering the wave forces estimated using Goda's method (Goda, 2000). Touch piles are preferred for sub-surfaces characterized by over-consolidated clays, soils above the water table with cohesion and free-draining soils such as sand. The consecutively driven adjacent piles support both gradual and drastic changes in elevations and they are practically impermeable (see Figs. 2.7(b) and 2.7(c)). The size, spacing and depth of these piles can be conveniently designed as per the site requirements. The penetration depths of such piles are estimated using the method proposed by Goda (2000) for wave force estimation. Sankarbabu *et al.* (2018) and Juma *et al.* (2018) have detailed the benefits of touch pile technology, emphasizing its renewable nature and potential for sustainability with proper management. It is lauded for being relatively light in weight yet possessing a favourable

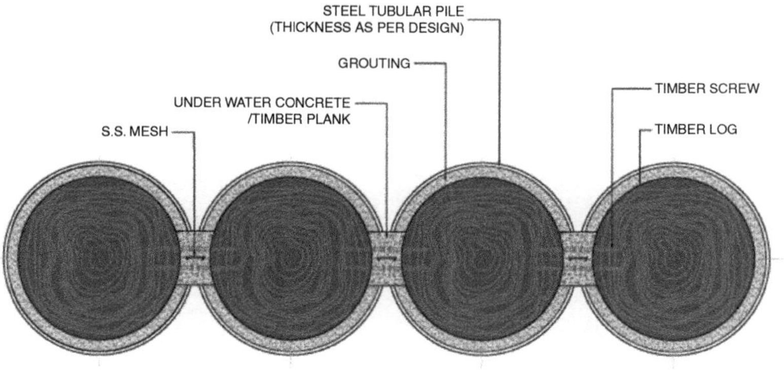

Fig. 2.7(b). Plan view of touch piles.
Source: Sankarbabu *et al.* (2018).

strength-to-weight ratio, boasting an appealing appearance and natural durability. Nonetheless, the expected drawbacks include its reliance on natural materials with inherent flaws and variability in properties, especially in terms of large sections, long lengths or high durability, which may only be available in limited quantities from specific species. It is only renewable over a relatively long timescale, making it difficult to demonstrate sustainability.

The primary reason for advocating for touch piles over conventional mound structures is that steel piles provide appreciable strength to the structure and it is rather practical to procure large quantities of steel and concrete rather than huge quantities of rubble stones. Additionally, these structures require minimum or no maintenance.

2.3.3 *Effectiveness as coastal protection*

Prior to 2004, the shoreline north of Chennai harbour situated along the southeast coast of India (13.0815°N, 80.2921°E) had been experiencing erosion over four decades over a stretch of about 9 km, with a width of about 300–400 m of beach being sacrificed to the Bay of Bengal due to erosion. Photographs of the damage are shown in Fig. 2.8. This erosion was due to the obstruction by the Chennai breakwaters to free passage of longshore sediments to a tune of about $0.8 \times 10^6\ \text{m}^3/\text{yr}$. To combat this problem, a groyne field of 10 groynes was constructed (from 13° 08′48.36″N, 80° 18′50.10″E to 13° 10′47.80″N, 80° 19′18.87″E). The effectiveness of the groynes has been detailed by Sundar (2005a,b), including the behaviour

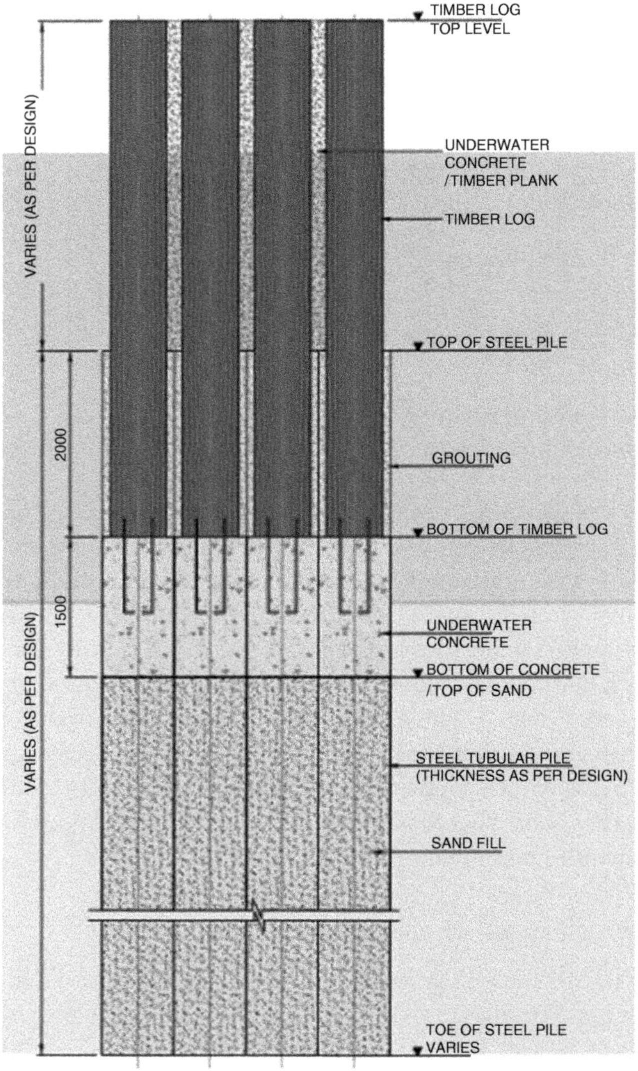

Fig. 2.7(c). Cross-section of typical touch piles.
Source: Sankarbabu *et al.* (2018).

of the shoreline before and after the 2004 tsunami. The groyne field effectively trapped the sediments, leading to recovery of the lost beach, images of which are shown in Figs. 2.9(a) and 2.9(b). Further north, in a densely populated region (13° 13′12″N and 80° 19′12″E) experiencing coastal flooding,

Fig. 2.8. Photographs of the erosion-affected stretch along the north of Chennai harbour, prior to 2004. Photo: Sundar.

Fig. 2.9(a). Effectiveness of the groyne field in recovering the lost beach **in the** north of the Chennai coastline (after 2004). Photo: Sundar.

a groyne field with a few T-groynes (Fig. 2.9(c)) has been serving its purpose as an effective mitigation measure. Sundar and Sannasiraj (2018) have reported in detail the success of a well-planned groyne field at about 40 stretches along the coast of Kerala, in the southwest of India. These examples prove the effectiveness of groyne fields as a coastal protection measure.

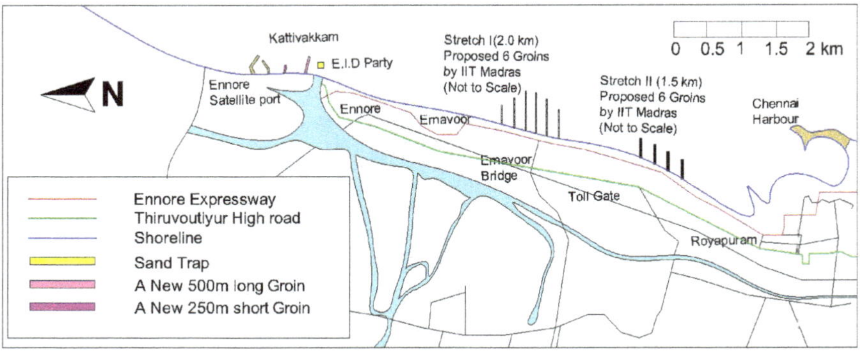

Fig. 2.9(b). Aerial view of the groyne field north of Chennai harbour (May 2020).

Fig. 2.9(c). Views of T-groynes as a protection measure against erosion and coastal flooding, north of Chennai harbour (May 2020).

2.4 Breakwaters

2.4.1 *General*

The main purpose of breakwaters is to help form artificial harbours, to protect an area in the sea against waves, to reduce dredging at the harbour entrance, to serve as a quay facility, to guide currents and to provide tranquillity conditions on its leeward side. The alignment of breakwaters depends on the prevailing wave climate (the height, periodicity and direction) with the wave direction being very important from a tranquillity point of view.

Breakwaters are broadly classified as mound type, vertical type, composite type, pile supported and special types. A mound-type breakwater does not produce substantial wave reflection, attributed to the incident wave energy breaking on a slope of rubble or concrete armour layer, which dissipates most of the incoming wave energy. Vertical breakwaters can either rest directly on the sea bed or be built on artificial rubble mound, known as composite breakwaters. On the other hand, vertical breakwaters reflect the incident wave energy as the breakwater is impervious/non-porous. Composite breakwaters are employed in locations where the tidal fluctuations are significant, so they function like mound types during the low tide and as vertical types during the high tide. Based on their location, breakwaters can be classified as detached, headland, near-shore, or shore-connected. They may be emergent, submerged below mean sea level (MSL), or floating. They may also differ in shape like the semi-circular breakwaters discussed earlier.

2.4.2 *Mound breakwaters*

2.4.2.1 *Concrete armour units*

The armour layer is the protective outer layer of the most commonly adopted mound-type breakwater, which is typically composed of rubble/quarry stones or plain unreinforced concrete units. For locations where the availability of large natural rocks is scarce, a wide variety of concrete armour units are developed for coastal and harbour engineering applications. The different available concrete armour units referred in CEM (2006) are illustrated in Fig. 2.10. These units should exhibit good interlocking ability and maintain the stability and durability of coastal structures over time. There has been a continuous effort to improve their shape as well. Usually, Hudson's formula (Equation (2.6)) is employed for assessing the weight of the armour stones, which is a function of the design wave height.

These armour units dictate numerous resultant hydrodynamic parameters such as reflection, transmission, run-up and overtopping when deployed in the field. Therefore, detailed parametric studies are necessitated prior to field implementation of any such novel armour unit design.

$$W = \frac{w_r H^3}{K_D (S_r - 1)^3 \cot\theta} \tag{2.6}$$

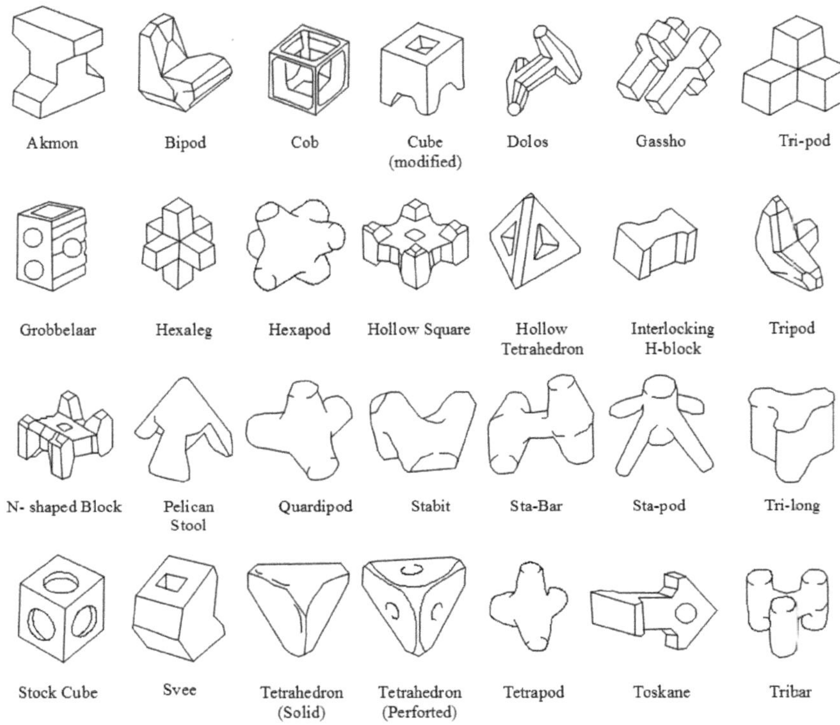

Fig. 2.10. Different types of concrete armour units.
Source: Sundar and Sannasiraj (2019).

where W is the weight of the individual armour unit (N), w_r is the unit weight of the armour unit in N/m^3, H is the design wave height (which could be the significant wave height), S_r is the specific gravity of the armour unit, W_w is the unit weight of water, θ is the angle of the structure's slope and K_D is the stability coefficient (depends on the type of the armour block, number of layers, n, type of placement and type of waves whether breaking or non-breaking).

2.4.2.2 *KOLOS*

KOLOS has been conceptualized by incorporating structural improvements to the existing armour unit, Dolos. The common failure mode for the Dolosse was found to be either at the shank due to torsion or at the fluke by bending-induced shear as claimed by Zwamborn *et al.* (1980). Higher structural stresses are largely due to the longer lever arm. The elongated shank length of the Dolosse was shrunk by 21.4% in the KOLOS. The saved material offsets the need for thicker legs on the KOLOS, which features a central shorter concrete member and two outer elongated concrete members connected on opposite sides of the central member. The outer members have parallel longitudinal axes extending normally to the longitudinal axis of the central member. These members have an octagonal cross-section and are designed so that their cross-sectional area decreases from the intermediate portion towards the opposite ends, as detailed by Chandramohan *et al.* (2012). The isometric view of the KOLOS is projected in Fig. 2.11. The initial field installation of these armour units, after a comprehensive laboratory examination to determine its stability coefficient, was adopted for the breakwaters of Krishnapatnam Port in Andhra Pradesh, India (14.2546° N, 80.1094° E) (see Fig. 2.12).

2.4.2.3 *Stability of KOLOS*

The stability coefficient, *KD* for the concrete armor units (CAUs) must be assessed and it is the main design parameter for the armour layer. Herein the assessment of *KD* for an innovative block, KOLOS has been discussed. The armour layer will subsequently expose the relatively lightweight underlayer and core to wave action, thus leading to doubts about the stability of the

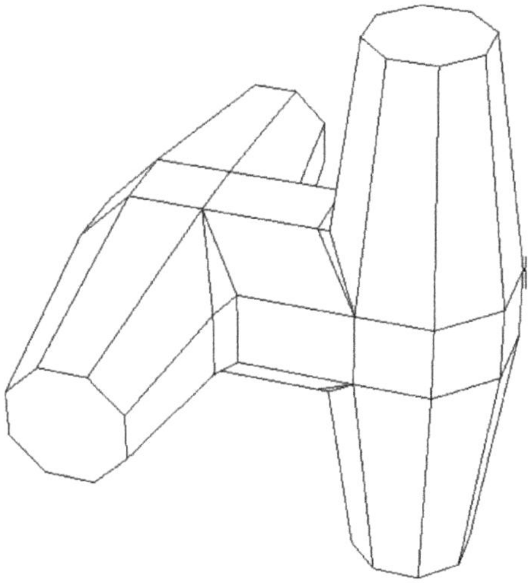

Fig. 2.11. An isometric view of Kolos.

Fig. 2.12. Field installation of KOLOS for the breakwaters of Krishnapatnam Port.
Photo: Sundar.

entire structure. The damage is thus related to the armour and defined as the displacement of an armour unit; according to CEM (2006), the term "displacement" is either the units sliding over the slope or being dislodged from the layer. The stability parameters of KOLOS units randomly placed in two layers were figured based on two-dimensional physical model tests with an adopted scale, under the action of at least 1000 regular as well as random waves. The scale ratio adopted was 1:21.7. The examination of the stability coefficient involved the test criteria of Owen and Allsop (1983) and changing the slope for both the head and trunk sections of the breakwater armoured with the KOLOS. The K_D value was evaluated using the well-known Hudson's formula both for the trunk and head sections.

The hydraulic stability of breakwater armour units is determined by wave height, unit size, specific density (of armour unit and seawater) and the shape of the unit. The stability coefficient (K_D) can be derived using Hudson's formula, written in terms of the stability number (N_s) as follows:

$$N_s = \frac{\boldsymbol{H}}{\Delta \boldsymbol{D_n}} = (\boldsymbol{K_D} \cot \alpha)^{1/3} \tag{2.7}$$

where Δ = relative density = $[(\rho_c/\rho_w) - 1]$ and D_n = the nominal diameter of the armour unit.

Prior to obtaining the value of K_D for the KOLOS, the damage level was fixed in line with the proposed guidelines on the classification of damage levels suggested by d'Angremond and van Roode (2004). For an armour slope of 1:1.5, no repair is needed for initial damage (no repair of armour) of up to 0.5% for the DOLOS, irrespective of the weight of the individual unit of the armour layer, and as the KOLOS is an improved version of the DOLOS, the same damage level has been incorporated in the stability study. The initial part of the study focused on the establishment of Hudson's stability coefficient of KOLOS, where the test sections had been subjected to the maximum wave height they can withstand without being damaged extensively. For additional details on the precise damage assessment aspects, please read Chapter 5 of Sundar and Sannasiraj (2019).

Detailed experimental investigations were conducted for breakwaters with the KOLOS armour layer (see a typical cross-section of a trunk in Fig. 2.13) over a model scale of 21.7 (see Fig. 2.14 for laboratory installation). The following results were obtained. Hudson's stability number was established for the KOLOS in order to compare it with other armour units. As shown in Table 2.1, the KOLOS can have a K_D value of up to 32, which is quite high, with a maximum damage percentage of 0.5%. Using such a high value for the design can be justified because random placement and double-layer configuration add to the stability of the armour and even at a damage percentage of 0.5%, further infliction of damage ceases to exist.

The stability number is one of the most important parameters that reveals the relationship between the structure and the wave. From the above-mentioned study, based on the wave height corresponding to the damage percentage considered, stability number Ns is obtained. The stability number corresponding to zero damage is 3.64 and for damage of 0.5% is 3.28. Although, the trunk and the head sections with slopes, 1:1.5, 1:2 and 1:2.5 were tested to establish the KD for KOLOS, the variation of KD with wave height for the steeper slope of the trunk and head section is only projected in Fig 2.15 and Fig 2.16, respectively. The recommended values of K_D to be adopted for the KOLOS are given in Table 2.1.

2.4.2.4 *Run-up and overtopping for KOLOS*

The acceptable rate of overtopping discharge over the crest depends on the stability of the leeward side armour, the ability of the crest to withstand high-velocity flows over the crest (CIRIA, 2007) and most importantly the level of tranquillity required for the harbour. Overtopping discharges can be reduced either by increasing the crest level beyond the maximum run-up height or by the construction of a concrete crown wall on the crest. Detailed information on wave run-up and overtopping over slopes formed by a variety of armour blocks has been provided by Sundar and Sannasiraj (2019);

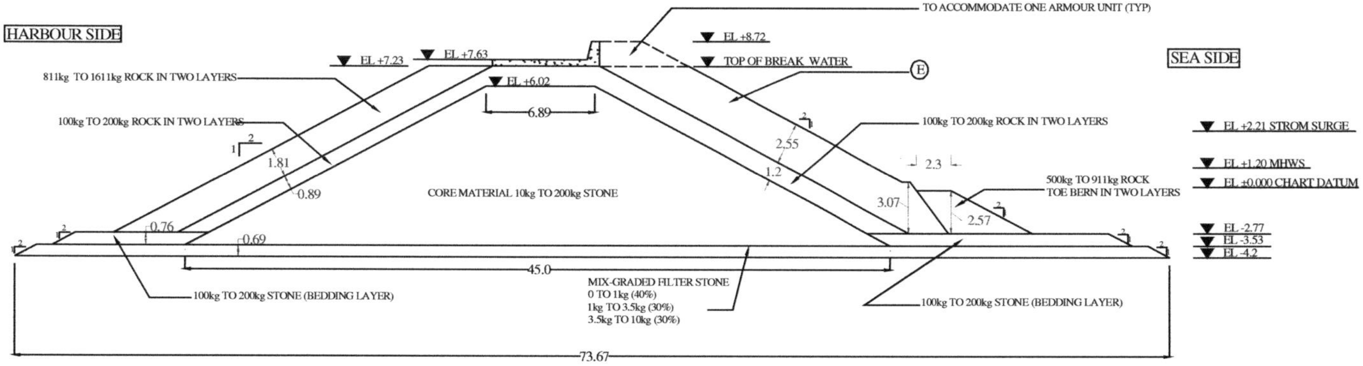

Fig. 2.13. Typical cross-section of a trunk with the KOLOS armour.
Source: Sundar and Sannasiraj (2019).

Fig. 2.14. Laboratory installation of a KOLOS armour layer.
Source: Sundar and Sannasiraj (2019).

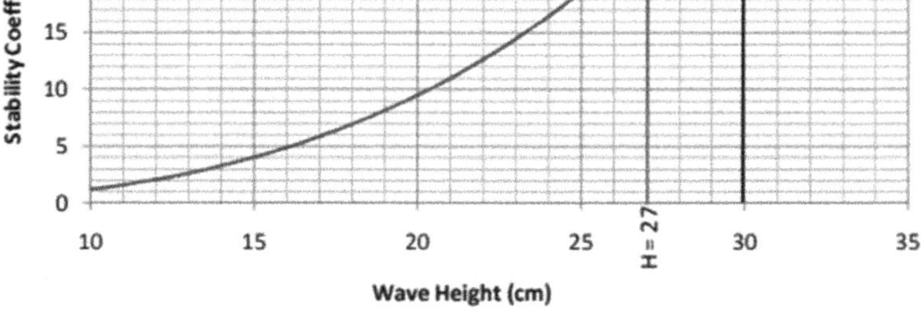

Fig. 2.15. Wave height vs. Hudson's stability coefficient graph (trunk slope: 1 in 1.5).

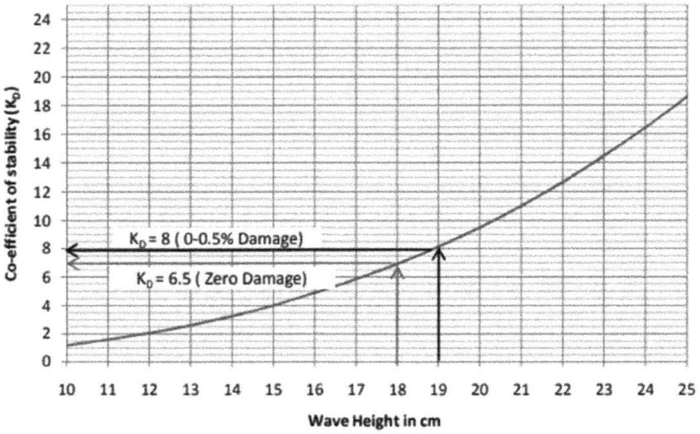

Fig. 2.16. Wave height vs. coefficient of stability (K_D) graph (head slope: 1:1.5).

Table 2.1. Recommended values of K_D for KOLOS (non-breaking waves).

	Hudson's stability coefficient for trunk (K_D)	
Slope	Zero damage	Up to 0.5% damage
1:1.5	23	32
1:2.0	19.5	27
1:2.5	14	23
	Hudson's stability coefficient for head (K_D)	
1:1.5	6.5	8
1:2.0	4.5	5
1:2.5	3	3.5

herein, the empirical relationships for the two parameters for KOLOS derived through a well-controlled experimental programme are presented.

The tests with KOLOS and natural rubble armoured breakwaters were carried out in four water depths, d of 50 cm, 55 cm, 60 cm and 65 cm with freeboards (Rc) as 34.5 cm, 29.5 cm, 24.5 cm and 19.5 cm, respectively. For complete details, refer to Arunjith *et al.* (2013a, 2013b). The *run-up* ($R_{u2\%}$) and overtopping discharge (q) were from the measurements. The variation of the relative run-up levels ($R_{u2\%}$) with the surf similarity parameter (ξ_{op}), considering the deep-water peak period (ξ_{op}), is illustrated for KOLOS in Fig. 2.17. This figure also superposes the relations for wave run-up over natural rubble (CEM, 2006) and Core-Loc (Melito and Melby, 2002) with the present measurements. The empirical relations for wave run-up over the KOLOS as armour units are estimated as follows:

$$\frac{R_{u2\%}}{H_s} = 0.72\xi_{\text{op}}^{0.56} \quad \text{for } \phi_{\text{op}} < 3.1, \tag{2.8}$$

$$\frac{R_{u2\%}}{H_s} = 0.72\xi_{\text{op}}^{0.56} \quad \text{for } \xi_{\text{op}} > 3.1. \tag{2.9}$$

The empirical relations for the estimation of *wave overtopping* over the KOLOS armoured slope are obtained as follows:

$$\frac{q}{\sqrt{gH^3}} = \frac{1}{245}e^{-2.65\frac{R_c}{H_s}}. \tag{2.10}$$

The wave run-up over the KOLOS is in general found to be less than that over armour units like rubble stones and the Dolosse. The KOLOS is found to perform better against overtopping compared with rubble stones.

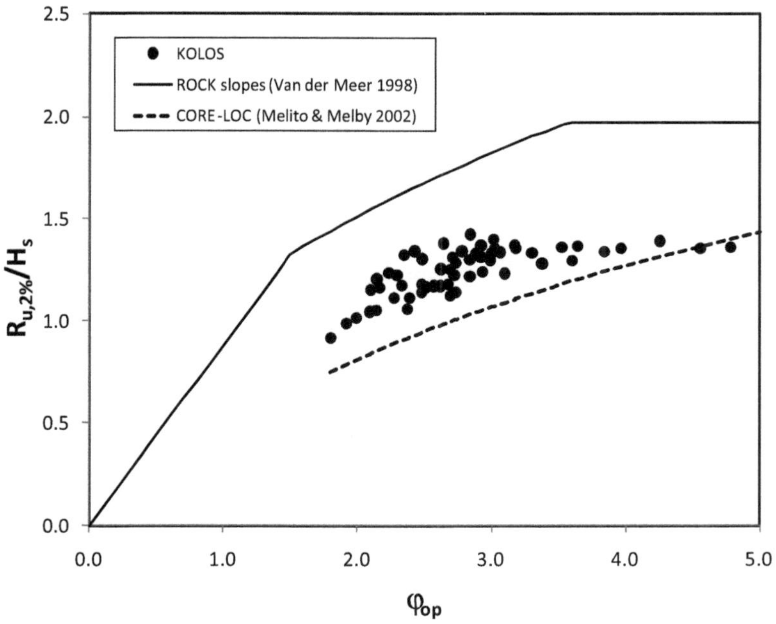

Fig. 2.17. Comparison of wave run-up of the KOLOS with different armour units.

2.4.3 *Piled breakwaters*

The demand for handling deep-draft vessels has been ever-increasing, forcing a search for alternative options for berthing vessels in deeper waters in which maintenance dredging, a persistent problem involving huge cost and risk, can be avoided. A rubble mound consisting of layers of graded stone with large concrete blocks as an armour layer is uneconomical for deeper waters. Further, in offshore locations with severe wave climates, the construction of rubble mound breakwaters is not practical as the stones are easily carried away by the waves. Concrete caisson-type breakwaters have inherent problems like scour and instability because of the excessive reflection of waves from their vertical front faces. The basic concept of a pile-supported breakwater is to serve as an impervious barrier near the free surface, where the water particle movements are greater, while leaving the lower region unobstructed. Construction of a pile-supported breakwater will allow the free passage of currents and allow the wave-induced littoral drift to pass through without leading to the instability of the shoreline that occurs in the case of shore-connected breakwaters. Further, at locations with soft soil and no hard stratum within a reasonable depth, a pile-supported breakwater can be an alternative.

The efficiency of such structures depends on the depth of immersion, its type, whether the skirt is permeable or impermeable, and the incident wave characteristics. Closely spaced piles and submerged skirts have been tried as alternatives for breakwaters in the past. The hydrodynamic characteristics of pile breakwaters consisting of just closely spaced piles have been discussed in detail by Karthik Ramnarayan *et al.* (2020a). The hydrodynamic performance characteristics of a quadrant front-face, pile-supported breakwater are such that the relative spacing between the piles (space between piles/pile diameter) does not affect the coefficient of reflection as mentioned by Sundar and Subba Rao (2003) and Suh *et al.* (2006). The performance of a pile breakwater consisting of closely spaced piles alone is not very encouraging and driving in very close proximity to the piles is very difficult. However, if we resort to a pile-supported breakwater, PSB with a skirt, deep-draft vessels can approach closer and berth on its leeward side for loading and unloading operations. In fact, the seaward side of the skirt can also be converted into an oscillating water column device for the extraction of energy from ocean waves, the details of which will be discussed later in this chapter. Photographs of a spaced pile breakwater and a sheet pile-supported breakwater are shown in Figs. 2.18(a) and 2.18(b), respectively.

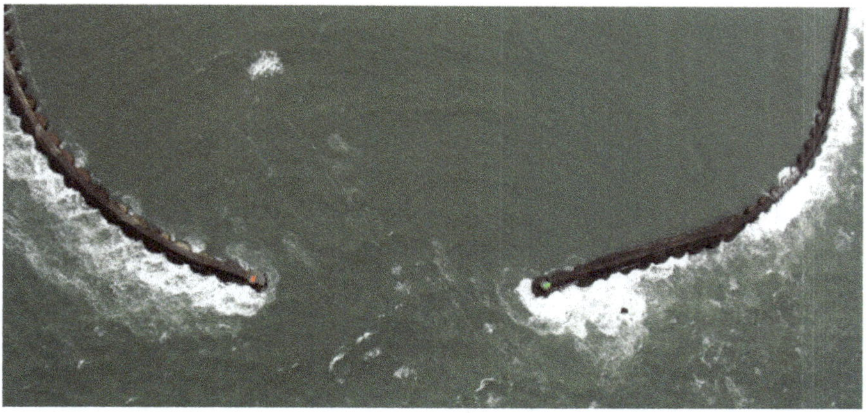

Fig. 2.18(a). Closely spaced pile breakwater.
Source: Karthik Ramnarayan *et al.* (2020a).

Fig. 2.18(b). Sheet pile-supported breakwater.
Photo: Schüttrumpf.

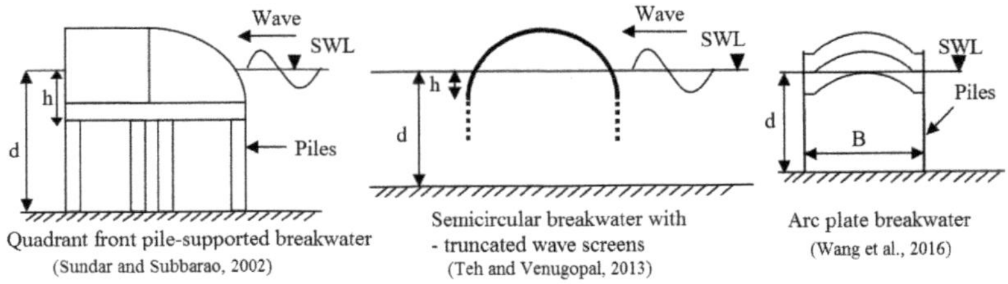

Fig. 2.19. Curved-front-type breakwaters.
Source: Karthik Ramnarayan *et al.* (2020a).

Different configurations for the seaward side face of pile-supported breakwaters have been investigated in the past by Takahashi (1996). A curved front or alternatively a porous barrier on the seaward side can reduce reflection. Interestingly, the curved-front breakwater models investigated to date are of a convex shape on the seaward side, while concave profiles have also received attention in the recent past. The different curved-front-type breakwaters that have been reported to experience less reflection are presented in Fig. 2.19.

The hydrodynamic coefficients of different types of pile-supported breakwaters are summarized in Table 2.2.

The curved-front-type pile-supported breakwaters discussed in Section 2.2.2 have convex profiles facing the sea. Based on the merits of these shapes of the seaward side face as discussed by Anand *et al.* (2011), Karthik Ramnarayan *et al.* (2020a, 2020b) conducted laboratory tests using a 1:25 scale model to evaluate pile-supported superstructures with a GS profile (referred to as GS-PSB) and CPS profile (referred to as CPS-PSB). Given that the primary function of the piles is to support the topside structure rather than dissipate the energy, a wide spacing between the piles was maintained. Schematic views of the GS-PSB and CPS-PSB models are projected in Fig. 2.20.

Two stainless steel wires of 1 mm Ø spaced at 10 mm on the concave profiles measured the wave run-up. Five pressure sensors were fixed on the concave profiles of both models to measure the wave-induced dynamic pressures. The elevation of the pressure sensors for both the models was maintained as shown in Fig. 2.21.

The non-dimensional parameters derived from rigorous physical modeling measurements were subjected to least squares multiple regression analysis based on the method of least squares. The transmission coefficient K_t and reflection coefficient K_r are expressed as a function of dimensionless

Table 2.2. Hydrodynamic performance of pile-supported breakwaters.

Literature	Breakwater type	Optimal water depth (m)	Hydrodynamic Coefficients K_t	Hydrodynamic Coefficients K_r
	Vertical front type			
Neelamani and Vedagiri (2002)	Twin vertical barrier	10–25	0.05–0.80	0.40–0.90
Neelamani and Rajendran (2002)	T-type breakwater	10–25	0.15–0.67	0.65–0.95
Rageh *et al.* (2009)	Caisson supported on piles	5–10	0.03–0.75	0.45–0.99
Rageh and Koraim (2010)	Vertical wall with horizontal slots	10–15	0.20–0.81	0.15–0.75
Koraim *et al.* (2011)	Double row of vertical walls with horizontal slots	10–15	0.05–0.70	0.20–0.99
	Porous type			
Nakamura *et al.* (2001)	Inclined plate array onseaward side	6.5–9.5	0.05–0.65	0.05–0.70
Brossard *et al.* (2003)	Semi-immersed breakwater with absorbing caisson		0.05–0.45	0.05–0.65
Wang *et al.* (2006)	Multiple layers of horizontal plates	12	0.10–0.95	0.10–0.60
Koraim *et al.* (2014)	C-shaped bars supported on piles	2.5–5	0.40–0.99	0.03–0.70
	Curved-front type			
Sundar and Subba Rao (2003)	Quadrant front-face breakwater	20–25	0.10–0.50	0.25–0.85
Teh and Venugopal (2013)	Semi-circular breakwater with truncated wave screens	20	0.05–0.90	0.12–0.48
Wang *et al.* (2016)	Arc plate breakwater	12	0.05–0.90	0.05–0.40

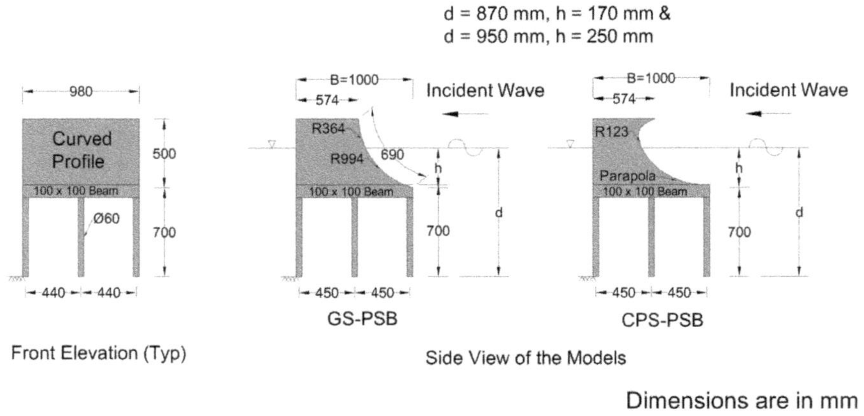

Fig. 2.20. Sectional view of the GS-PSB and CPS-PSB models.

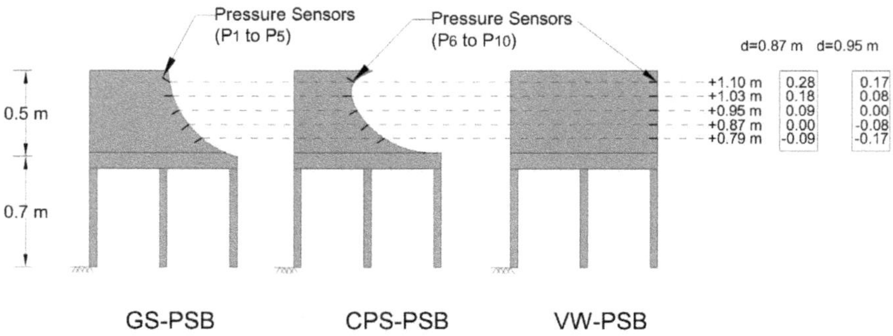

Fig. 2.21. Locations of the pressure sensors.
Source: Karthik Ramnarayan *et al.* (2020a).

parameters, relative breakwater width B/L and relative depth of submergence h/d. The non-dimensional pressure $P_{\text{sh}}/\gamma H_i$ is expressed as a function of d/L and z/d. The relative run-up R_u/H_i is expressed as a function of d/L and h/d. Herein, P_{sh} is the shoreward peak pressure, γ is the specific density of water, H_i is the wave height, d is the water depth, R_u is the wave run-up, L is the wavelength, z is the pressure port location negative below still water level, h is the depth of submergence of the curved-face superstructure and B is the width of the breakwater. The equations for the aforementioned different parameters for the GS-PSB and CPS-PSB as presented in Equations (2.11) to (2.14) display better performance than VW-PSB as discussed by Karthik Ramnarayan *et al.* (2020b).

For transmission

$$\textbf{GS-PSB: } K_t = 0.003 \times (B/L)^{-1.68} \times (h/d)^{-1.24}. \tag{2.11a}$$

$$\textbf{CPS-PSB: } K_t = 0.005 \times (B/L)^{-1.58} \times (h/d)^{-1.11}. \tag{2.11b}$$

For reflection

$$\textbf{GS-PSB: } K_r = 0.91(B/L) + 2.01(h/d) - 0.11. \tag{2.12a}$$

$$\textbf{CPS-PSB: } K_r = 0.98(B/L) + 1.69(h/d) - 0.29. \tag{2.12b}$$

For wave-induced pressures

$$\textbf{GS-PSB: } P_{\text{sh}}/\gamma H_i = -1.61(d/L) + 2.31(z/d) + 1.43;\ (\text{For } z \leq 0). \tag{2.13a}$$

$$\textbf{CPS-PSB: } P_{\text{sh}}/\gamma H_i = -1.65(d/L) + 3.20(z/d) + 1.51;\ (\text{For } z \leq 0). \tag{2.13b}$$

For run-up

$$\textbf{GS-PSB: } R_u/H_i = 1.54(d/L) + 0.67(h/d) + 0.75. \tag{2.14a}$$

$$\textbf{CPS-PSB: } R_u/H_i = 1.01(d/L) - 3.71(h/d) + 2.01. \tag{2.14b}$$

The above-mentioned equations are valid for a d/L range of 0.13–0.61 (B/L of 0.14–0.64). The energy dissipation offered by the curved profiles does not affect the wave transmission past the structure in deep-water conditions, but results in lesser transmission in intermediate water depths. The energy dissipation affects wave reflection more than transmission. Wave reflection is the least for CPS-PSB which is more receptive to the incoming waves due to the flat slope at lower elevations. However, wave-induced pressures are greatest for CPS-PSB followed by GS-PSB and VW-PSB. At lower elevations, the induced pressures due to short waves are the least for CPS-PSB due to the flatness of the profile. Though the run-up over CPS-PSB is high, there is no overtopping due to the presence of the recurve. Overtopping occurred over GS-PSB, thus necessitating an additional energy-dissipating mechanism on the seaward side. For the complete discussion, the reader can refer to Karthik Ramnarayan *et al.* (2020a, 2020b).

Jemi Jeya *et al.* (2021) investigated the hydrodynamic characteristics of vertical face pile-supported breakwaters (VPSBs) and quadrant face pile-supported breakwaters (QPSBs) under oblique waves and reported that the K_r for both models decreases with an increase in the angle of wave attack. The presence of a convex shape on the seaside face of the QPSB model reduces K_r significantly. For any angle of the wave attack, K_r increases with an increase in the water depth for both models. In summary, it was reported that the QPSB satisfies the requirement of less reflection and transmission of incident wave energy and less pressure exerted on the superstructure compared to the VPSB. However, the run-up alone was found to be higher and hence it was claimed that the QPSB could be accommodated in locations of permissible occasional overtopping or higher crest elevation could be included for the QPSB.

2.5 Design Philosophy

2.5.1 *General*

The requirements that fulfil the main design criteria are the life of the seawall, probability of risk of failure, acceptable limit, and the required rehabilitation method and time, which are governed by the characteristics of waves and tides. The design process involves the serviceability of the structure and duration in years for how long each component of the structure can serve its purpose. The life of the structure needs careful consideration if found to be less as the cost towards its maintenance is bound to increase significantly. On the other hand, if the structure is designed to sustain for a longer period, it will require more investment. Hence, a compromise is essential by taking into consideration both the above-stated aspects. The availability and sourcing distance of materials, type of labor, and the significance of the problem play crucial roles in determining costs. For instance, if total avoidance of over-topping is required, the crest elevation along the entire coast must be increased, leading to higher costs. Conversely, if occasional over-topping is acceptable, the crest elevation can be reduced, thus lowering the costs.

The design life of a coastal protection measure is typically 50 years, but its status should be checked periodically, particularly after a severe coastal hazard. Its rehabilitation should be assessed prior to the planning process and should be implemented immediately. This will again depend on the materials adopted for the construction of the coastal protection measure. As per the Code of Practice for Maritime Structures, the minimum design life for shore protection works and breakwaters is 60 years and for sea defence works is 100 years. The typical design life and degrees of protection (CIRIA, 2007) are mentioned in Table 2.3.

2.5.2 *Risk of exceeding design criteria*

An integrated approach in designing structures exposed to phenomena such as winds, waves, tides, and surges involves selecting extreme values that

Table 2.3. Design life and level of protection.

Type of use of seawall structure	Design life (years)	Level of protection (return period of event in years)
Temporary and short-term measures	1–20	5–50
Majority of coast protection or sea defence walls	30–70	50–100
Occasional works subject to specific greater risk	30–70	100
Special structures of high capital cost and/or high risk (including conventional power stations and barrier works)	—	Up to 1000
Nuclear power stations	—	10000 and greater

Source: CIRIA (2007).

are very rare. By choosing such rare extreme values, designers ensure their structures can withstand even the most severe conditions. This approach accounts for the inherent uncertainties associated with natural phenomena and provides a safety margin against unforeseen events. However, determining absolute maximum values for these phenomena is often challenging, except in certain cases, such as when the wave height is limited by depth.

When a realistic maximum cannot be established, the solution is to select extremely rare extreme values. Consequently, there exists a finite probability that the design conditions will surpass these values over the lifespan of the structure. To simplify matters, this likelihood of exceeding the design conditions is typically referred to as the "Return Period." On average, an event with a return period of N years is anticipated to occur once every N years. The appropriate return period should be chosen based on the potential consequences of exceeding the design conditions, such as overtopping and subsequent flooding. This is because the return period represents the average time interval between occurrences of an extreme event that is expected to be exceeded on average once within that time frame.

On the other hand, design life refers to the intended lifespan or durability of a structure or system. If the return period and design life are set to be the same, the extreme event is expected to occur once within the design life. However, due to statistical probabilities, there is a 63% chance that this extreme event will occur before the design life ends. This discrepancy arises because return periods are based on a statistical analysis of historical data and assumptions about future conditions. However, design life considers factors such as maintenance, degradation over time and changing environmental conditions. By setting the design life time shorter than the return period, as shown in the following, engineers can ensure that structures are safe:

$$R = 1 - \left(1 - \frac{1}{nT_R}\right)^{Ln}, \tag{2.15}$$

$$T_R = \frac{1}{n}\left(\frac{1}{1-(1-R)^{1/Ln}}\right), \tag{2.16}$$

where R = risk of exceedance, L = design life, T_R = return period and n = number of events per year.

It is crucial to consider the specific design aspects and their potential consequences when determining the return period. One can ensure that appropriate safety measures are implemented by tailoring the return period to match the severity of potential damages. In the case of a seawall, which serves as a protective barrier against coastal flooding, overtopping may result in short-term damage such as erosion or minor flooding. While this can cause inconvenience and require maintenance, it may not immediately threaten life or property. Therefore, a relatively shorter return period could be assigned to this design aspect.

On the other hand, if the body of the seawall was to be completely destroyed during an extreme event, it could lead to catastrophic consequences. The loss of protection would expose nearby areas to severe flooding and endanger lives and infrastructure. To prevent such devastating outcomes, a longer return period should be considered for designing against complete destruction by recognizing that different design aspects have varying levels of vulnerability and potential impacts.

2.5.3 *Problem of Risk and return period*

A seawall with a design life of 30 years is built as a protective barrier against coastal flooding. A 15% acceptable risk of exceedance can be considered during the life of the seawall.

If considering extreme high waters with $n = 307$ number of tides/year, the approriate return period can be calculated as follows:

$$\begin{aligned} T_R &= \frac{1}{n}\left(\frac{1}{1-(1-R)^{1/Ln}}\right), \\ &= \frac{1}{307}\left(\frac{1}{1-(1-0.15)^{1/30\times 307}}\right), \\ T_R &= 185 \text{ years.} \end{aligned}$$

Another example considers 30 years of the design life of a seawall built against a design wave condition with a return period of 100 years.

For a wave condition that lasts three hours, the number of events per year (n) can be calculated as

$$n = \frac{365 \times 24}{3} = 2920.$$

The risk of exceedance can be calculated as

$$R = 1 - \left(1 - \frac{1}{2920 \times 100}\right)^{30 \times 2920},$$

$R = 0.26$, that is to say, there is a 26% chance that design conditions will already have been exceeded by the time the structure reaches the end of its design life.

2.5.4 *Probabilistic design*

The probabilistic approach in the design of seawalls recognizes that the forces acting on them, such as waves, currents and water levels, are inherently uncertain and variable. Instead of assuming fixed values for these factors, one can evaluate them probabilistically by considering their statistical distribution and likelihood of occurrence. By adopting a probabilistic perspective, one can better account for the inherent variability in environmental conditions. Once the probabilities are established, deterministic methods can be applied to design the seawall structure. Then, the complementary strengths and dimensions of the seawall, like the strengths of materials and the thickness of the armour layers, can be calculated. These deterministic calculations ensure that the seawall can withstand or accommodate the anticipated forces within an acceptable level of risk. The combination of probabilistic evaluation and deterministic design ensures that seawalls are robust enough to handle any eventuality.

The development of advanced numerical models allows engineers to simulate various scenarios and assess the performance of different design alternatives. These models consider uncertainties in material properties, environmental conditions and other influential variables to represent the seawall's response under different loading conditions accurately. By incorporating probabilistic analysis into the design process, one can better understand potential outcomes and account for uncertainties. This approach leads to a more robust design that can withstand a wider range of extreme events and ensures long-term structural integrity. However, despite these advancements, full probabilistic design remains feasible only for major seawall projects.

References

Anand, K.V., Sundar, V. and Sannasiraj, S.A. (2011). Hydrodynamic characteristics of curved-front seawall models compared with vertical seawall under regular waves. *Journal of Coastal Research*, 27(6), 1103–1112.

Arunjith, A., Sannasiraj, S.A. and Sundar, V. (2013a). Wave overtopping over crown walls and run-up on rubble mound breakwaters with KOLOS armour under random waves. *International Journal of Ocean and Climate Systems*, 4(2), 125–132.

Arunjith, A., Sannasiraj, S.A. and Sundar, V. (2013b). Pressures on the crown wall of breakwater formed by new armor blocks 'KOLOS' due to regular waves. *ASCE's Journal of Waterway, Port, Coastal, and Ocean Engineering*, 518–526. https://doi.org/10.1061/(asce)ww.1943-5460.0000203.

Bilskie, M.V., Hagen, S.C., Medeiros, S.C. and Passeri, D.L. (2014). Dynamics of sea level rise and coastal flooding on a changing landscape. *Geophysical Research Letters*, 41, 927–934.

Blaeu, J. (1652). Vlissingen in Toonneel der Steden van de Vereenighde Nederlanden. Met HareBeschrijvingen. http://objects.library.uu.nl/reader/index.php?obj=1874-237713&lan=en#page//52/21/31/52213138426633615601451169782851892476.jpg/mode/1up (Accessed on 29 June 2018).

Brody, S.D., Highfield, W.E. and Kang, J.E. (2011). *Rising Waters: The Causes and Consequences of Flooding in the United States.* Cambridge: Cambridge University Press.

Brossard, J., Druaux, A., Marin, F. and Aoul, E.H. (2003). Fixed absorbing semi-immersed breakwater. *Coastal Engineering*, 49, 25–41.

Bruun, P. (1953). Coastal Protection. The Dock and Harbour Authority, Vol. XXXIV, November-December 1953.

CEM. (2006). Coastal Engineering Manual. U.S. Army Coastal Engineering Research Centre (Part VI).

Chandramohan, P.V., Sundar, V., Sannasiraj, S.A. and Arunjith, A. (2012). Development of 'KOLOS' armor block and its hydrodynamic performance. In *Proceedings of 8th International Conference on Coastal and Port Engineering in Developing Countries (PIANC-COPEDEC VIII)*, Chennai, India, February, pp. 1344–1354.

CIRIA/CUR/CEFMEF. (2007). The Rock Manual. The Use of Rock in Hydraulic Engineering, 2nd edn., p. 1200.

D'Angremond, K. and van Roode, F.C. (2004). *Breakwaters and Closure Dams.* London: SPON Press.

Dassanayake, D.T. and Oumeraci, H. (2012). Engineering properties of geotextile sand containers and their effect on hydraulic stability and damage development of low-crested/submerged structures. *The International Journal of Ocean and Climate Systems*, 3(3), 135–150. DOI: 10.1260/1759-3131.3.3.135.

Galili, E., Benjamin, J., Eshed, V., Rosen, B., McCarthy, J.K. and Horwitz, L.K. (2019). A submerged 7000-year-old village and seawall demonstrate earliest known coastal defence against sea-level rise. *PLoS One*, 14(12), e0222560. DOI: 10.1371/journal.pone.0222560.

Goda, Y. (2000). *Random Seas and Design of Maritime Structures.* World Scientific, Singapore. http://dx.doi.org/10.1142/3587.

Jackson, L.A., Tomlinson, R., Corbett, B. and Strauss, D. (2012). Long-term performance of a submerged coastal control structure: A case study of the Narrow neck multi-functional artificial reef. *Coastal Engineering Proceedings,* 1, 54.

Jemi Jeya, T.J., Sriram, V. and Sundar, V. (2021). Hydrodynamic characteristics of vertical and quadrant face pile supported breakwater under oblique waves. *Journal of Engineering for the Maritime Environment.* https://doi.org/10.1177/14750902211031353.

Juma, I.M., Abdulrahim, A., AlHarmoudi, Hokal, N.M. and Kizhisseri, A.S. (2018). Real time coastal monitoring for effective management of Dubai coast. In *Proceedings of 36th Conference on Coastal Engineering,* No. 36, Baltimore, Maryland.

Kamikubo, Y., Murakami, K., Irie, I. and Hamasaki, Y. (2000). Study on practical application of a non-wave overtopping type seawall. In *International Conference on Coastal Engineering, Proceedings of the 27th International Conference Held in Sydney Australia,* Vol. 126, pp. 2215–2228.

Karthik Ramnarayan, S., Sundar, V. and Sannasiraj, S.A. (2020a). Hydrodynamic characteristics of concave front pile-supported breakwaters with a tubular wave screen. *Journal of Waterways, Oceans, Ports and Coasts,* 148, 1. https://ascelibrary.org/doi/10.1061/%28ASCE%29WW.1943-5460.0000682.

Karthik Ramnarayan, S., Sannasiraj, S.A. and Sundar, V. (2020b). Hydrodynamic characteristics of curved and vertical front face pile-supported breakwaters in regular waves. *Ocean Engineering,* 216, 108105, 1–11.

Kim, H.J., Park, T.W., Dinoy, P.R., Kim, H.S., and Kim, H.J. (2018). Design and consolidation analysis of geotextile tubes for the Saemangeum project in Korea. *Geosynthetics International,* 25(5), 507–524. https://doi.org/10.1680/jgein.18.00015.

Kirkgoz, S. (1991). Impact pressure of breaking waves on vertical and sloping walls. *Ocean Engineering,* 18, 45–99.

Koraim, A.S., Heikal, E.M. and Rageh, O.S. (2011). Hydrodynamic characteristics of double permeable breakwater under regular waves. *Marine Structures,* 24, 503–527.

Koraim, A.S., Iskander, M.M. and Elsayed, W.R. (2014). Hydrodynamic performance of double rows of piles suspending horizontal c shaped bars. *Coastal Engineering,* 81, 81–96.

Kortenhaus, A., Pearson, J., Bruce, T., Allsop, N. and Van der Meer, J.W. (2003). Influence of parapets and recurves on wave overtopping and wave loading of complex vertical walls. In *Coastal Structures,* pp. 369–381.

Kudale, M.D., Mahalingaiah, A.V. and Tayade, B.R. (2014). Use of sand-filled geotextile tubes for sustainable coastal protection-case studies in Indian scenario. *Indian Journal of Geo-Marine Sciences,* 43, 1241–1246.

Melito, I. and Melby, J.A. (2002). Wave runup, transmission and reflection for structures armoured with CORE-LOC. *Coastal Engineering*, 45(1), 33–52.

Muller, G.U. and Whittaker, T.J.T. (1993). An investigation of breaking wave pressures on inclined wall. *Ocean Engineering*, 20, 349–358.

Murakami, K., Irie, I. and Kamikubo, Y. (1996). Experiments on a non-wave overtopping type seawall. In *International Conference on Coastal Engineering, Proceedings of the Twenty-Fifth International Conference Held in Orlando, Florida*, Vol. 122.

Nakamura, T., Kohno, T. and Morita, Y. (2001). Performance of a double-walled barrier with a front wall of inclined plate array. In *Proceedings of 11th International Offshore and Polar Engineering Conference*, Stavanger, Norway, June, 499–505.

Neelamani, S. and Rajendran, R. (2002). Wave interaction with T-type breakwaters. *Ocean Engineering*, 29, 151–175.

Neelamani, S. and Vedagiri, M. (2002). Wave interaction with partially immersed twin vertical barriers. *Ocean Engineering*, 29, 215–238.

Oumeraci, H. and Kortenhaus, A. (2011). Core made of geotextile sand containers for rubble mound breakwaters and seawalls: Effect on armour stability and hydraulic performance. *Ocean Engineering*, 38, 159–170.

Owen, M.W. and Allsop, N.W.H. (1983). Hydraulic modelling of rubble mound breakwaters. In *Proceedings of Conference on Breakwaters, Design and Construction*. ICE, London.

Rageh, O.S. and Koraim, A.S. (2010). Hydraulic performance of vertical walls with horizontal slots used as breakwater. *Coastal Engineering*, 57, 745–756.

Rageh, O.S., Koraim, A.S. and Salem, T.N. (2009). Hydrodynamic efficiency of partially immersed caissons supported on piles. *Ocean Engineering*, 36, 1112–1118.

Saathoff, F., Oumeraci, H. and Restall, S. (2007). Australian and German experiences on the use of geotextile containers. *Geotextiles and Geomembranes*, 25, 251–263.

Sankarbabu, K., Juma, I.M., Abdulrahim, A.A. and Yang. Z. (2018). Hydrodynamics of extended Dubai creek system. In *Coastal Engineering Proceedings: No. 36, Proceedings of 36th Conference on Coastal Engineering*, Baltimore, Maryland, 2018

Schoonees, T. (2014). Impermeable recurve seawalls to reduce wave overtopping. Master's thesis, Stellenbosch University, Stellenbosch.

Suh, K.D., Shin, S. and Cox, D.T. (2006). Hydrodynamic characteristics of pile-supported vertical wall breakwaters. *Journal of Waterway, Port, Coastal, and Ocean Engineering*, 132(2), 83–96.

Sukanya, R., Sundar, V. and Sannasiraj, S.A. (2016). Stability and settlement analysis of compound geo-synthetic materials in coastal protection measure. In *Proceedings of 9th International Conference on Coastal and Port Engineering in Developing Countries (PIANC-COPEDEC IX)*.

Sukanya. R., Sundar. V. and Sannasiraj. S.A. (2022). Hydrodynamic performance assessment of a geo-synthetic seawall. *Journal of Ocean and Coastal Engineering*, 2250001-1–2250001-30. June 2022. https://www.worldscientific.com /doi/epdf/10.1142/S2529807022500014.

Sukop, M.C., Rogers, M., Guannel, G., Infanti, J.M. and Hagemann, K. (2018). High temporal resolution modelling of the impact of rain, tides, and sea level rise on water table flooding in the Arch Creek basin, Miami-Dade 666 County Florida USA. *Science of the Total Environment*, 616–617, 1668–1688.

Sundar, V. (2005a). Behaviour of shoreline between groin field and its effect on the tsunami propagation. In *Proceedings of Solutions to Coastal Disasters Conference of ASCE*, 8–11 May 2005, Charleston, South Carolina, U.S.A.

Sundar, V. (2005b). Behaviour of shoreline between groin field and its effect on the tsunami propagation. In *Proceedings of 5th International Symposium on Ocean Wave Measurement and Analysis*, WAVES2005, Madrid, Spain, 3–7 July, paper No: 323.

Sundar, V. (2010). Feasibility of geo-bags as core for rubble mound breakwaters International workshop on geosynthetics and modern materials in coastal protection and related applications, jointly organized with committee on Coastal and Maritime Hydraulics of IAHR, IIT Madras 6–7 August, pp. 245–250.

Sundar, V. and Anand, K.V. (2010). Dynamic pressure and run-up on curved seawalls compared with vertical wall under Cnoidal waves. *Indian Journal of Geo-Marine Sciences*, 39(4), 579–588. December 2010, EID: 2-s2.0-84976523273.

Sundar. V. and Sannasiraj. S.A. (2018). Coastal protection along vulnerable stretches along the south Indian peninsula. In *Handbook of Coastal and Ocean Engineering* [Chapter 46], Vol. 2 (edition 2), Young C. Kim (ed.), World Scientific, Singapore.

Sundar, V. and Sannasiraj, S.A. (2019). *Coastal Engineering-Theory and Practice.* Advanced Series on Ocean Engineering, Vol. 47, April, World Scientific, Singapore. https://www.worldscientific.com/worldscibooks/10.1142/11148.

Sundar, V. and SubbaRao, B.V.V. (2003). Hydrodynamic characteristics of quadrant face pile supported breakwater. *Journal of Waterway, Port, Coastal, and Ocean Engineering* (division of ASCE), 129(1), 22–33. January/February.

Sundar, V., Maiti, D.K., Sannasiraj, S.A. and Venkatraman, M. (2009). Geosynthetic application for coastal protection at Shankarpur, West Bengal, India. In *Proceedings of 5th International Conference on Asian and Pacific Coasts (APAC 2009)*, 23–25 September 2009, Singapore, pp. 58–64.

Takahashi, S. (1996). Lecture Notes on Design of Breakwaters. Port and Harbour Research Institute, Japan International Cooperation Agency.

Teh, H.M. and Venugopal, V. (2013). Performance evaluation of a semi-circular breakwater with truncated wave screens. *Ocean Engineering*, 70, 160–176.

USACE. (2006). Coastal Engineering Manual (CEM). US Army Corps of Engineers.

Wang, Y.X., Wang, G.Y. and Li, G.W. (2006). Experimental study on the performance of the multiple-layer breakwater. *Ocean Engineering*, 33(13), 1829–1839.

Wang, G., Ren, B. and Wang, Y. (2016). Experimental study on hydrodynamic performance of arc plate breakwater. *Ocean Engineering*, 111, 593–601.
Weber, C. (1934). Seawall, United States Patent Office, S.No. 681005.
Zwamborn, J.A., Bosman, D.E. and Moes, J. (1980). Dolsse past, present, furture. In *Proceedings of 17th International Conference on Coastal Engineering*, Sydney, Australia, March 23–28, pp. 1948–1967.

Chapter 3

Rubble Mound Structures

3.1 Introduction

Rubble mound sections are commonly adopted in the coastal environment as design elements of breakwaters, seawalls and groynes. The usage of rubble mound structures primarily stems from the necessity of accommodating diverse functionalities along extensive coastlines, coupled with the continued allure of rubble stones as a favourable design material. The design guidelines follow the recommendations of the Coastal Engineering Manual (2004) and Rock Manual (CIRIA, 2008). The selection of the type of breakwaters is in general based on geotechnical investigations and the availability of materials, particularly for rubble mound structure. Forces due to waves, cyclones, tidal variations, ambient currents, storm surges and design water levels are the major factors governing the design of breakwaters. The main criterion in the design is the stability of the armour blocks for a given design wave climate. Although, the proper structural design of the rubble mound section will not guarantee the functional achievements of the structures, the sectional design still requires much attention. This is because many rubble mound sections have failed due to defective designs. Hence, this chapter details the structural design of rubble mound sections.

3.2 Structure of Rubble Mound Section

A typical rubble mound section is shown in Fig. 3.1. An armour layer at the top protects the other layers beneath it from washing away. Thus, the armours have to be designed to withstand severe environmental forces. And, an inner core layer of smaller-sized stones prevents any sediment from seeping through the rubble mound section. Depending upon the requirements, there may be few (none to two) under-layers between the armour and core layers. The stones in the under-layer are chosen in such a way that they will not fit through the voids of their immediate overlaid layer.

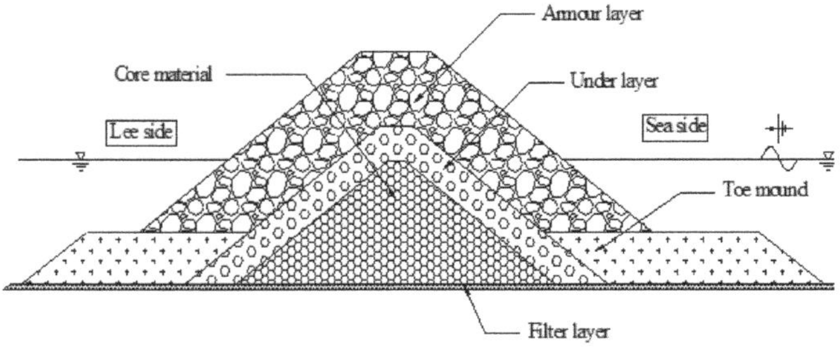

Fig. 3.1. Conventional rubble mound section.

The rubble mound sections are often provided with rubble toe protection that serves as a scour blanket to prevent undermining and thereby a reduction in lateral stability. The toe also serves to protect the section against slip failure due to the poor soil conditions. The toe may be provided on one side or both depending upon the functionality and wave climate in addition to the scour and slip failure requirements. In terms of functionality, for harbour breakwaters, the toe is essential on the seafront, while for the groynes, the toe would be provided on both sides of the section. This is because the wave attack could be from any direction for a groyne section. The entire cross-section will be placed over the filter layer blanket, which is laid on the seabed. The filter layer evenly distributes the weight of the entire structure into the seabed, serving as the foundation for the rubble mound superstructure. It also prevents seabed materials from seeping into the core layer and stops larger stones from settling into the soft sands.

3.3 Hydrodynamic and Structural Design

3.3.1 *General*

The construction of a rubble mound breakwater requires a large quantity of natural rock. The density and shape of the rocks are important because of the following:

- Correlation between density and durability.
- Density of rock considered as input for armour stability and geotechnical analyses.

In the design, the basic assumption is that rocks are randomly shaped with rough faces. A smooth surface reduces friction between adjacent stones and forfeits the main design criteria. The rocks required for the construction of the breakwaters are in general sourced from existing quarries, and hence specific gravity and water absorption are also important design elements. In general, water absorption by percentage should not exceed 0.01.

The following steps are involved in the detailed design process of the breakwater:

- A basic design is created including the details of materials and forces.
- A preliminary design of sections is created.
- The physical model is tested for stability and hydraulic response.
- A detailed final design is made incorporating observations from the physical model tests.

3.3.2 *Forces acting on rubble mound sections*

3.3.2.1 *Wave forces*

Rubble mound sections should preferably be used at exposed sites because of their ability to withstand severe wave loads and decrease wave reflection. Moreover, the risk of scouring and the formation of strong rip currents along the rubble section is reduced. Most rubble mound sections are designed with quarry stones as the armour which are heavy enough to be stable under a selected design wave height.

3.3.2.2 *Current forces*

Currents can exert forces on rubble mound sections both as longshore currents flowing over low sections and as seaward-flowing rip currents along a section's flank. However, forces from current-induced flow are usually small when compared with the forces due to waves. Normally, the stone weight necessary for stability against currents will be lesser than the stone weight necessary for stability against wave action.

3.3.2.3 *Buoyancy forces*

When the rubble stone is submerged in water, it experiences an upward buoyant force. The buoyant force is equal to the weight of the water displaced by the submerged volume of the stone, as per Archimedes' principle. It means that the stone effectively becomes lighter by the weight of the water that it displaces.

The reduction in weight due to buoyancy makes the object feels lighter when submerged in water. For rubble stones in the marine environment, the effective weight of stones on itself or on the seabed would be less when they are underwater compared to in air. This is crucial for various engineering calculations, such as the stability of submerged structures, the design of breakwaters, or the assessment of underwater foundations.

3.3.2.4 *Frictional resistance*

Frictional resistance is induced parallel to the slope, either upwards or downwards, but contrary to the direction of the wave force. Instability occurs if the friction is insufficient to neutralize the other forces parallel to the slope.

3.3.2.5 *Other forces*

A section might experience impact forces due to wave-carried debris and small craft collisions. The magnitude of these forces is difficult to predict because the cause of the impact and the mass of the impacting body are not known a priori. If debris is suspected to be a problem, appropriate levels of conservation should be included in the design.

Designing structural sections to withstand construction loads is a critical aspect of engineering that ensures the safety and integrity of a structure during its erection. By accounting for the temporary loads from construction equipment and surcharge due to temporary fills, engineers can prevent construction-phase failures and ensure a smooth transition from construction to the structure's operational phase. Proper planning and design adjustments during the construction phase play a vital role in the overall success and safety of the project. Further, these forces may be critical and exceed forces due to other more routine causes such as waves and currents.

3.3.3 ***Hydraulic response***

The hydraulic response of the breakwater is measured in terms of the following:

- Wave overtopping,
- Wave run-up and run-down,
- Wave reflection and wave transmission.

The detailed methods to estimate the hydraulic response are discussed in the following sections.

3.3.3.1 *Wave overtopping*

The overtopping rate should be estimated as per the recommendations of the EurOtop Manual (2018) or CEM (2004) to ensure the stability of leeward side armour units and the traffic on the rear side. For the design of the breakwater, the average overtopping rate should not exceed $0.1\,m^3/s/m$ run of the breakwater. For estimation of overtopping, the methods proposed by Pedersen and Burcharth (1992), Pedersen (1996), van der Meer and Janssen (1995) and the EurOtop Manual (2018) can be used.

3.3.3.2 *Wave run-up*

The breakwaters for major facilities should be designed as a non-overtopping section, i.e. the crest level should be such that the waves do not overtop the section. To avoid overtopping, the crest level has to be sufficiently high but should also be as low as possible. Wave run-up governs the design parameters and its estimation is important. Wave run-up can be calculated using the EurOtop Manual (2018) and should also be based on the method proposed by Van Gent (2001), which takes into account the effect of surface roughness of the armour, wave period and direction of wave attack.

3.3.3.3 *Wave reflection and transmission*

In general, the rubble mound section provides high porosity and, hence, it is expected that the wave reflection is minimal, which would avoid amplification in front of the structure. Similarly, with proper design of the core layer, wave transmission past the structure will be minimal unless the section is designed as an overtopping section.

3.4 Armour Layer

The outermost armour layer protects the entire structure from wave action. It dissipates the wave energy through its porosity. This armour layer can be formed either using natural rock debris or concrete blocks depending upon the size of the armour units required to withstand the wave action, the availability of rocks and their quality. The concrete blocks can be simple cube forms or highly interlocking tetrapods, accropodes and core-locs.

3.4.1 *Stability criteria*

There are hardly any standards available for the design of armour units except for European standards. However, the design of armour units is based on the results of physical model studies and field observations. Some empirical formulae have been developed from experimental work to find suitable sizes of units which allow the block to withstand wave-induced forces. The main governing parameter of the armour layer stability is the stability number (N_s) which is defined as follows:

$$N_s = \frac{H}{(\gamma_s/\gamma_w - 1)D_{n50}}, \tag{3.1}$$

where H is the design wave height, γ_s is the unit weight of the armour stone, γ_w is the ratio of the unit weight of water and D_{n50} is the nominal median diameter of the stones. The nominal median diameter is related to the median weight W_{50} by the following:

$$W_{50} = \gamma_s\ D_{n50}^3. \tag{3.2}$$

From Equation (3.1),

$$W_{50} = \frac{\gamma_s H^3}{(\gamma_s/\gamma_w - 1)^3 N_s^3}. \tag{3.3}$$

The stability (amount of movements) of the armour unit is related to the value of N_s. Stone motion is, in general, not expected on conventional rubble mound sections. For stable sections, the value of N_s ranges between 1 and 4. The armour unit size can then be estimated using either Hudson's formula (Hudson, 1958) or Van der Meer's findings (1988).

3.4.2 *Hudson's formula*

The size of the armour stones is commonly estimated using the widely accepted Hudson formula and the same is recommended by the CERC (1984). Hudson (1958) established the stability coefficient (K_D) for different types of armour units from physical model tests under regular waves. These values are given as a function of the damage. In the design of rubble mound sections, a damage percentage of 0–5% is acceptable to ensure no movement of the stones. The stability coefficient thus can be correlated to the weight of the armour units and the slope (angle, θ with the horizontal) in which the units are laid. The suggested K_D values for different structure slopes, in general, satisfy the stability criteria of N_s in the range of 1–2:

$$N_s^3 = K_D \cot\theta. \tag{3.4}$$

In this method, the weight of the individual stones (W) is proportional to the cubic power of the design wave height (H_D) for a given structure

Table 3.1. Recommended K_D values for the estimation of armour unit size.

Armour units	N	K_D		Slope Cot θ	Porosity (P) %
		BW	NBW		
Rough angular quarry stone	1	NR	2.9	1.5–3.0	
	2	2	4	1.5–3.0	37
	>3	2.2	4.5	1.5–3.0	40
Graded rough angular stone	—	1.3	1.7	1.5–5.0	37
Tetrapod	2	7.0	8.0	1.5–3.0	50
Dolos	2	15.8	31.8	2.0–3.0	63
Cube (modified)	2	6.5	7.5	1.5–3.0	47

Note: BW – breaking wave; NBW – non-breaking wave.
Source: CEM (2004).

slope (θ) and the type of stones/blocks. This is given by Equation (3.3) as follows:

$$W = \frac{\gamma_s H_D^3}{K_D \left(\gamma_s/\gamma_w - 1\right)^3 \cot\theta}. \tag{3.5}$$

Table 3.1 provides the stability coefficient for different types of stone/concrete blocks for use in a trunk section. The corresponding coefficients for head sections are more conservative because head sections directly face the wave action from all directions.

3.4.3 *Van der Meer's formula*

Van der Meer (1988) proposed stability criteria that incorporate a broader range of parameters compared to Hudson's formula, based on physical model tests using irregular waves. This would make it more difficult to apply. A stability criterion is estimated using the damage level S, relating the erosion area in a cross-section (A, m^2) and the mean diameter of stones (D_{n50},m) (Van der Meer and Heydra, 1991) as follows:

$$S = \frac{A}{D_{n50}^2}. \tag{3.6}$$

For a rubble quarry stone, the stability criteria under the action of breaking waves are given as follows:

For plunging waves ($\xi < \xi_c$):

$$\frac{H_s}{\left(\gamma_s/\gamma_w - 1\right) D_{n50}} = 6.2\, P^{0.18} \left(S/\sqrt{N}\right)^{0.2} \frac{1}{\sqrt{\xi}}. \tag{3.7}$$

For surging waves ($\xi > \xi_c$):

$$\frac{H_s}{(\gamma_s/\gamma_w - 1)\, D_{n50}} = 1.0\, P^{-0.13} \left(S/\sqrt{N}\right)^{0.2} \cot\theta\, \xi^{P}. \tag{3.8}$$

In these formulae, many variables have been introduced such as the number of waves N, the permeability (P) of the surface and the surf similarity parameter ξ, which relates the slope to the wave steepness H/L as follows:

$$\xi = \frac{\tan\theta}{\sqrt{H/L}} \tag{3.9}$$

and

$$\xi_c = (6.2P^{0.31}\sqrt{\tan\theta})^{\frac{1}{P+0.5}}. \tag{3.10}$$

The condition of "no damage" introduced by S in Van der Meer's formula leads to a value of N_S nearly 1–2. Hudson's formula and Van der Meer's formula differ only in extreme wave conditions. The stability coefficient K_D is explicitly defined in terms of the permeability coefficient in Hudson's formula and wave steepness in the Van der Meer's formula. As the wave intensity increases from surging to plunging, the importance of slope was given due consideration by increasing its power of influence.

The armour layer slope should be equal to or greater than 1V:1.5H. Hudson's formula (Equation (3.5)) is applicable for armour units of nearly uniform size, and the overtopping of waves is not allowed. For graded riprap armour stone, the modified Hudson's formula uses W_{50} representing the weight of the median size (50% size) in the gradation. In the graded armour layer, the maximum weight of graded rock is 1.25 W_{50} and the minimum weight is 0.75 W_{50}. The stability coefficient has to be modified on the conservative side and the recommended values are 1.3 and 1.6 for depth at the toe of a structure less than 6.0 m and greater than 6.0 m, respectively. These coefficients were chosen based on the allowable 5% damage criteria. However, graded stones are not recommended if the design wave height exceeds 1.5 m.

3.5 Under-Layer

The purpose of one or more under-layers in between the armour and core layers is to prevent smaller stones in the core from seeping into the voids of the armour units. This criterion is met if the first under-layer stone weighs $W_{50}/10$ to $W_{50}/15$ where W_{50} is the median weight of the armour stone.

This criterion assumes that the stone in the under-layers has approximately the same unit weight as the armour stone. By this criterion, the second under-layer stone should weigh approximately $W_{50}/100$ to $W_{50}/150$. For the rubble mound section system, one under-layer is recommended. To prevent smaller stones in the under-layer from escaping through the pores of the armour layer, the following filter design criterion needs to be followed for graded stones:

$$D_{15}(\text{cover}) \leq 5D_{85}(\text{under}), \tag{3.11a}$$

where D_{15} (cover) is the diameter exceeded by the coarsest 85% of the layer immediately above the under-layer and D_{85} (under) is the diameter exceeded by the coarsest 15% of the under-layer.

Permeability criteria must be checked to prevent the development of hydraulic gradients. The equation used .to verify these criteria is as per Equation (3.11b).

$$D_{15}(\text{armour}) > 4D_{85}(\text{under}). \tag{3.11b}$$

3.6 Core Layer

The core layer supports the protective armour cover and any other additional under-layers. It prevents sediments from passing through the section. The ideal core of a rubble mound breakwater should have a wide gradation of rocks so that fine material cannot be drawn out by wave actions. This will ensure low permeability with respect to wave transmission and will help maintain tranquillity conditions on the leeward side of the breakwater. The size of stones in the core layer can be $W_{50}/200$ to $W_{50}/4000$ (CERC, 1984) following a single under-layer. If there is more than one under-layer, the weight of the core stones may further be reduced. The percentage of fines below the lower bound shall not exceed 1% of the total mass of core materials.

3.7 Toe Layer

Breakwaters exposed to breaking waves should have their primary armour layers supported by a quarry stone toe berm. The stone weight needed for stable toe protection can be determined from the stability against the scour arising under the armour stones. This requirement dictates that the weight of the toe-layer stones should be equal to the weight of under-layer stones, i.e. $W_{50}/10$. The minimum width and height of the toe berm are about

$3k_\Delta D_{n50}$ and 2 $k_\Delta D_{n50}$, respectively. Here, k_Δ is the layer coefficient and it can be taken as approximately 1.0.

The forces causing loss of foundation materials from beneath a rubble mound structure are severe at the structure's toe. Wave pressure fluctuations and groundwater flow may produce quick sand conditions at the toe and currents may carry the suspended soil away. A shallow scour hole may remove support for the cover layers, allowing them to slump down the face. A deeper scour hole may destabilize the support, causing the entire face to slip. To prevent seabed erosion in the vicinity of the toe, an apron is provided which can be an extension of the bedding layer. Normally, the toe apron must be 1.0–1.5 m thick and extend well beyond the toe of the structure. The material used will be quarry run with stones up to 0.3 m in size.

A scour apron in addition to the above-mentioned toe layer width is required if the wave's downrush reaches the toe layer. An additional toe berm should be provided if bearing failure is possible.

3.8 Filter Layer

The massive rubble stone structure should be stable against disintegration due to excessive settlements caused by leaching, undermining or scour because of wave- and current-induced turbulence and quick soil conditions, particularly on sandy beds. A filter or bedding layer is required to retain the rubble mound structure while passing large volumes of water through it. The filter layer should satisfy the following:

$$D_{15}(\text{filter material}) \leq 5D_{85}(\text{foundation soil}). \tag{3.12}$$

The filter layer thickness should not be less than 300 mm to ensure that the bottom irregularities are completely covered. Geo-textiles can be considered instead of filter materials. The lower limit of filter materials is specified by their median size to avoid placing very small stones comparable to the sediment.

The stone size gradation of each layer as a percent of the mean stone size is given in Table 3.2.

3.9 Thickness of Armour/Under-Layer

The thickness of a single armour or under-layer t is equal to the mean diameter of stones in that particular layer. However, if there are n number

Table 3.2. Stone size gradation in a graded rubble design.

Layer	Stone size	Gradation (%)
Armour layer	W_{50}	75–125
Under-layer	$W_{50}/10$ to $W_{50}/15$	70–130
Toe layer	$W_{50}/10$ to $W_{50}/15$	70–130
Core	$W_{50}/200$ to $W_{50}/4000$	30–170
Filter layer	100mm to $W_{50}/6000$	30–170

of layers, then the thickness is as follows:

$$t = nk_{\Delta} D_{n50}, \tag{3.13}$$

where k_{Δ} is the layer coefficient in the range of 1.0 (for dolosse) to 1.15 (for rough quarry stones).

The placing density of rubble stones is expressed in terms of the required number of armour units (N_r) for a given surface area, A:

$$\frac{N_r}{A} = \frac{nk_{\Delta}(1 - P/100)}{D_{n50}^{2/3}}, \tag{3.14}$$

where P is the average porosity of the armour layer.

3.10 Crest Width

The crest width (r) is the maximum of the following aspects: (a) depends upon the mimum number of stones (n_s), as per CERC (1984) a minimum of 3 stones are recommended. (b) degree of overtopping and the sections are designed as a non-overtopping structure to avoid sediment bypassing. (c) functional aspects during the constructions and later requirements. CERC (1984) suggests a minimum of 3 stones. Next, the width depends on the degree of overtopping and the sections are designed as a non-overtopping structure to avoid sediment bypassing. The last criterion demands the functional aspect during the construction and after use. If access on the top of the sections is not mandatory after construction, the minimum required crest width will be equal to the wheelbase of the crane used for the construction to place the stones on the slope:

$$r = \max(n_s k_{\Delta} D_{n50}, \text{ construction requirement}). \tag{3.15}$$

3.11 Crest Elevation

The rubble mound section height or the crest elevation is an important parameter in the section design. It depends on the maximum possible wave height (H_{max}) that one would come across during the lifetime of the section, wave run-up (R_u), the freeboard (R_c) requirement and the permissible overtopping of water.

Hence, the crest elevation above the maximum high-water line =

$$H_{\max} + R_u + R_c. \tag{3.16}$$

The freeboard is an additional safety provided to the section to avoid overtopping of waves if the temporary water level setup submerges the section during storms or cyclones. A typical value of 1.5 m may be assumed for sections. If suitable storm data were available, R_c should be increased accordingly.

3.11.1 *Wave run-up*

The wave run-up over the section slope is calculated from the design wave height (H) and the structure slope (θ):

$$R_u = H\sqrt{\frac{\pi}{2\theta}}. \tag{3.17}$$

The wave run-up according to Van der Meer and Stam (1991) includes the frequency of the wave:

$$R_u = 0.83\, H\,\xi. \tag{3.18}$$

3.12 Design Criteria for Roundheads

The roundhead is the deepest portion of the breakwater and is exposed to wave attacks from all directions. The head section of the breakwater requires special attention in terms of stability during storms. This can be achieved by increasing the weight of the armour rock or flattening the slope at the head section. For the design of the roundhead using Hudson's method, the stability coefficient has to be chosen for the roundhead section.

3.13 Geotechnical Stability of Breakwaters

The geotechnical stability of breakwaters should be analyzed based on the available soil data at the location.

The stability of the breakwater should be checked for the following depending on the type of soil:

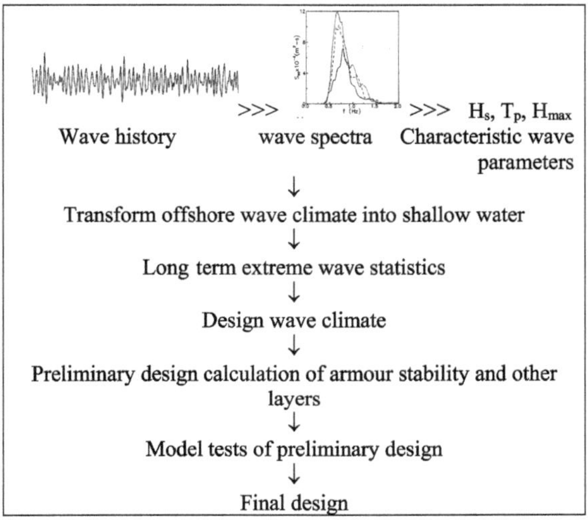

Fig. 3.2. Design procedure of rubble mound sections.

- Liquefaction,
- Slope stability,
- Safe bearing capacity,
- Subsidence and settlement.

3.13.1 *Design procedure*

Figure 3.2 presents an overview of the design procedure of sections. For the preliminary section design under the stability criterion, the following steps need to be followed:

a. Determine the water level range for the site.
b. Determine the wave heights.
c. Select suitable section configurations.
d. Select suitable armour unit type and size (rubble mound/concrete units and toe protection).
e. Determine the potential run-up to set the crest elevation.
f. Determine the amount of overtopping expected which should not be more than allowed.
g. Develop a cost estimate for each alternative.

The most critical design elements are a secure foundation to minimize settlement and toe protection to prevent undermining. Both these are potential causes of failure of such walls, apart from the main cause of stability of armour units against waves.

3.14 Design of Rubble Mound Section

3.14.1 *General*

The designs of various structural components such as armour stones, toe mounds and filter bed layers are detailed in the following. The core layer and under-layers, if any, are also been suggested based on the design of the armour stones.

The main parameters in the design of the rubble mound cross-section are the water depth and the design wave height. The various cross-sections need to be designed in different water depths from the shoreline to the tip of the breakwater or groyne. Generally, the armour slope of a head section (offshore tip) is flatter than a typical trunk section. The following sections present typical designs of a rubble mound groyne in water depths of 4.0 m and 6.5 m.

For a water depth of 4.0 m

Design wave

The design water depth can be calculated from the mean water depth, tidal level and water level setup during storms:

$$\text{Design water depth} = 4\,\text{m} + 0.5\,\text{m for}$$

$$\text{Storm and tide corrections} = 4.5\,\text{m}.$$

The maximum possible sustainable wave height in a particular water depth is as follows:

$$\begin{aligned}\text{Maximum wave height, } H_{\max} &= 0.78 \times \text{water depth}\\ &= 0.78 \times 4.5 = 3.51\,\text{m}.\end{aligned}$$

In the calculation of armour weight, the design wave height, represented by the significant wave height, needs to be established:

$$\begin{aligned}\text{Significant wave height, } H_s &= H_{\max}/(1.6\text{–}2.0)\\ &\cong 1.95\,\text{m}.\end{aligned}$$

3.14.1.1 *Armour unit (rubble)*

The formula of Hudson *et al.* (1979) is used for the estimation of the stable weight of armour rubble stone W_{50} from Equation (3.5). The following rubble characteristics are assumed. The unit weight of sea water γ_w is 1.025 t/m^3. The stones available in the nearby quarries are rough and angular with a unit weight γ_s of 2.65 t/m^3. Hence, the stability coefficient K_D adopted is 2.0 from Table 3.1. This coefficient is for severe wave climate existing at the site, i.e. waves breaking on the section. For the trunk section, an armour slope of 1V:1.5H ($\cot\theta = 1.5$) is suggested.

From Equation (3.5),

$$W_{50} = 1650 \text{ kg.}$$

The nominal median diameter of stones corresponding to the armour rubble stone weight can be calculated using Equation (3.2) as follows:

$$\begin{aligned} D_{n50} &= (W_{50}/\gamma_s)^{1/3} \\ &= 0.86 \text{ m say } 1.0 \text{ m.} \end{aligned}$$

Then, the thickness of the armour layer (t) can be calculated using Equation (3.13). The suggested number of armour layers is two ($n = 2$) and the layer coefficient is 1.0:

$$t = 2.0\,\text{m}.$$

Hence, 1.25–2.0 t armour stones are provided in two layers of thickness of 2.0 m. A typical section in a water depth of 3 m is shown in Fig. 3.3.

3.14.1.2 *Under-layer*

The under-layer stone weights should be one-tenth of the armour stone weight ($W_{50}/10$). Hence, the under-layer is provided with stones of 125–200 kg mean weight. From Equation (3.13), for two layers of similar under-layers, the layer thickness is 1.0 m. A typical section in a water depth of 5 m with an under-layer is shown in Fig. 3.4.

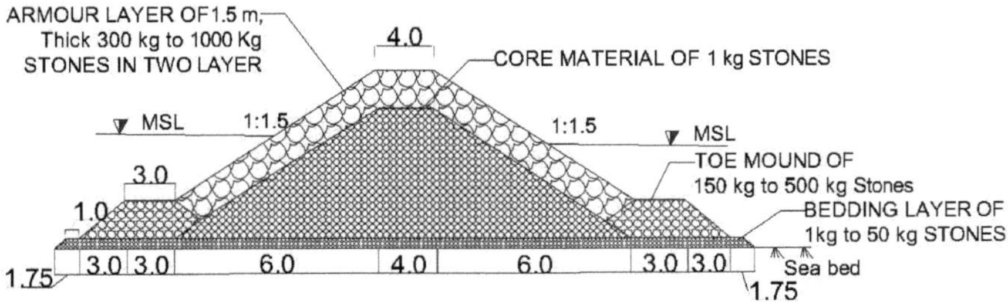

Fig. 3.3. Typical rubble mound cross-section at a water depth of 3.0 m.

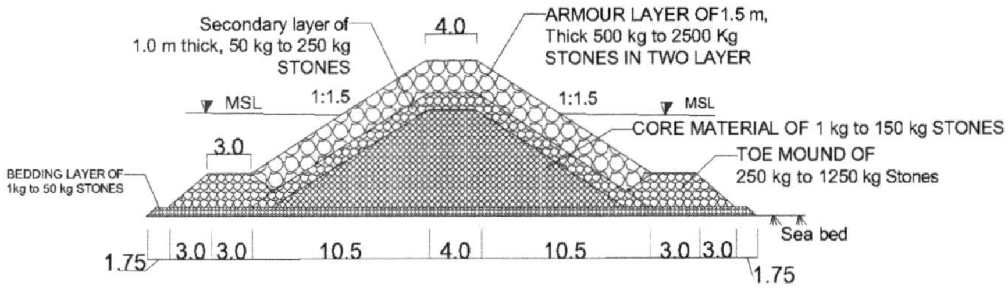

Fig. 3.4. Typical rubble mound cross-section at a water depth of 5.0 m.

3.14.1.3 *Core*

The core layer comprises the core part of the section. Only one under-layer is proposed for the sections. Hence, the core layer below the under-layer should be one-tenth of the under-layer's stone weight so that the rubble in the core do not penetrate through the top layers. A stone weight of $W_{50}/100$ can be used, which is about 12.5–20 kg. However, for practical limitations as well as to ensure that silt does not pass through the sections, a wider range is usually suggested.

3.14.1.4 *Filter layer*

A filter layer or bedding layer acts as a foundation to support the entire structure. Stones of 10 mm to 50 kg are provided for a bedding layer of 0.5 m thickness.

3.14.1.5 *Toe mound*

A toe mound is to be placed on both sides of the groynes to prevent the sliding of the stones and to protect the groyne from slope failure. A toe mound of 3.0 m top width and 2.0 m height with 1:1.5 side slopes is provided. The toe mound can be built using stones similar to the under-layer as the armour is supported by the toe mound. Table 3.3 provides design details for various water depths up to 5.5 m.

For a water depth greater than 6.5 m and up to 8.0 m, concrete cubes as armour stones were first suggested as shown in Fig. 3.5. In recent times, based on construction difficulties and operational expenses, rough angular stones have been preferred. Figure 3.6 presents a redesigned section in a water depth of 7.5 m.

Table 3.3. Typical design details for trunk section up to a water depth of 5.5 m.

Trunk section	0–1 m water depth	1–3 m water depth	3–5.5 m water depth
Wave Height	up to 0.7 m	Up to 1.0 m	Up to 2.5 m
Crest elevation (m)	+3.5	+3.5	+3.5
Crest width (m)	4	4	4
Armour layer	300–500 kg (2 layers of thickness 1.1 m)	300–1000 kg (2 layers of thickness 1.5 m)	500–2500 kg (2 layers of thickness 2.0 m)
Slope	1:1.5	1:1.5	1:1.5
Under-layer	Nil	Nil	50–250 kg (thickness 1.0 m)
Core layer	1–150 kg	1–150 kg	1–150 kg
Filter layer (thickness 0.3 m)	1–50 kg	1–50 kg	1–50 kg
Toe	50–100 kg	150–50 kg	250–1250 kg
Toe width (m)	3.0	3.0	3.0
Toe height (m)	1.0	2.0	2.0

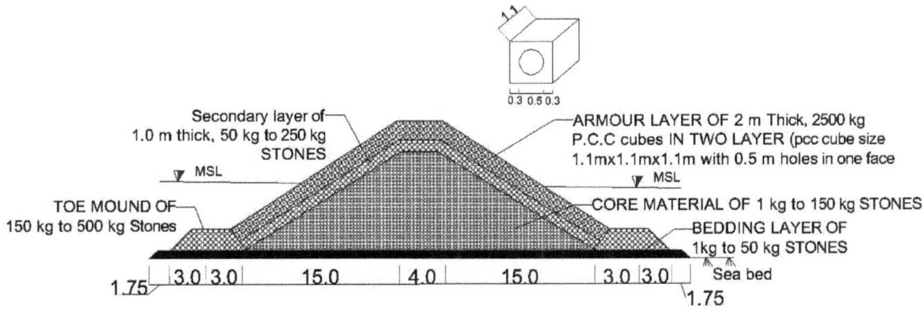

Fig. 3.5. Typical rubble mound cross-section using concrete cubes at a water depth of 8.0 m.

Figure 3.7 depicts a typical head section. The head section directly faces the wave action and, hence, the slope is flattened compared to the trunk sections. Table 3.4 presents the design details of a head section using concrete cubes and random rubbles as armour units. In the revised design using rubble stones, the slope is flattened to 1:2.5 compared to concrete cubes at a 1:1.5 slope.

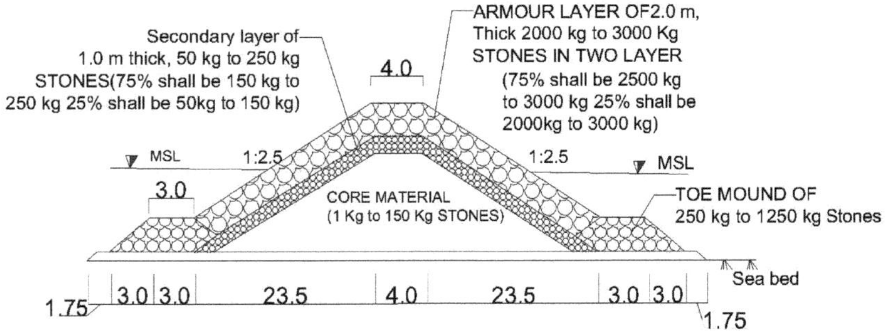

Fig. 3.6. Modified rubble mound cross-sectional details at a water depth of 6.5–7.5 m.

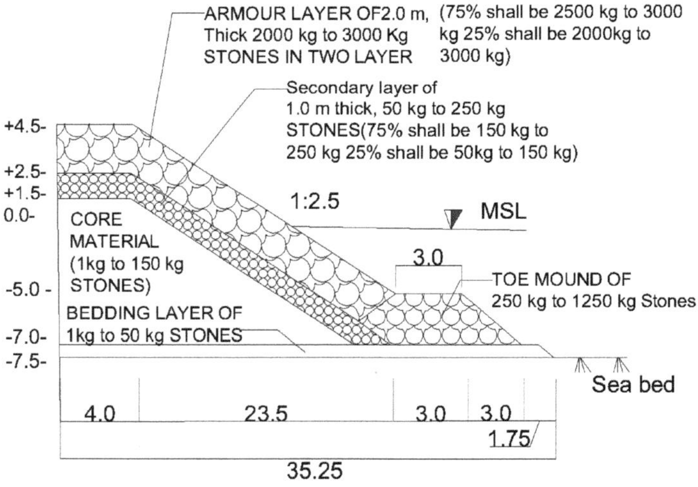

Fig. 3.7. A typical head section at a water depth of 7.5 m.

3.14.2 *Design example*

Design a breakwater for a water depth of 8 m.

Given: HHWL = 2.4 m, Storm surge = 0.5 m, and $H_s = 5$ m for 50-year wave climate.

Solution:

$$\begin{aligned} d_{\text{total}} &= d_{cd} + \text{HHWL} + \text{Storm surge} \\ &= 8 + 2.4 + 0.5 \\ &= 10.9\,\text{m}. \end{aligned}$$

Table 3.4. Comparison of design details of a head section using concrete cubes and rubble stones as armour units.

Head section	Concrete cubes	Rubble stones
Crest elevation (m)	+4.5	+4.5
Crest width (m)	4.0	4.0
Armour layer	2500 kg in 2 layers of thickness 2.0 m	2000–3000 kg in 2 layers of thickness 2.0 m
Slope	1:1.5	1:2.5
Secondary layer	50–250 kg of thickness 1.0 m	50–250 kg of thickness 1.0 m
Core layer	1–150 kg	1–150 kg
Filter layer	1–50 kg of thickness 0.30 m	1–50 kg of thickness 0.30 m
Toe	250–1250 kg	250–1250 kg
Toe width (m)	3.0	3.0
Toe thickness (m)	2.0	2.0

$$\begin{aligned}\text{Maximum sustainable wave height} &= 0.78 \times d\\ &= 0.78 \times 10.9\\ &= 8.5\,\text{m}.\\ \text{Significant wave height} &= H_{\text{max}}/1.8 = 8.5/1.8\\ &= 4.72\,\text{m} < 5\,\text{m}.\end{aligned}$$

Breaking of wave occurs.

(*Herein, it is checked if the maximum sustainable wave height/breaking wave height is less than the design wave height. In case the breaker wave height exceeds the design wave height, it would be wise to use the breaker wave height to minimize the damage to the armour layer.*)

The structure should be designed for breaking conditions with a design wave height of 4.72 m.

Armour design: (Hudson formula)

$$N_s = \frac{H_{\text{des}}}{\Delta_a D_a} = (K_D \cot\theta)^{\frac{1}{3}}.$$

$$\gamma_s = 2.65; \gamma_w = 1.025.$$

$$\Delta_a = \left(\frac{\gamma_s}{\gamma_w} - 1\right).$$

$$\Delta_a = \left(\frac{2.65}{1.025} - 1\right) = 1.585.$$

$K_D = 2$ for breaking condition for a rubble mound structure.
$\cot\theta = 2$ for random rubble masonry.
Weight of the armour stone, $W = \frac{\rho_a H_{\text{des}}^3}{K_D \Delta_d^3 \cot\theta}$.

$$\mathrm{W} = \frac{2.65 \times 4.72^3}{2 \times 1.585^3 \times 2} = 17.49 \text{ tonnes.}$$

Adopt 18 tonnes.

$$\begin{aligned} \text{Gradation} &= (75 \text{ to } 125\%)W, \\ &= 13.5 \text{ to } 22.5 \text{ T.} \end{aligned}$$

Provide two layers of armour.

$$T_{\text{armour}} = \sqrt[3]{\frac{W}{\rho_s}} K_\Delta n.$$

K_Δ is taken as 1.
$n =$ number of under-layers.
$\rho_s = 2.65$ tonnes/m^3.

$$\begin{aligned} T_{\text{armour}} &= \sqrt[3]{\frac{18}{2.65}} \times 1 \times 2. \\ &= 3.8\,\text{m.} \end{aligned}$$

$$\text{Diameter of armour stone} = \sqrt[3]{\frac{W}{\rho_s}} = \sqrt[3]{\frac{18}{2.65}} = 1.9\,\text{m.}$$

$$\begin{aligned} 1_{\text{st}}\ \text{Under} - \text{layer} &= \frac{W}{10} - \frac{W}{15}, \\ &= 1.8 \text{ to } 1.2 \text{ T.} \end{aligned}$$

Thickness of under-layer:

$$\begin{aligned} T_{\text{underlayer}} &= \sqrt[3]{\frac{1.8}{2.65}} \times 1 \times 2, \\ &= 1.7\,\text{m.} \end{aligned}$$

Diameter of under-layer stone$= \sqrt[3]{\frac{W}{\rho_s}} = \sqrt[3]{\frac{1.8}{2.65}} = 0.879\,\text{m.}$

$$\begin{aligned} \text{Core} &= \frac{W}{100} - \frac{W}{225} \\ &= 180 \text{ to } 80\,\text{kg.} \end{aligned}$$

Provide a filter layer of 300 mm thickness.

For the filter layer, the size of a stone to be 10 mm and the max weight of a stone = 10 kg.

$$\text{Toe layer} = \frac{W}{10} - \frac{W}{15},$$
$$= 1.8 \text{ to } 1.2\,\text{T}.$$

$$\text{Diameter of toe stone} = \sqrt[3]{\frac{\text{W}}{\rho_s}} = \sqrt[3]{\frac{1.8}{2.65}} = 0.9\,\text{m}.$$

Width of toe = D_{n50} = 2.6 m.

Provide 3 m width.

Thickness of toe layer = D_{n50} = 1.8 m.

For wave run-up, = 0.0075.

From graph,

$$\frac{R_u}{H_o} = 0.8.$$
$$R_u = 3.8\,m.$$

Fig. 3.8 presents the design section in a water depth of 8 m.
Crest elevation:

$$= d_{cd} + \text{HHWL} + \text{Storm surge} + \text{Runup} + \text{Free board}$$
$$= 8 + 2.4 + 0.5 + 3.8 + 0.5$$
$$= 15.2\,m.$$

Fix crest elevation at 15.5 m from the sea bed.

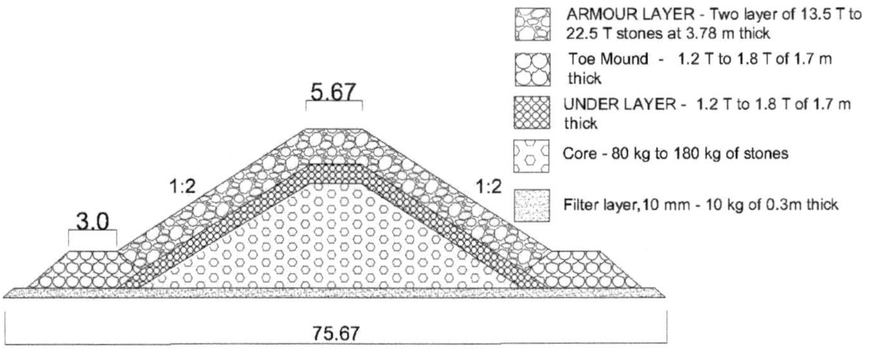

Fig. 3.8. A typical trunk section at a water depth of 8 m.

3.15 Stability of Rubble Mound Structure/ Porous Structure

The stability of rubble mound breakwaters involves a large number of parameters. These parameters were listed by Hudson *et al.* (1979). Sharp (1985) pointed out that Hudson's (1958) grouping of dimensionless products into the stability number is one of several possible combinations that can be formed from dimensional analysis considerations. Based on past success, Hudson *et al.* (1979) suggested that the appropriate length scale range for rubble mound stability tests is between 1:5 and 1:70. Most tests conducted by Hudson at the Waterways Experiment Station were at scales between 1:40 and 1:50. Oumeraci (1984) stated that rubble mound structure models have linear scales ranging between 1:10 and 1:80, with 1:50 being the most common.

Further, in coastal engineering, one can use 2D or 3D models. The 2D models are basically investigated in a wave flume and 3D models in a wave basin. The 3D models are expensive and carried out at a reduced horizontal scale (depending upon the problem or requirements). However, after the general structure shape has been optimized in a 3D model, an analysis on a two-dimensional model is often conducted on a larger scale to examine armour and toe stability, quantify overtopping, etc. Further, the larger scale reduces the potential for scale effects and provides an optimum cost-effective design that fulfils design criteria. Thus, in many situations, a 2D model is all that is necessary to provide a functional design.

The stability of the rubble mound breakwater or any coastal structure requires a systematic physical model study. Hughes (1993) pointed out the following different conditions that a model needs to undergo after installation or construction in a facility:

(a) Tests with increasing wave impacts,
(b) Tests for design conditions,
(c) Tests for reproducing individual storms,
(d) Long-duration tests,
(e) Tests for accumulated storm impacts,
(f) Tests for residual stability,
(g) Tests for structures under construction.

These tests will provide more insight into the functionality and the physical processes of the wave–structure interactions. Apart from these tests, studies can be conducted to verify specific designs or develop general design guidance, determine the hydrodynamic forces exerted on monolithic structures by wave action, and optimize the structure type, size and geometry to meet performance requirements and budget constraints. Studies can also investigate structure characteristics such as wave run-up, run-down, overtopping, reflection, transmission, absorption and static/dynamic internal pressures for different structure types, geometries and/or construction methods; develop and/or test methods of repairing damage on existing structures or improving the performance of an existing structure; determine the effects a proposed modification might have on an existing structure's stability and performance; and examine alternate construction sequences under different wave conditions.

Defining Damage of the armour layer in Rubble Mound Models: One of the important aspects of assessing stability is damage assessment. The assessment of damage to any breakwater cross-section can be carried out through two primary methods: measuring the damaged portion or counting the number of displaced units. Both approaches are detailed in the following:

Damage Assessment via Measuring the Damaged Portion: This method involves evaluating the damaged area of the breakwater cross-section. The damage to the armour layer can be quantified as a percentage of displaced blocks relative to a specific area, which can encompass the entire layer or a designated portion of it.

Damage Assessment via Counting Displaced Units: Alternatively, damage assessment can be conducted by counting the number of displaced units. This involves defining the relative damage, denoted as N_o, which represents the actual number of units displaced within a certain width along the longitudinal axis of the structure, based on a nominal diameter D_n.

The subscript in N_o can be extended to differentiate between various scenarios.

Displacement out of the layer: Units rocking within the layer, that may occur once or multiple times. These methods provide comprehensive insights into the extent of damage sustained by the breakwater cross-section, facilitating effective analysis and subsequent maintenance or repair efforts. The actual number is related to the width of one D_n.

$$N_{\text{od}} = \text{units displaced out of the armour layer (hydraulic damage)},$$
$$N_{\text{or}} = \text{rocking units},$$
$$N_{\text{omov}} = N_{\text{od}} + N_{\text{or}}.$$

3.15.1 *Number of units method*

The approach for assessing damage through this method aligns with the procedure outlined by Owen and Allsop (1983). Damage to the groyne section is quantified by the number of armour units that have become completely dislodged from the armour layer. This assessment is typically expressed as a percentage relative to the total number of units present on the armour layer. The evaluation of damage considers the movement of armour units, which can be categorized into four distinct categories as outlined in the following:

P — Unit seen to be rocking, but not permanently displaced,
Q — Unit displaced by up to 0.5 D,
R — Unit displaced by between 0.5 and 1.0 D,
S — Unit displaced by more than 1.0 D,

where D is the equivalent diameter of an individual armour unit. The order of damage for the armour layer has been assessed using the following formula:

$$\% \text{ of damage} = \frac{\text{Number of armour units dislodged}}{\text{Total number of units in armour layer}} * 100. \qquad (3.19)$$

If the sum of P, Q, R and S (P + Q + R + S) is less than 5%, the section is deemed to be safe. In the case of breakwater sections, damage assessment using the number of units method is implemented. This involves continuous monitoring of units during wave propagation over the Coastal Armor Unit (CAU). The stability analysis of breakwater sections and subsequent evaluation are conducted through visual observations of their physical behaviour. As outlined in this section, the assessment of damage for rubble mound sections is consistently based on the number of units method. A comprehensive description of the assessment of armour layers for breakwaters is discussed in Chapter 5 of Sundar and Sannasiraj (2019).

3.16 Concluding Remarks

The basis of designing rubble mound structures lies in assessing topography, hydrography, still-water characteristics and wave dynamics. Key structural considerations encompass the attributes of armour units, including shape, dimensions, specific weight, individual unit weight, porosity, thickness and interlocking qualities. Additionally, determining the still-water level involves accounting for tidal elevation, storm surges and the resulting wave setup at the site.

The capacity of a structure to endure environmental forces hinges primarily on factors such as wave climate, rock density and armour stone size. Reduction in the size of individual pieces due to abrasion or breakage can jeopardize the integrity of the structure, potentially leading to failure. Studies have shown that the loss of volume in armour stone due to abrasion over the lifespan of a rubble mound section can be quantified, and material properties can be correlated with wave climate and structural considerations to develop a damage model. However, the challenge of armour stone breakage caused by in-service motion remains unresolved, and further research is needed to understand stone movement, velocity and the probability of breakage.

Finally, while sectional design is conducted using stability criteria, notably employing the widely accepted Hudson's formula, it is recommended to validate the stability of the armour layer through physical model studies. None of the empirical models can substitute for the rigorous stability testing offered by scaled-down models.

References

Coastal Engineering Manual (CEM). (2004). Vols. 1–6, US Army Corps of Engineers, Washington, DC.

Coastal Engineering Research Centre (CERC). (1984). Shore Protection Manual, Vol. I & II. Department of Army, US Army Corps of Engineers, Washington, DC, USA.

Construction Industry Research and Information Association (CIRIA). (2008). Rock Manual.

EurOtop Manual. (2018). EurOtop Wave Overtopping of Sea Defenses and Related Structures. An Overtopping Manual Largely Based on European Research, But for Worldwide Application, p. 320.

Hudson, R.Y. (1958). Design of Quarry-Stone Cover Layers for Rubble-Mound Breakwaters; Hydraulic Laboratory Investigation. Research Report No. 2-2, US Army Engineer Waterways Experiment Station, Vicksburg.

Hudson, R.Y., Herrmann, F.A., Sager, R.A., Whalin, R.W., Keulegan, G.H., Chatham, C.E. and Hales, L.Z. (1979). Coastal Hydraulic Models, Special Report No. 5, US Army Engineer Waterways Experiment Station, Vicksburg.

Hughes, S.A. (1993). *Physical Models and Laboratory Techniques in Coastal Engineering*. Advanced Series on Ocean Engineering, World Scientific, Vol. 7. https://doi.org/10.1142/2154.

Oumeraci, H. (1984). Scale effects in coastal hydraulic modelling. In *Proceedings of International Association for Hydro- Environment Engineering and Research (IAHR), Symposium on Scale Effects in Modelling Hydraulic Structures*, Esslingen, Germany.

Owen, M.W. and Allsop, N.W.H. (1983). Hydraulic modelling of rubble mound breakwaters. *Proceedings of Conference on Breakwaters, Design and Construction*. ICE, London.

Pedersen, J. (1996). Wave forces and overtopping on crown walls of rubble mound breakwaters: An experimental study, Ph.D Thesis, Aalborg Universitet, https://core.ac.uk/download/pdf/60299542.pdf.

Pedersen, J. and Burcharth, H.F. (1992). Wave forces on crown walls. *Coastal Engineering*, American Society of Civil Engineers, pp. 1489–1502.

Sharp, J.J. (1985). Scaling the weight of breakwater armor units. *Journal of Waterway, Port, Coastal, and Ocean Engineering*, 111(3). https://doi.org/10.1061/(ASCE)0733-950X(1985)111:3(594).

Sundar, V. and Sannasiraj, S.A. (2019). *Coastal Engineering Theory and Practice*. Advanced Series on Ocean Engineering, Vol. 47. World Scientific. https://doi.org/10.1142/11148.

van der Meer, J. (1988). Rock slopes and gravel beaches under wave attack. Delft Hydraulics Laboratory, Ph.D. Dissertation.

van der Meer, J.W. and Heydra, G. (1991). Rocking armour units: Number, location and impact velocity. *Coastal Engineering*, 15, 21–39.

van der Meer, J.W. and Janssen, W. (1995). Wave run-up and wave overtopping at dikes. *Wave Forces on Inclined and Vertical Wall Structures*, American Society of Civil Engineers.

van der Meer, J.W. and Stam, C.-J.M. (1991). Wave runup on smooth and rock slopes. Publications no. 454, Delft hydraulics, Netherlands.

Van Gent, M.R.A. (2001). Wave runup on dikes with shallow foreshores. *Journal of Waterway, Port, Coastal, and Ocean Engineering*, 127(5). October. https://doi.org/10.1061/(ASCE)0733-950X(2001)127:5(254).

Chapter 4

Principles of Physical Modelling

4.1 Introduction

Prior to advancements in computers and computing capabilities, researchers primarily relied on physical models to address various challenges in coastal, harbor, and offshore engineering. Even in today's technological age, physical modelling remains the primary and dependable approach for comprehending intricate phenomena in coastal engineering. Key issues such as wave–structure interaction, coastal processes, sediment dynamics and near-shore characteristics persist as widespread problems in this field. The inherent randomness of sea states, sediment dynamics, vortex formation and flow through porous media makes precise modelling using numerical techniques challenging, despite significant advancements in this area over the years. Physical modelling provides engineers, researchers, scientists and planners with a real-time visualization of the various phenomena occurring in the vicinity of proposed structures. This chapter delves into a comprehensive discussion on various aspects of physical modelling.

4.2 Dimensional Analysis

4.2.1 *General*

The initial step in physical modelling involves conducting a dimensional analysis that encompasses all variables relevant to defining the problem. The primary goal of dimensional analysis is to reduce the number of variables while emphasizing the foundation for physical modelling and the principles of similarity in physics. This process includes identifying crucial independent variables to derive one or more dimensionless parameters that aid in describing the problem.

The fundamental representation of all variables can be distilled into three basic units: mass, length and time. Additional physical variables defining a problem in coastal engineering may encompass the height and period of the wave, sediment size and structural characteristics. Table 4.1 provides

Table 4.1. Unit representation and scale factor for various parameters.

Variable	Unit	Definition
Geometry		
Length	L	A dimension representative of the features of the object
Area	L^2	Area exposed or projected onto a flat surface
Volume	L^3	For any segment of the object
Kinematics and dynamics		
Moment of Inertia Area	L^4	
Moment of Inertia Mass	ML^2	Considered with respect to a fixed point
Centre of Gravity	L	Determined from a reference point
Time	T	The identical reference point (e.g. initial time) is treated as zero time
Acceleration	LT^{-2}	Rate of change of velocity
Velocity	LT^{-1}	Rate of change of displacement
Displacement	L	Zero position is designated when at rest
Angular Acceleration	T^{-2}	Rate of change of angular velocity
Angular Velocity	T^{-1}	Rate of change of angular displacement
Angular Displacement	None	Reference is set at zero degrees
Spring Constant (Linear)	MT^{-2}	Force per unit length of extension
Damping Coefficient	MT^{-1}	Viscous resistance opposing oscillation
Damping Factor	None	The ratio of the damping coefficient to the critical damping coefficient
Natural Period	T	Period at which inertia force equals restoring force
Momentum	MLT^{-1}	Mass times linear velocity
Angular Momentum	ML^2T^{-1}	Product of mass moment of inertia and angular velocity
Torque	ML^2T^{-2}	Product of tangential force and distance
Work	ML^2T^{-2}	Product of force applied and distance moved
Power	ML^2T^{-3}	Rate of work
Impulse	MLT^{-1}	Product of constant force and its short duration of time
Force, Thrust, Resistance	MLT^{-2}	The influence of one body on another to alter or induce a change in the state of motion of the acted-upon body
Statics		
Stiffness	ML^3T^{-2}	Product of modulus of elasticity and the moment of inertia, EI
Stress	$ML^{-1}T^{-2}$	Force on an element per unit area
Moment	ML^2T^{-2}	The product of the applied force and its distance from a fixed point

(*Continued*)

Table 4.1. (*Continued*)

Variable	Unit	Definition
Shear	MLT^{-2}	Force per unit cross-sectional area parallel to the force
Section Modulus	L^3	Area moment of inertia divided by the distance from the neural axis to the extreme fibre
Hydraulics		
Kinetic Energy	ML^2T^{-2}	The potential of a body to perform work based on its configuration
Pressure Energy	ML^2T^{-2}	Energy resulting from pressure head
Potential Energy	ML^2T^{-2}	The ability of a body to perform work based on its configuration
Loss to friction	ML^2T^{-2}	Energy or work dissipation caused by friction
Scour		
Grain size of sediments	L	For the same prototype material
Free fall velocity	LT^{-1}	Final velocity of a freely falling particle in a medium
Wave mechanics		
Wave Height	L	Distance between the crest and the trough
Wave Period	T	Time between two successive crests crossing a point
Wave Length	L	Distance between two successive crests at a given time
Celerity	LT^{-1}	Speed of the wave
Particle Velocity	LT^{-1}	Particle velocity is the velocity of particle of media at any instant as a progressive wave travels through media
Particle Acceleration	LT^{-2}	The rate of change (or time derivative) of velocity
Particle Orbit	L	Path of a water particle (closed or open)
Wave Elevation	L	Variation of the free surface about a level (MSL)
Wave Pressure	$ML^{-1}T^{-2}$	Force exerted by a water particle per unit area
Stability		
Displacement (Volume)	L^3	Volume of water moved by a submerged body
Righting & Overturning Moment (Hard Volume)	ML^2T^{-2}	Moment about a fixed point due to displaced weight and static (dead) weight, respectively
Natural Period	T	Natural period of oscillation in calm water resulting from an initial disturbance
Metacentre	L	Instantaneous centre of rotation
Centre of Buoyancy	L	Distance of C.G. of displaced volume from a fixed point

(*Continued*)

Table 4.1. (*Continued*)

Variable	Unit	Definition
Material properties		
Density	ML^{-3}	Mass per unit volume
Modulus of Elasticity	$ML^{-1}T^{-2}$	Ratio of tensile or compressive stress to strain
Modulus of Rigidity	$ML^{-1}T^{-2}$	Ratio of shearing stress to strain

Note:

1. The theory of dimensional analysis is entirely algebraic in nature.
2. Problems are articulated through dimensionally homogeneous equations.
3. One can utilize the governing differential equation of a phenomenon or any preliminary theory to identify all variables that may exert influence on the same.

a detailed breakdown of the basic unit representation for the majority of parameters associated with Maritime hydraulics.

4.2.2 *Rayleigh's method*

Rayleigh's method is a technique used in mathematical analysis and optimization to determine a parameter for a dependent variable based on the maximum values of three or four independent variables. The method is particularly applicable when dealing with situations where the number of independent variables is limited, and using more variables would complicate the analysis. Let us consider X to be a dependent variable that is a function of three independent variables, X_1, X_2 and X_3. As per the Rayleigh's method,

$$X = f(X_1, X_2, X_3). \tag{4.1}$$

It can be rewritten as

$$X = K \quad X_1^a, X_2^b, X_3^c. \tag{4.2}$$

Herein, "K" is a non-dimensional constant and a, b, c are arbitrary powers that are obtained by comparing the powers of fundamental dimensions (dimensional homogeneity).

Rayleigh's method is often used in optimization problems where the goal is to find the optimal values of the independent variables to maximize or minimize a certain objective function represented by the dependent variable. Keep in mind that the method assumes that the relationship between the variables is well behaved and can be adequately modelled by the given mathematical expressions.

4.2.3 *Buckingham's Pi theorem*

Buckingham's Pi theorem is a powerful tool in dimensional analysis to obtain a relationship between the variables involved in a given physical problem. This theorem provides a method for determining the dimensionless, non-redundant and independent dimensionless groups (π-groups) that describe the relationships between physical variables in a given problem. Dimensional analysis is particularly useful when seeking to understand the behaviour of physical systems and when there is uncertainty about the form of the functional relationship between variables. The theorem states that "*If there are 'n' variables in a problem and these variables contain 'm' primary dimensions (for example M, L, T) the equation relating all the variables will have (n-m) dimensionless groups.*"

The dimensionless groups are represented by π and are called π-groups. According to this theorem, the final expression is given by

$$\pi_l = f(\pi_2, \pi_3, \ldots, \pi_{n-m}).$$

The goal of constructing dimensionless groups (π-groups) is to obtain a set of independent dimensionless ratios that characterize the behaviour of a physical system. These groups are constructed in such a way that each π-group is independent of the others, and no π-group can be formed by multiplying the powers of other π-groups. Compared to Rayleigh's method of solving simultaneous equations, this method is a lot more advantageous and simpler.

Example: Group the variables of a vegetal belt response subjected to both wave and current actions. The following variables of vegetal response are of interest for both currents and waves.

$$\text{Vegetal Response} = f(\rho, g, h, H, T, B_s, D, D_b, BG, f_1, l, SP, E, L, \beta, R_u, V), \quad (4.3)$$

where ρ – mass density of water; g – gravitational acceleration; h – depth of water in the flume; H – wave height; T – wave period; B_s – width of the structure; D – diameter of the vegetation; D_b – diameter at the root of the vegetation; BG – width of green belt; f_1 – frequency of first mode of the vegetal stem; l – height of the vegetation; SP – centre-to-centre spacing between vegetation; E – modulus of plant stiffness; L – wavelength; β – beach slope; R_u – run-up.

The variables seen in Equation (4.3) are grouped as per Buckingham's Pi theorem, in which $U_{\max}$ is the velocity V_f and L_o is the deep-water wavelength. The non-dimensional quantities investigated in the study are Darcy's f and Manning's n in steady uniform flow and wave run-up R_u/H, pressures and forces on structures. They are designated as follows:

$$f = \left(\frac{8H_f gh}{LV_f^2}\right). \tag{4.4}$$

$$n = \left(\frac{R^{(2/3)} * S^{(1/2)}}{V_f}\right). \tag{4.5}$$

$$F* = \left(\frac{F_{\max}}{0.5 * \rho g H^2 B_s}\right). \tag{4.6}$$

$$P* = \left(\frac{P_{\max}}{H}\right). \tag{4.7}$$

4.3 Model Analysis

4.3.1 *General*

Prior to embarking on an engineering project, creating scaled-down models of the prototype can offer valuable insights into the upcoming endeavour. This process aids in explaining the fundamental physics associated with the problem. However, it is important to note that model studies for certain issues are constrained by a few limitations.

The application of general modelling principles to fluid mechanics problems was initiated by Reynolds and Froude, who conducted a series of experiments to establish the criteria for viscous and inertia effects. The modelling approach in ocean engineering has been expanded beyond fluid mechanics, considering the influence of varying environmental parameters.

Insufficiency of Analytical Methods: Sometimes, traditional analytical methods may not be adequate to solve complex problems or provide accurate predictions. In such cases, adopting model studies, which involve mathematical or computational models, can offer additional insights and solutions.

Determining Empirical Coefficients: Models are often used to determine empirical coefficients for known equations. By fitting experimental data into mathematical models, researchers can derive coefficients that enhance the accuracy and applicability of these equations.

Validation of Analytical Tools: When new analytical tools or techniques are introduced, it is crucial to validate their results. Model studies can be employed to compare and validate the outcomes of the new tools, ensuring their reliability and effectiveness in capturing the underlying phenomena.

Substantiating Results: Model studies can be utilized to substantiate or verify experimental results obtained through other means. This helps in building confidence in the accuracy and reliability of the findings, providing a more comprehensive understanding of the studied phenomena.

Investigating Failure Patterns and Mechanisms: Models are valuable in investigating failure patterns and mechanisms, especially in fields like engineering and materials science. By simulating conditions leading to failure, researchers can gain insights into the underlying causes and design more robust and reliable systems.

In summary, adopting model studies is a versatile approach that can address various challenges in research and problem-solving, ranging from insufficient analytical methods to the validation of new tools and the exploration of failure mechanisms.

The terminologies often used in model analysis and their definitions are given in the following:

Prototype: The actual system or field problem under investigation.

Model: A scaled-down substitute system designed to replicate the prototype.

Homologous: Corresponding conditions or parameters in the model and its prototype, ensuring a meaningful relationship between the two.

These terminologies are fundamental in the context of model analysis, especially in fields such as engineering, physics and other scientific disciplines where models are employed to gain insights into real-world systems or phenomena.

4.3.2 *Complete similarity*

The primary objective in many model studies is to determine numerical values, such as the destructive or ultimate load, which are dependent on various variables as discussed in the following.

Dimensional Analysis: To obtain accurate results, dimensional analysis is employed. This involves grouping the variables to form a complete set of dimensionless products, typically denoted as $\pi 1, \pi 2, \ldots, \pi n$.

Complete Similarity: Attaining perfect similarity is a theoretical scenario in which the dimensionless products (independent variables) of both the model and the prototype exhibit identical values. However, realizing complete similarity proves to be a formidable task in practical applications.

Influential Forces and Variables: Some forces, like surface tension and surface roughness, may influence the model but not the prototype. These forces are considered in the analysis, and dimensionless variables with secondary influences may be neglected.

Scale Effects: Scale effects, discussed later in the chapter, refer to the unavoidable differences in magnitude that occur in models. To mitigate scale effects, larger models are recommended.

In summary, the foregoing discussion highlights the challenges in achieving complete similarity between a model and its prototype, emphasizing the need for careful consideration of influential forces and variables. The use of dimensional analysis and larger models is suggested to improve the accuracy of model studies and minimize scale effects.

4.3.3 *Applications of model analysis*

Most marine structures such as breakwaters, jetties, groins, training walls, seawalls, piers, and intake and outfall structures can be tested for their interaction with waves as well as their effects on the coast. The behaviour and stability of coastal structures can be verified in the laboratory by applying a suitable model scale, the results of which can be extended in the field. The extent of sedimentation in a harbour approach channel, the influence of waves and tides, the behaviour of estuaries, the bottom stability of breakwaters and other gravity structures, and the behaviour of shorelines can be investigated through physical model studies.

4.4 Principles of Similitude

4.4.1 *General*

The breakdown of the key points emphasizing the importance of the principles of similarity, also known as principles of similitude, in designing, testing and predicting the behaviour of a model in relation to its prototype is discussed in the following:

Definitive Relation with Prototype: The behaviour of a model, when subjected to external forces, is expected to have a clear and definitive relationship with that of its prototype. This implies that the model's response should accurately reflect the expected behaviour of the actual system.

Principles of Similarity (or Similitude): The design, construction and testing of a model should adhere to certain principles known as the principles of similarity or similitude. These principles guide the process of creating a model that faithfully represents the relevant characteristics of the prototype.

Applicability to Testing and Prediction: The principles of similarity are not only relevant during the design and testing phases but also extend to the extrapolation of model results. This suggests that the principles guide the application of model data to predict the response of the actual prototype.

In summary, this passage underscores the significance of the principles of similarity throughout the entire process of model development, testing and the application of results to predict the behaviour of the prototype. Adhering to these principles ensures that the model provides meaningful and reliable insights into the real-world system it represents.

4.4.2 *Similitude in hydrodynamic problems*

The principle of similitude adopted for hydrodynamic problems needs to be treated separately from conventional similitude. The three major types of similitude, viz., geometric similitude, kinematic similitude and dynamic similitude, are explained in detail in the following. A convention of using "p" as a subscript for prototype and "m" as a subscript for model is followed.

The following are types of similitude:

Geometric Similitude: Geometric similarity pertains to the resemblance in physical structure or configuration between the prototype and the model. It involves maintaining the same dimensional proportions when scaling down from the prototype to the model. An undistorted model is one where both the horizontal and vertical scales are the same. In contrast, a distorted model has different horizontal and vertical scales. Departing from geometric similarity introduces distortion in the model. While geometric similarity is generally desired, definitive objectives can be achieved by departing from geometric similarity. Departures may be necessary due to constraints in laboratory testing facilities or other factors. A few problem areas are associated with departing from geometric similarity, including confining to the boundaries of laboratory testing facilities, the surface roughness of the model, scaling thickness and depth of water. An example that illustrates the preference for a distorted model is *warranted* in the study of an offshore structure deployed at greater depths is the investigation of a deep-sea oil platform. Designers need to study the behaviour and stability of a deep-sea oil platform in a depth of 1000 m exposed to various environmental forces due to waves and currents. The solution is to resort to a distorted model for the physical simulation. Herein, the model is scaled down but with different scales for vertical and horizontal dimensions. For instance, the horizontal dimensions might be scaled by a factor of 1:100, while the vertical dimensions are scaled by a factor of 1:50. This distortion helps to simulate the forces due waves more accurately than a uniform scale model. This approach demonstrates why distorted models are preferred in such studies, as they provide more detailed and practical insights into the complex interactions in deep-sea environments.

Kinematic Similitude: Kinematic similarity refers to the similarity of motions between a model and its prototype. It implies both geometric similarity and similarity of time intervals. Geometric similarity involves having the same length scale ratio between the model and prototype, while kinematic similarity requires the same time scale ratio. Therefore, the motions in both the model and prototype should exhibit similar characteristics in terms of length and time scales.

Dynamic Similitude: Dynamic similarity is accomplished when both the model and prototype exhibit congruent ratios in terms of length scale, time scale and force scale (mass scale). This means that not only are the geometrical and temporal aspects (kinematic similarity) matched but the forces involved are also scaled appropriately. To achieve dynamic similarity, three key ratios must be the same: length scale ratio, time scale ratio and force scale (mass scale) ratio. This ensures that the forces acting in the two systems are scaled proportionally. This underscores that within dynamically similar systems, the proportion of any two forces in one system must match the proportion of the corresponding forces in the other system. This equivalence of force ratios is crucial for dynamic similarity. While kinematic similarity can be achieved without considering any properties of the fluid, dynamic similarity involves a more comprehensive matching of properties, including force scales. Achieving dynamic similarity is considered more challenging, especially when fluid properties are involved. Kinematic similarity

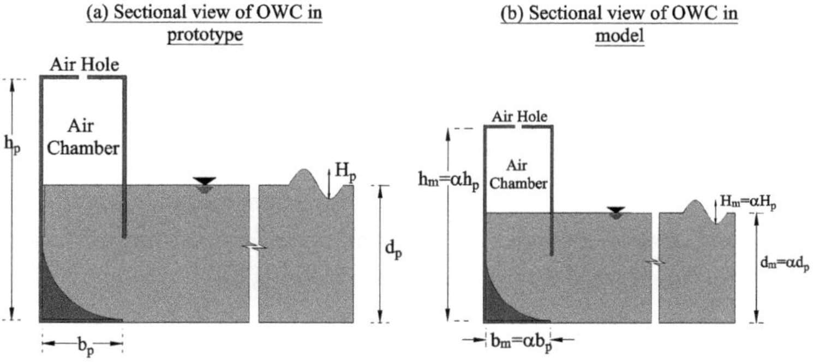

Hydrodynamic Parameters	Dimension	Froude scale
Mass	[M]	α^3
Force	$[MLT^{-2}]$	α^3
Pressure and stress	$[ML^{-1}T^{-2}]$	α
Energy and work	$[ML^2T^{-2}]$	α^4
Power	$[ML^2T^{-3}]$	$\alpha^{3.5}$

Fig. 4.1. Geometric, kinematic and dynamic similitude achieved by a model of an oscillating water column, OWC.
Source: Ashlin *et al.* (2017).

can be achieved without considering any properties of the fluid, which is unlike the dynamic similarity:

$$\frac{F_m}{F_P} = \frac{M_m a_m}{M_p a_p} = \frac{\rho_m L_m^3}{\rho_p L_p^3} \times \frac{\lambda_L}{\lambda_T^2} = \lambda_\rho \lambda_L^2 \left(\frac{\lambda_L}{\lambda_T}\right)^2 = \lambda_\rho \lambda_L^2 \lambda_u^2.$$

Figure 4.1, depicting a model of an Oscillating Water Column (OWC), emphasizes the use of Froude's scaling law to achieve similitude between the model and the prototype in the context of wave energy conversion studies.

4.4.3 *Scale effects*

Model scales are often derived from dimensional analysis or governing equations. This involves scaling down certain parameters to create a model that represents the prototype. Achieving complete similarity between a model and a prototype is impossible due to certain quantities, such as gravity, that cannot be easily scaled down. Gravity, fluid viscosity and density are noted as examples of parameters that remain the same in both the model and prototype. As a result, these parameters remain constant in both the model and the prototype. It is in fact challenging to simulate both gravity-driven forces (e.g. waves and currents) and viscous effects concurrently in model studies. The preference is often given to simulating gravity accurately, which may lead to challenges in reproducing viscosity effects properly. The non-similarity arising from the inability to scale certain fundamental parameters is referred to as "scale effect". This implies that certain aspects of the model behaviour may deviate from what is observed in the prototype due to these inherent non-similarities. The trade-offs made in modelling can lead to "scale effects", where certain phenomena may not be accurately reproduced in the model compared to the prototype.

Model laws: The principles guiding the design of models for dynamic similarity are denoted as model laws or laws of similarity. These model laws are categorized based on their specific applications, namely Reynolds' model law, Froude's model law, Euler's model law, Weber's model law and Mach's model law. The specifics of various model laws are covered in several foundational texts and are thus omitted here. However, interested readers can consult Sundar and Sannasiraj (2019) for more in-depth information.

Typical case study: A sustainable seawall cross-section comprising different geo-synthetic products was proposed to be erected along Pallana Beach (19°17′55:19″ N and 76° 23′18:55″ E), located in Alleppey district of Kerala along the southwest coastline of India. Since the subsoil at the location is poorly graded, it was decided to replace conventional materials like rock boulders and concrete armour units with geo-synthetic products.

A comprehensive physical model study was conducted to assess the hydrodynamic performance (i.e. reflection, run-up and pressure distribution) of the geo-synthetic seawall cross-section for a wide range of random wave characteristics and two water depth conditions.

In line with the newly adopted materials and techniques in coastal protection, an innovative cross-section of a seawall comprising different geo-synthetic units, such as geo-synthetic reinforced soil-retaining walls (GRW/barrier force system), geo-bags (small bags and mega bags), geo-cells, geo-grids, and geo-textiles (woven and non-woven), was proposed as shown in Fig. 4.2. The cross-section is shaped by filling up the geo-synthetic units with locally available soil/sediments; the geo-bag components are to be filled up to 80% of their volume (Den Adel *et al.*, 1996), and owing to the packing density and multiple layers of geo-synthetic materials, the structure will be weakly permeable.

The physical model testing for the proposed seawall design at Pallana Beach involved a scale model of 1:8.5 to fit within the existing flume facility.

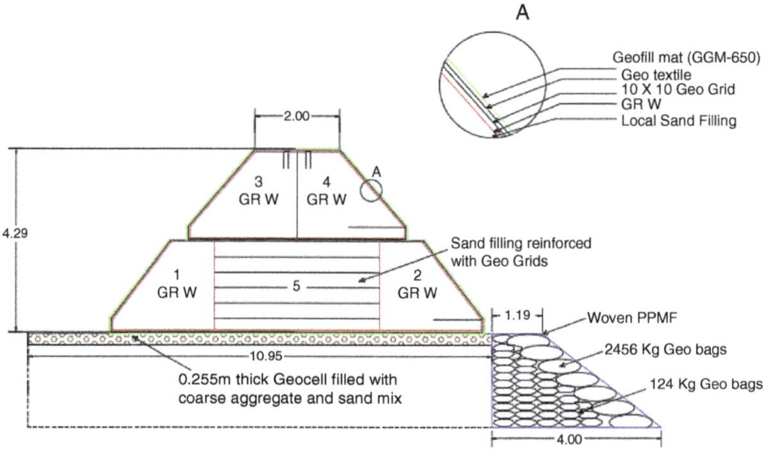

Fig. 4.2. Cross-section of the sea wall prototype (Sukanya *et al.* (2022)).

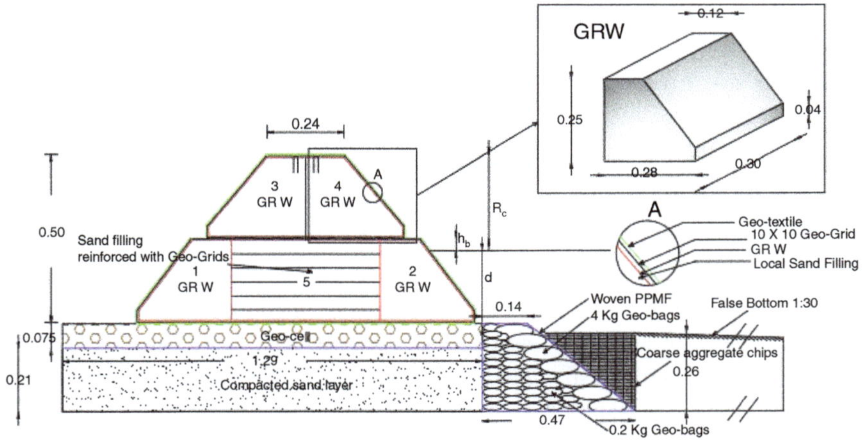

Fig. 4.3. Cross-section of the scaled-down seawall model (Sukanya *et al.* (2022)).

The GRW units are interconnected and stitched using high-denier polyester thread.

Each GRW unit is 0.25 m high, 0.28 m wide at the bottom and 0.12 m wide at the top. Two types of geo-bags are used in the model – 0.2-kg bags made from non-woven geo-textile and 4 kg bags made from woven geo-textile. The geo-bags constitute the excavated toe of the structure.

Geo-cells, with a thickness of 0.75 m, serve as the foundation layer of the seawall. Geo-grids, with the required length, are used to reinforce the structure. Both woven and non-woven geo-textiles are used in the design. These materials contribute to the overall stability and permeability of the structure.

The model seawall's cross-section, as shown in Fig. 4.3, includes the interconnected geo-synthetic units filled with locally available soil/sediments. The construction sequence adopted in the laboratory is not explicitly mentioned but may involve assembling the geo-synthetic components in a specific order to mimic the proposed seawall design accurately.

The model testing considered variations in water levels to analyze the structure's performance under different conditions. The small-scale model provides valuable insights into the hydrodynamic behaviour of the innovative seawall design, offering a practical approach to assess its effectiveness in coastal protection.

The model was subjected to the action of predefined spectral characteristics. The complete details of this study are discussed by Sukanya *et al.* (2022).

4.4.4 *Scale effects on impact dynamics*

The design of coastal structures is intricately influenced by numerous factors, with impact force emerging as a pivotal parameter. As elucidated in Chapter 1, the approach of Cuomo *et al.* (2010) showcases commendable efficacy within the impact zone, whereas the Froude method exhibits superior applicability in the uprushing zone or non-impact region. Nonetheless, the current quandary involves determining the optimal method for extrapolating impact force from pressure measurements.

Subsequent analysis revealed that employing the Froude method for upscaling impact pressure in the impact zone proved impractical while utilizing the Cuomo method in the non-impact zone led to an underestimation of pressure. Consequently, a rational approach for determining force through pressure integration entails applying the Cuomo method in the impact zone and the Froude method in the uprushing zone. Nonetheless, this hypothesis warrants meticulous examination.

The proposed methodology of Ravindar *et al.* (2021), coined the Cuomo–Froude method, to accurately predict impact forces from pressure measurements was subject to rigorous examination.

The cumulative horizontal force on the structure is measured as follows using Equations (4.7a), (4.7b) and (4.8) and the pressure measurements:

For large scale,

$$F_h = \sum_j P_j \times \Delta z_j. \tag{4.7a}$$

For small scale (Cuomo method),

$$F_h = \sum_k \lambda_s \times P_k \times \Delta z_k. \tag{4.7b}$$

For small scale (Cuomo–Froude method),

$$F_h = \sum_i \lambda_s \times P_i \times \Delta z_i + \sum_l \lambda \times P_l \times \Delta z_l. \tag{4.8}$$

Herein,

P_j: Pressure transducer j readings in large scale, with $j = 1, 2 \dots 16$.
P_k: Pressure recorded by the transducer k in small scale, with $k = 1, 2 \dots 7$.
P_i: Pressure recorded by the transducer i in small scale only in the impact zone, $(i = 1, 2, \dots 4)$.
P_l: Pressure recorded by the transducer l in small scale only in the uprushing zone, $(l = 5, 6, 7)$.
$\Delta z_j, \Delta z_k, \Delta z_i, \Delta z_l$: Distance between two adjacent transducers on the seawall.
λ_s: Cuomo scaling factor calculated for each test case.
λ: Froude scaling factor.

In calculating the horizontal force on the recurved parapet segment, pressure measurements undergo multiplication by the angle to isolate the

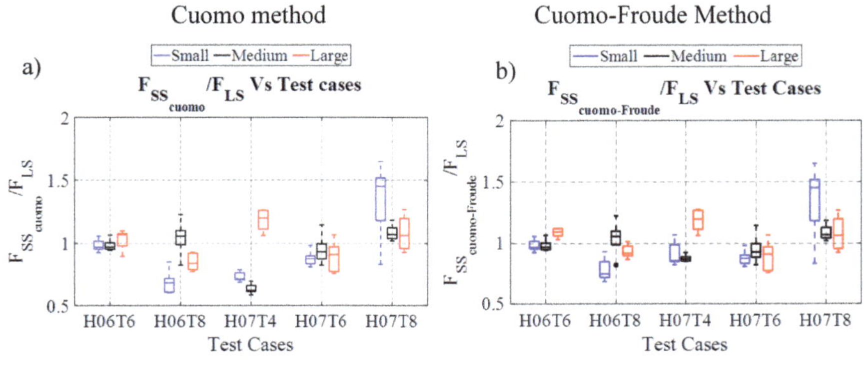

Fig. 4.4. Quartile representation of local force peak ratio for upscaled model and prototype for all three recurved parapet types using (a) Cuomo method and (b) Cuomo–Froude method (Ravindar *et al.* (2021)).

horizontal component. As a result, the derived value F_h signifies the force, specifically directed shoreward, that acts exclusively on the entirety of the seawall recurved parapet structure. Figure 4.4 depicts the quartile representation of the local force peak ratio between small-scale and large-scale scenarios utilizing both the Cuomo method (Fig. 4.4(a)) and the Cuomo–Froude method (Fig. 4.4(b)) across three distinct parapet types: small, medium and large recurved parapets (denoted as 480, 700 and 900, respectively). Upon examination of the figure, it is evident that the force ratio demonstrates variations within ±0.5 units for the majority of parapet cases when comparing small-scale and large-scale scenarios.

Upon comparing Figs. 4.4(a) and 4.4(b), it becomes apparent that the Cuomo–Froude method exhibits a slight improvement over the Cuomo method alone. This is underscored by the median of force ratios, which tends to centre on 1 for the majority of recurved parapet types. However, it is important to acknowledge that the disparities between the two methods are not notably pronounced. This minimal variation in results is attributed to the primary force contribution, particularly for maximum or peak force, originating from the impact zone.

4.4.5 *Why physical model?*

One of the pertaining questions that arises is the need for physical model studies. The objective of physical model studies is to simulate unknown physical prototype behaviour in the past, present and future. It can be used for prospective analysis and design as well as for retrospective analysis. In the case of prospective analysis, it will be helpful to predict and understand the future behaviour of the prototype system. In the latter case, one can conduct hindcast and understand the past behaviour of the prototype system. In both cases, the goal is to reduce or minimize error sin understanding the physical system under consideration.

Even though physical modelling has a lot of benefits and advantages, it does have limitations. The differences between the up-scaled model and prototype, according to the literature, are caused by the following effects, which are divided into model and laboratory effects, scale effects and non-repeatability effects, as seen in Table 4.2. The disparities caused by wave generation, type of water used for measuring, materials used for model design, choice of instruments used and their locations, and measurement results (Schüttrumpf and Oumeraci, 2005) must be considered in model and laboratory effects (Yalin, 1971; Ivicsics, 1978; Kobus, 1980; Novak, 1984;

Table 4.2. Classification of effects between small scale and large scale in an impact pressure study.

Scale effects	Model effects (including Measurement effects)	Laboratory effects	Non-repeatability effects
Yalin (1971), Le Méhauté (1990), Hughes (1993), Martin and Pohl (2000), and Heller (2007), Hudson (1958), Hudson *et al.* (1979)	Yalin (1971), Ivicsics (1978), Bretschneider (1980), Novak (1984), Hughes (1993), Schüttrumpf and Oumeraci (2005), and Kortenhaus *et al.* (2004)		Streicher *et al.* (2019)
1. scaling law 2. compressibility 3. surface tension 4. viscosity 5. fluid and structure properties	1. material properties used in model construction (surface roughness) 2. measurement setup 3. instrumentation choice	1. flume dimensions 2. wave generation and wave absorption systems – paddle leakages, control system 3. water properties 4. neglected processes like wind or currents	1. 3D effects of turbulent bore front or interactions 2. air entrainment 3. air entrapment – cushioning effect Bullock *et al.* (2007)

Hughes, 1993). Scaling rules, compressibility, surface friction, viscosity and water properties are discussed in scale effects (Yalin, 1971; Le Méhauté and Hanes, 2005; Hughes, 1993; Martin and Pohl, 2000; Heller, 2007). The 3D effects of turbulent bore front, air entrapment and air entrainment, i.e. the cushioning effect (Bullock *et al.*, 2007), must be considered in non-repeatability effects (Streicher *et al.*, 2019).

Scale effects are the variations in the reaction between the prototype and the model that result from the model's failure to replicate all applicable forces at the appropriate scale as dictated by the scaling (similitude) criterion (Hughes, 1993).

Laboratory effects are variations in prototype and model response caused by restrictions in laboratory facilities such as physical constraints of flow limits, re-reflection, unintended nonlinear effects caused by the use of mechanical wave and current generation, and water form selection. The differences in material properties used during model design, the measuring precision of the instruments used, the orientation inaccuracy of the wave probes, model manufacturing or fabrication inaccuracy, slope construction and the choice of water used all contribute to model results (Hughes, 1993). Differences in impact pressure/force calculation due to the stochastic action of the wave impact mechanism, resulting in non-repeatability of wave impact measurements, are known as non-repeatability effects (Streicher *et al.*, 2019).

For additional reading, refer to Chakrabarti (1994) and Den Adel *et al.* (1996). Irrespective of model size, the errors involved decrease with a large scale. Sometimes, physical models are more expensive to operate than numerical models. Thus, the following two requirements must be met (Hughes, 1993; Kamikubo *et al.*, 2000):

(a) Under the constraints of the study, the model should be equivalent to the prototype.
(b) The model data must be properly interpreted.

In coastal engineering, long-wave approximations (representing long periods compared to depth) are used. When one needs to model these long waves and their interaction with structures from the prototype to model scale, some necessary conditions need to be met. These conditions can be obtained by non-dimensionalizing the governing equations. The detailed derivation can be found in Hughes (1993). Only, the scaling requirements for long-wave models are provided in the following:

1. Long-wave hydrodynamic physical models may be either geometrically distorted or undistorted.
2. Viscous effects can only be modelled in a long-wave model when both Froude and Reynolds criteria are met.
3. Long-wave models maintain a similitude of vertical viscous shear stresses only when the model is geometrically undistorted and the Reynolds criterion has been met.

The types of models that are in practice are illustrated in the block diagram in Fig. 4.5. In this chapter, we discuss the measurement models. Young researchers think that model studies are easy. However, this is not the case; in fact, the following was pointed out by Svendsen (1985):

- If there is a discrepancy between a theory and the experiment carried out to verify it, it is likely to be due to inaccuracies in the experiment.

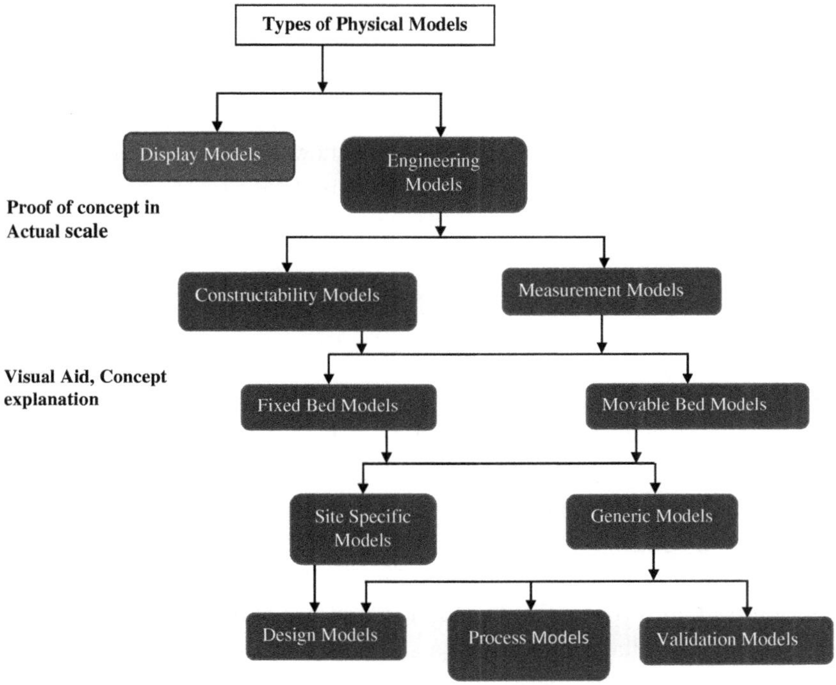

Fig. 4.5. Types of physical models.

- It is far more difficult to conduct good experiments than it is to make good theories.

These two statements are true to date even after advancements in measurement technologies and numerical models or theories.

4.5 Modelling of Mitigation Measures against Extreme Flows

4.5.1 *Background*

In coastal regions, the repercussions of extreme events are starkly evident through the catastrophic damage inflicted upon the social, environmental and economic fabric. Over recent years, the recurrence of these extreme events has been increasingly attributed to climate change (IPCC, 2021). Various methodologies proposed by researchers aim to mitigate the energy of these extreme flows, thereby reducing their impact on coastal structures. Additionally, proactive measures such as timely evacuation protocols and the installation of inundation-delaying structures are imperative. Protective structures deployed on site should align with economic feasibility relative to vulnerability levels. Introducing roughness elements, such as buffer blocks strategically positioned along the flow path, can effectively mitigate flow dynamics, resulting in moderated outflow downstream. Laboratory-generated extreme flows often employ the dam break mechanism for experimentation. This section explores two recent case studies elucidating these principles.

4.5.2 *Case study 1 – Energy dissipation using buffer blocks*

4.5.2.1 *Experimental setup*

Experiments were conducted in a flume of 31 m × 2.7 m × 2 m with a reservoir of length 14.43 m at the Department of Ocean Engineering, IIT Madras, India. Within the experimental setup, a horizontal false bottom is elevated to a height of 0.85 m above the flume's bottom on its downstream side, facilitating the installation of buffer blocks and aiding in water recirculation back into the reservoir. Upon the release of two pneumatic pistons, the gate opens radially, simulating a surge akin to an extreme event hitting the shore. Buffer blocks were strategically positioned 10 m away from the gate opening to engage with the flow dynamics effectively.

The prototype buffer block's height was limited to 80 cm to ensure comfortable seating during calm sea conditions. Following the guidelines of the USBR (Peterka, 1984), the width and length of the buffer blocks are determined based on ratios relative to the height, set at 1.5 for width and 0.5 for length. Thus, the prototype's dimensions are 80 cm × 120 cm × 40 cm (Percy *et al.*, 2023). The spacing in the lateral direction was the same as the width of the buffer block. The reservoir's water depth was varied from 20 cm to 40 cm for free runs without buffer blocks. From the experimental investigation, the maximum flow depth that can be attained at the buffer block position is around 12 cm. A Froude scale of 8 was opted, considering the buffer block height and inundation heights of the extreme flows. Therefore, the model buffer block dimensions are 10 cm × 15 cm × 5 cm. The ratio of maximum flow depth at the buffer block position to its height was taken as the dimensionless number known as the flow depth to buffer block height ratio (R). Free run cases were conducted for different ratios of flow depth to buffer block height (R) of above and below unity. Then, the buffer blocks were placed at the 10 m from the gate opening position and the experiments were repeated for same ratios of flow depth to buffer block height (R). The details of the testing flume with a single row of buffer blocks are

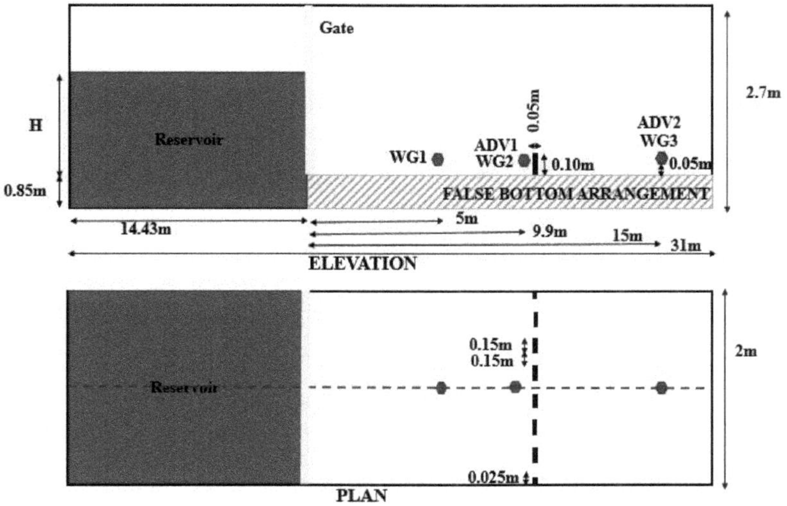

Fig. 4.6. Experimental flume with the buffer blocks (not to scale).

shown in Fig. 4.6. The wave gauges are positioned at 5 m, 9.9 m and 15 m from the gate opening positions at the centre of the flume, namely WG1, WG2 and WG3, respectively. Similarly, instruments for acoustic doppler velocimetry, ADV1 and ADV2, are placed slightly parallel to the WG2 and WG3 positions, respectively. The details of the test setup are discussed by Percy *et al.* (2024).

4.5.2.2 *Flow characteristics*

Flow depth and flow velocity values at WG2 and WG3 were taken from the wave gauges and acoustic doppler velocimetry instruments, respectively. From the flow depth and flow velocity values, the Froude number and momentum flux are calculated. The analysis focuses on the region where the Froude number lies within the range of 1–2, as shown in Fig. 4.7. Similarly, the momentum flux that gives the combined effect of flow depth and flow velocity is plotted in Fig. 4.8. It is to be noted that for the majority of extreme events, the flow occurs in this Froude regime (Fritz *et al.*, 2006). This region represents the quasi-steady phase, which is important for designing tsunami loads (Harish *et al.*, 2021).

Once the buffer blocks are positioned in the flow path, the gate is opened, allowing water to surge into the flume, colliding with the buffer blocks amidst a significant spray of air and water. This collision induces

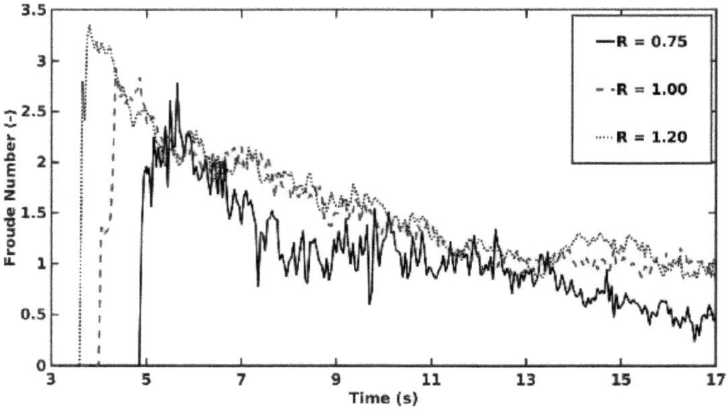

Fig. 4.7. Time series of Froude number for different flow depth to buffer block height ratios (R) at WG2.

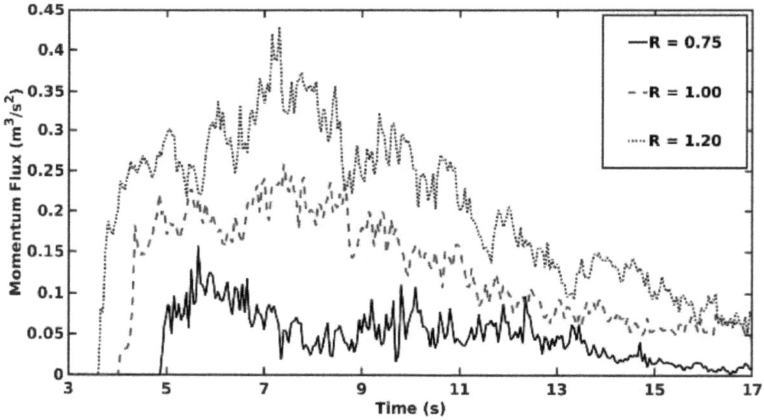

Fig. 4.8. Time series of momentum flux for different flow depth to buffer block height ratios (R) at WG2.

turbulence around the buffer blocks due to their obstruction, disrupting the flow field and causing a reduction in velocity in the surrounding areas. Consequently, water gets diverted through gaps between the blocks. This decrease in velocity in the buffer block vicinity leads to an increase in flow depth, effectively conserving momentum. Ultimately, this phenomenon diminishes flow properties downstream.

The reduction in flow property can be represented in a percentage reduction plot as shown in Fig. 4.9 at the WG3 position. As the flow depth

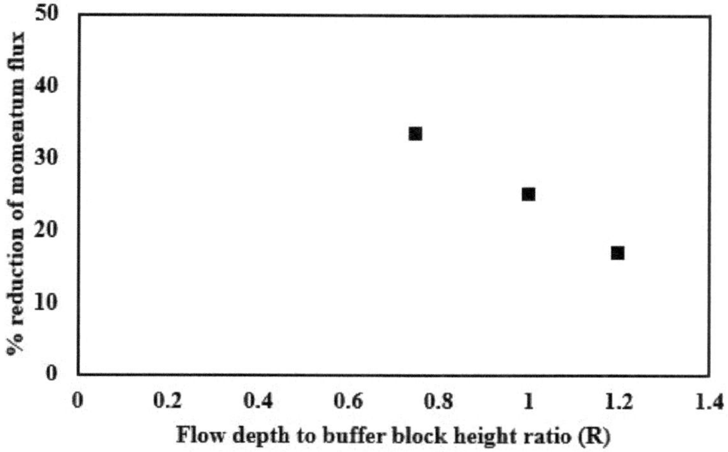

Fig. 4.9. Percentage reduction of momentum flux for different flow depth to buffer block height ratios, R.

to buffer block height ratio increases, the percentage reduction decreases. This emphasizes the importance of the maximum flow depth reaching the buffer block position. The higher the flow depth, the lower the reduction property of the buffer block on the leeward side. So, the addition of rows in a staggered manner can enhance the performance of the buffer blocks.

4.5.3 *Case study 2 – Forces on shorefront structures*

4.5.3.1 *General*

Many coastal structures are being destroyed due to loads generated by extreme flows. So, to understand extreme flow interaction with structures, many researchers have conducted experiments and arrived at design standards. Extreme flows are reproduced in the laboratory by generating (a) solitary waves using a piston-type wavemaker (Ramsden, 1996), (b) bores and surges using a dam break mechanism and controlling the valves of pumps and periods of bore generation (Chanson, 2006; Harish *et al.*, 2021), and (c) long-period waves using pumps (Goseberg *et al.*, 2013). In this study, a pump-type bore generator that discharges water from the underground reservoir was adopted.

The experiments were conducted by changing the valves of the pump and the periods of bore generation. The bore generation mechanism at

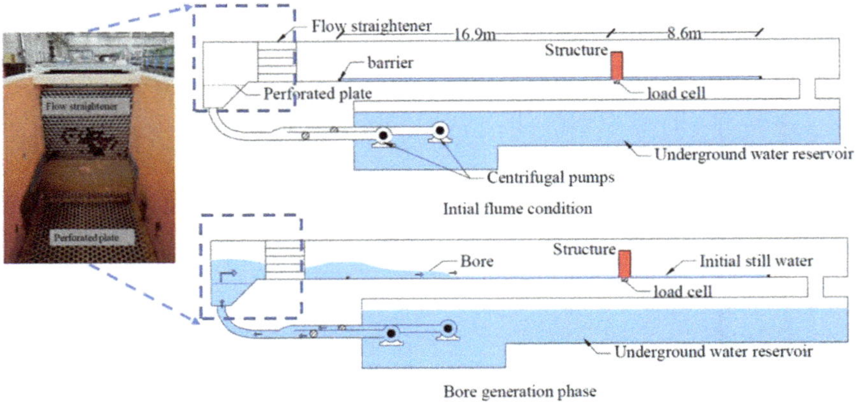

Fig. 4.10. Bore generation facility in IWW, RWTH Aachen University.
Source: Harish (2023).

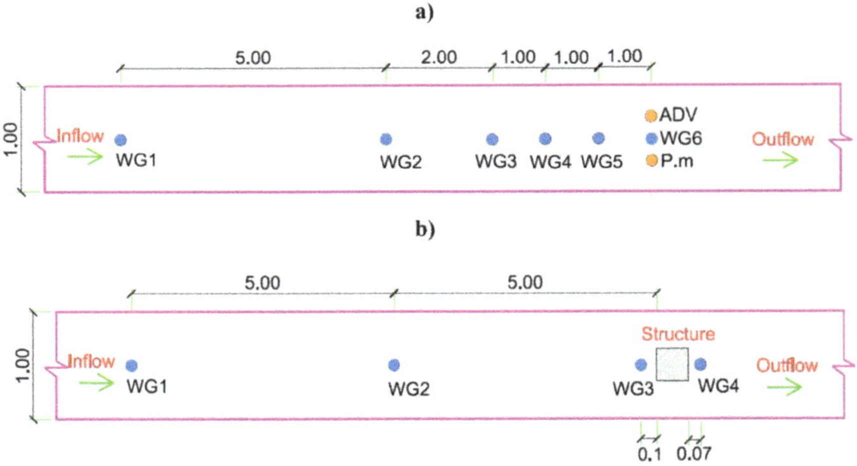

Fig. 4.11. Plan view of the flume showing the location of the instruments (a) without the structure and (b) with the structure.
Note: P.m = Propeller meter, ADV = Acoustic Doppler Velocimetry and WG = wave gauge (Ultrasonic depth sensors). All dimensions are in m.
Source: Harish (2023).

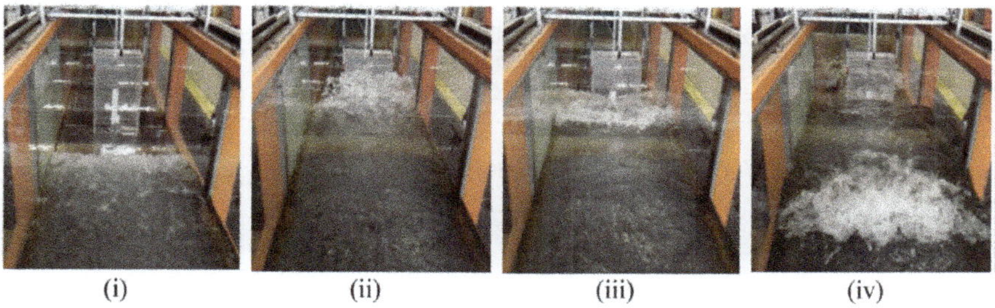

Fig. 4.12. Picture showing the bore's interaction with the structure at different phases of force: (i) just before interaction, (ii) impulsive pressure phase, (iii) bore pressure phase or run-up phase and (iv) quasi-steady phase.
Source: Adapted from Harish (2023).

IWW, RWTH Aachen University, is shown in Fig. 4.10 with a perforated plate and a flow straightener at the inlet point to reduce turbulence. Figure 4.11 shows the location of wave gauges, ADVs, the propeller meter and the instrumented structure in the flume. The bore's interaction with the structure at different phases of force is shown in Fig. 4.12. The entire analysis was carried out for quasi-steady phase, where most of the shorefront structure tends to fail. This study was conducted to develop an equation to estimate forces on coastal structures considering the effect of b/W and Fr, where b is the structure width, W is the width of the flume and Fr is the Froude number. The force was attained from the assumption of the hydrostatic pressure at the upstream and downstream sides of the structure (Harish, 2023). The bore height at the structure front (h_f) was evaluated from the conservation of mass and momentum equations (CMM). From the experimental investigation, the authors arrived at the following equation for C_d (Harish *et al.*, 2022):

$$C_d = \left(1.25\left(\frac{b}{W}\right) + 0.94\right) + \frac{(2(b/W) + 1.03)}{\mathrm{Fr}^2}. \tag{4.9}$$

4.6 Wave Generation in the Laboratory

4.6.1 *General*

One of the important tasks in conducting the physical model is the generation of waves that are of interest or required for the project. It can be achieved by a wave-generating mechanism. Different types of waves, such as regular, random, solitary, cnoidal, wave groups, freak/rogue waves, tsunami-like waves and nonlinear waves, can be generated in a laboratory

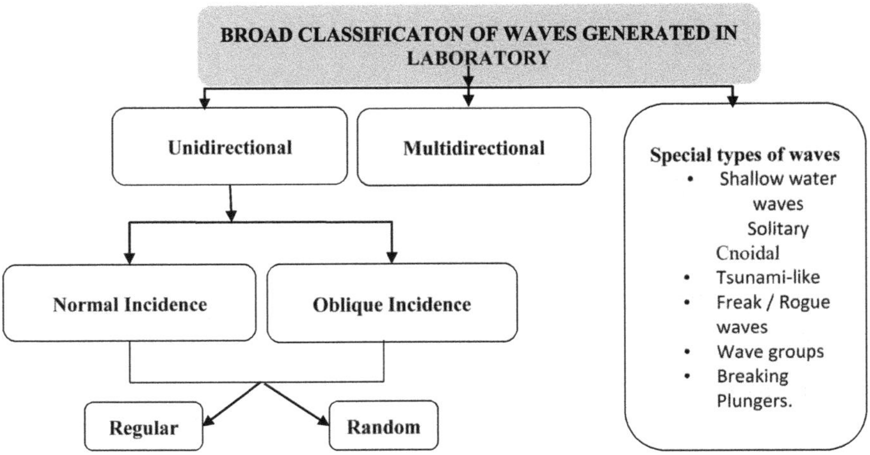

Fig. 4.13. Broad classification of laboratory-generated waves.

wave flume or basin. Among these, the most frequently observed types are depicted in Fig. 4.13. A detailed discussion on the laboratory-generated regular, random and directional waves is reported by Goda (1970) and Sundar (2021). The details of the generation of only a few of the important types of waves are discussed in this chapter.

4.6.2 *Generation of oblique waves*

The generation of oblique waves is possible using the multi-element wavemaker. A number of paddles installed at one end of a wave tank are operated at the same frequency and stroke, but the phase lag between the paddles is modulated such that a long-crested wave is generated at an angle to the wavemaker. See Fig. 4.14 for the explanation.

Let θ be the wave propagation angle made between the wave crest and the wave paddle elements:

$$\sin\theta = \frac{L}{NB}, \tag{4.10}$$

where N is the number of elements of width B in one oscillation along the wave paddle alignment.

Hence, the phase shift ϕ between adjacent elements can be written as

$$\phi = \frac{2\pi}{N}. \tag{4.11}$$

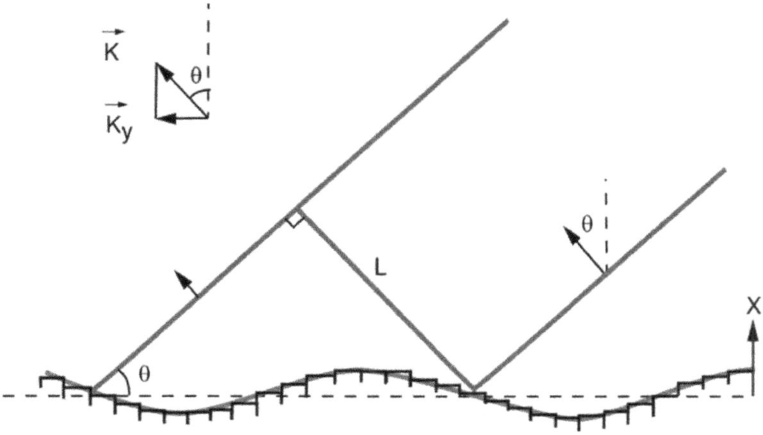

Fig. 4.14. Generation of regular oblique waves.

Hence,

$$\sin\theta = \frac{\phi L}{2\pi B} \tag{4.12}$$

$$\Rightarrow \quad \phi = \frac{2\pi B}{L}\sin\theta.$$

For a given paddle width B and the wavelength, the phase shift between adjacent paddles increases with the obliqueness of the wave. The phase shift decreases with the increase in the L/B ratio, as shown in Fig. 4.15.

Let the wave surface elevation at any point in the wave basin be expressed as

$$\eta(x, y, t) = \frac{H}{2}\cos(kx\cos\theta + ky\sin\theta - \omega t). \tag{4.13}$$

The paddle motion from its mean position along the wave paddle alignment ($x = 0$) is expressed as

$$x_p(y, t) = \frac{S}{2}\sin(ky\sin\theta - \omega t). \tag{4.14}$$

To generate a wave of height H and frequency f with a propagating angle of θ, the paddle displacement has to be increased by a factor of

$(1/\cos\theta)$:

$$\frac{H}{S} = \frac{TF_{\text{normal}}(f)}{\cos\theta}. \tag{4.15}$$

The transfer function of oblique wave generation becomes

$$TF_{\text{obl}}(f,\theta) = \frac{TF_{\text{normal}}(f)}{\cos\theta}. \tag{4.16}$$

However, the phase difference of the nth paddle should be considered. The movement of the nth paddle (number of paddles from the coordinate origin) can be defined as

$$x_p(n,t) = \frac{H}{2}\frac{1}{TF_{\text{obl}}(f,\theta)}\sin(kx\cos\theta - \omega t + \phi). \tag{4.17}$$

Substituting for ϕ,

$$x_p(n,t) = \frac{H}{2}\frac{1}{TF_{\text{obl}}(f,\theta)}\sin(kx_r\cos\theta - \omega t - k(nB - y_r)\sin\theta), \tag{4.18}$$

where n indicates the number of positions of the paddle and (x_r, y_r) are the coordinates of the reproduction point in the wave basin.

4.6.3 *Generation of freak/focusing and breaking waves*

Freak waves are defined as waves with height greater than two times the significant wave height (Kjeldsen, 1990). They are characterized as single, remarkably horizontally asymmetric and extremely high waves with a

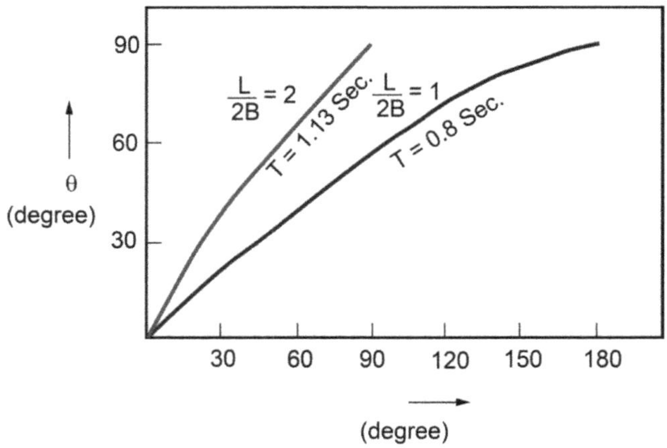

Fig. 4.15. Phase shift as a function of ratio of wavelength to paddle width.

broadening of the spectral bandwidth (Yasuda *et al.*, 1992). There are many pieces of evidence that show that freak waves exist in nature (Kjeldsen, 1984). It is believed that the phenomenon occurs both in deep and shallow waters. Much research has been carried out to find the exact cause of these freak waves, but it is still not clear. Sand *et al.* (1990) proposed wave focusing due to waves from various directions and superposition cause these waves. These waves are highly nonlinear in nature. Moreover, the wave heights are expected to be higher than those given by Raleigh's distribution (Dean, 1990).

Freak waves are simulated by considering several components of waves that are configured such that the maxima of all these waves appear at the same point – the point where the freak wave is desired. This ensures that the maximum wave height will occur at that point. Using linear progressive wave theory, the wave height and orbital velocity are calculated. A sufficient number of constituent waves are taken so that, in the time series, there is only one large peak – the freak wave. The following factors need to be considered:

a. The wave generator's range of time periods and wave heights,
b. Restriction in movement of the wave generator paddles,
c. Peaks are located based on the positive maximum and negative minimum, and waves between zero and crossing points are taken as one wave.

Input data include the following:

- Wave data file containing the number of component waves, and for each wave, the wave height, the period and the direction.
- The number of waves and the position of the freak wave.

The frequency and the directions of the waves can also be specified, and the programme generates the wave data with the given information.

By using this principle, wave breaking can also be generated in a physical model. Wave breaking in the deep ocean may result from direct wind forcing (Banner and Phillips, 1974), wave–current interactions (Longuet-Higgins and Stewart (1964), Skyner and Easson (1998)) and wave–wave interactions (Longuet-Higgins, 1974; Melville, 1983). Though wind forcing plays a role in provoking breaking in the field, it is difficult to create this field condition in the laboratory. We simulate breaking through wave–wave interactions in the laboratory. The temporally constructive interference proposed by Longuet-Higgins (1974) provided the principal methodology in the generation of a localized, unsteady breaking wave, resembling breaking in the field (Rapp and Melville, 1990). This methodology of wave focusing was widely used for the simulation of an isolated and repeatable two-dimensional breaker within a wave packet (Longuet-Higgins, 1974; Greenhow and Vinje, 1982; Chan and Melville, 1988; Rapp and Melville, 1990; Kway *et al.*, 1998). This had been extended to the generation of a three-dimensional breaker by adding directionality, i.e. spatial focusing and diffraction, to the wave's evolution (Nepf *et al.*, 1998).

The fundamental experiments for studying two-dimensional wave breaking, apart from wind-generated waves, fall into three main categories: (1) the focusing of two-dimensional waves in the lateral direction (Ramberg and Griffin, 1987); (2) the towing of a submerged object, such as a hydrofoil, to produce steady breakers (Duncan, 1981, 1983); and (3) the focusing of variable-length waves from a modulated or chirped wavemaker to produce unsteady breakers (Duncan *et al.*, 1994; Rapp and Melville, 1990; Skyner *et al.,* 1990; Baldock *et al.,* 1996; Greenhow and Vinje, 1982) or the overturning of an irregular wave train (Ochi and Tsai, 1983; Bonmarin, 1989) to produce unsteady breakers.

One of the most extensive studies on unsteady deep-water wave breaking is the work of Rapp and Melville (1990). The dispersive characteristic of deep-water waves was used to focus a wave packet and generate a single unsteady breaking event at a controlled location in the wave channel.

Several methodologies have been used in the past to simulate wave breaking in the laboratory. Steady spilling breakers have been produced by the relative motion of a submerged hydrofoil and the surrounding fluid (Duncan, 1981; Battjes and Sakai, 1981) and by the steady motion of a surface-piercing hull (Longuet-Higgins, 1974). On the other hand, focusing the wave energy in space can generate unsteady breaking waves. For two-dimensional waves, the energy may be focused by a laterally converging channel wall (Van Dorn and Pazan, 1975) or by a longitudinal focusing caused by frequency dispersion (Longuet-Higgins, 1974). Given the complexity of wave breaking, it is advantageous to study wave breaking in the absence of wind and bottom boundary influences. This can be achieved through the methodology of wave focusing (Chan and Melville, 1988). This technique is, in principle, similar to that described by Longuet-Higgins (1974) where wave energy is focused longitudinally by linearly decreasing the wavemaker's frequency and thus increasing the group velocity of the generated waves, leading to a focusing of energy. The modulated wave packets were generated by summing up sinusoidal wave components of discrete frequencies. Making use of the frequency dispersion given by linear theory, the relative phases were chosen so that all wave crests arrive simultaneously at a point in space and time, leading to constructive interference at the desired location. Using this method, the resultant wave is extremely repeatable in time and space.

The free surface displacement, $\eta(x,t)$, from linear theory, was given by

$$\eta(x,t) = \sum_{n=1}^{N} a_n \cos(k_n x - 2\pi f_n t - \phi_n), \tag{4.19}$$

where N = number of wave components, a_n = amplitude of the nth frequency component, k_n = wave number, f_n = nth frequency component, ϕ_n = phase angle, and k_n and f_n are related by the dispersion relation

$$(2\pi f_n)^2 = k_n g \tanh(k_n d). \tag{4.20}$$

The phase of each component was computed by setting

$$\cos(k_n x_b - 2\pi f_n t_b - \phi_n) = 1, \tag{4.21}$$

where x_b = theoretical wave-breaking location and t_b = time of occurrence of wave breaking.

This yielded

$$\phi_n = k_n x_b - 2\pi f_n t_b + 2\pi m \ (m = 0, \pm 1, \pm 2, \ldots), \tag{4.22}$$

and the surface displacement becomes

$$\eta(x,t) = \sum_{n=1}^{N} a_n \cos[k_n(x - x_b) - 2\pi f_n(t - t_b)]. \tag{4.23}$$

The discrete frequency f_n, for each wave component can be equally spaced across a bandwidth. The amplitude in each wave component, a_n, could be simplified depending on the spectral distribution adopted. By defining the mean paddle position as x = *0*, the desired surface displacement at the paddle was

$$\eta(0,t) = \sum_{n=1}^{N} a_n \cos[-k_n x_b - 2\pi f_n(t - t_b)]. \tag{4.24}$$

The theoretical surface displacements generated based on the above-provided equations were converted into input signals by a conversion factor called the Amplitude Transfer Function (ATF), which is a function of frequency. A typical free surface elevation is shown in Fig. 4.16.

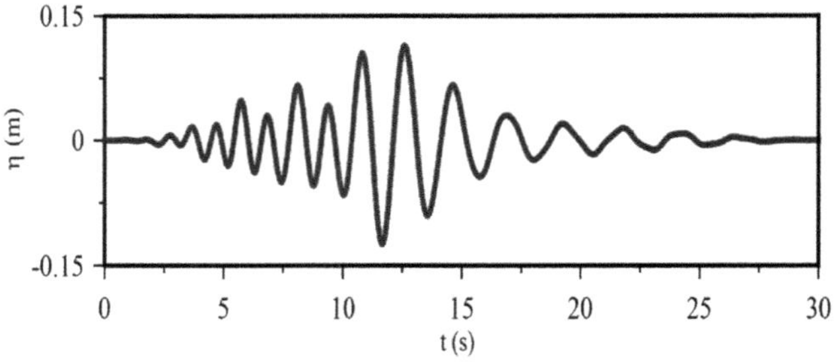

Fig. 4.16. Typical theoretically simulated breaking wave time history.

To eliminate any abrupt paddle movement, the first and last two seconds of the input signal were tapered with a half-cosine bell function. The desired wave packet can be generated using any computer-controlled wave simulation system. The success, however, depends on the careful incorporation of the wavemaker transfer function. In addition, the variable t_b has to be chosen systematically in order to ensure that the paddle motion starts with a relatively small stroke. The effect of changing t_b is delays or advances the entire wave signal with no influence on the theoretical focal point, x_b.

Although the breaking distance x_b was specified in the input to the algorithm, a deviation in the breaking location from that predicted by the linear theory was expected due to nonlinear effects, which became important as the amplitude increased. Another point to note was that because the breaking distance x_b had to be adjusted to observe different stages of breaking within the facility, it was realized that wave packets with different breaking distances had to be fine-tuned separately.

4.6.4 *Generation of wave groups*

4.6.4.1 *General*

In previous studies, researchers have discovered that maintaining constant spectral energy is not the sole consideration in designing structures; the sequence of waves within a wave train also plays a crucial role. One significant type of wave train is a group of waves, commonly referred to as wave groups. Despite ocean wave surface elevations typically assumed to follow normal processes in studies, the tendency for normal processes to exhibit groupiness is often limited.

Until the late seventies, structural testing under random wave conditions often overlooked the influence of wave groups. However, the failure of the Sines breakwater in Portugal brought attention to this oversight.

It was found that the breakwater's failure was partly due to the impact of wave groups, causing continuous oscillation of the armour unit and eventual instability of the primary layer, leading to their dislodging. Grouped waves can induce significant low-frequency motion in moored structures such as floating breakwaters or semi-submersible bodies, particularly when the natural period of the structure aligns closely with the modulation period of the natural waves.

The literature adequately covers the simulation of irregular waves with a specified target spectrum, but studies focusing on wave groups remain limited. Goda (1970, 1976, 1983) extensively discussed the importance of investigating wave groups and their impact on structures. An initial laboratory simulation of wave groups was conducted by Funke and Mansard (1980) using the SIWEH (Smooth Instantaneous Wave Energy History) method. However, this method was primarily relevant for specifying the height rather than the length of the wave group. Despite this limitation, several works (such as those by Mase *et al.*, 1983; Nolte and Hsu, 1972) have reported on the simulation of wave groups. Xu *et al.* (1993) proposed a method for simulating wave groups that considered both their length and height.

A procedure to simulate the presence of wave groups in a wave train is discussed here. The wave groups are simulated using the envelope approach with predefined spectral characteristics. The wave envelope is computed from the time series of wave elevation using the Hilbert transform technique. The spectral characteristics of waves and wave envelope time histories form the basis for the methodology. In order to identify the wave group, the group height factor (GHF), the relative height of the group and the group length factor (GLF), the relative length of the group is defined.

4.6.4.2 *Envelope Time History*

The envelope series can be obtained from the wave elevation time history using the Hilbert transform. The displacement of a narrow-banded sea surface is represented by

$$\eta(t) = \mathrm{Re}\{a(t)\, e^{i\phi(t)}\}, \tag{4.25}$$

where

$$a(t) = \sqrt{\{\eta^2(t) + \xi^2(t)\}}, \tag{4.26}$$

$$\phi(t) = \tan^{-1}\left\{\frac{\xi(t)}{\eta(t)}\right\}, \tag{4.27}$$

where $a(t)$ and $\phi(t)$ are real functions of time, representing the envelope and phase of $\eta(t)$, respectively, and $\xi(t)$ is the Hilbert transform of $\eta(t)$, defined as

$$\xi(t) = \frac{1}{\pi} P \int_{-\infty}^{\infty} \frac{\eta(t)}{t-\tau}\, d\tau, \tag{4.28}$$

where P denotes the Cauchy principal value of the integral evaluated at $\tau = t$. It has been proved that the Fourier transform (F) of $[\eta(t) + i\xi(t)]$ can be written as

$$F\{\eta(t) + i\xi(t)\} = \begin{cases} 2F\{\eta(t)\} & \omega \geq 0 \\ 0 & \omega < 0 \end{cases}, \tag{4.29}$$

where ω is the wave angular frequency. By using the Hilbert transform, it is easy to evaluate $\xi(t)$, and then $a(t)$ and $\phi(t)$ from $\eta(t)$.

In general, the sea surface displacements deviate from the normal distribution, and from a single spectrum, it is possible to generate a number of wave trains with markedly different groupiness.

4.6.4.3 *Characteristic factors of wave groups*

A simple way to represent the length of a wave group is by the run factors and run length. A group of high waves is referred as a run, and its length is measured in terms of the number of successive high waves that exceed a certain magnitude.

The characteristics of a wave group can be effectively specified using two factors, i.e. group height and group length. Funke and Mansard (1979) presented a groupiness factor (GF) to describe the degree of wave groups:

$$\text{GF} = \sqrt{\frac{m_{a,o}}{m_{\eta,o}}}. \tag{4.30}$$

The group height factor (GHF) characterizes the relative height of wave groups and is defined following the notations of Funke and Mansard (1979) as

$$\text{GHF} = \frac{m_{a,o}}{m_{\eta,o}}. \tag{4.31}$$

If $\eta(t)$ is assumed to be a normal process, it can be shown that

$$\text{GHF} = 2 - (2/\pi) \cong 0.43. \tag{4.32}$$

Thus, for the normal process, the GHF is a constant. But it is reasonable to consider the GHF to be a variable rather than a constant for a sea surface with marked wave groups. The values of the GHF for real wave groups, however, should not differ much from 0.43, because the sea surface does not deviate far from a normal process.

Similarly, the group length factor (GLF) characterizes the relative length of wave groups and is defined in terms of peak wave frequency, as this represents a narrow-banded process that depends largely on its spectral peak frequency:

$$GLF = \frac{f_p}{f_{a,p}}, \tag{4.33}$$

where f_p and $f_{a,p}$ are the peak frequencies of wave and envelope spectra, respectively.

4.6.4.4 *Wave group simulation procedure*

The procedure to simulate a wave group with the predefined group characteristics is as follows:

Step 1: The time series of $\eta(t)$ and $\eta_a(t)$ are generated using the method of superposition or Fast Fourier Transform from the specified target wave spectrum $S_\eta(\omega)$ and envelope spectrum $S_a(\omega)$, respectively. The time series should be made zero means.

Step 2: Compute zeroth spectral moments $\mathrm{m}_{\eta,o}$ and $\mathrm{m}_{a,0}$ from $S_\eta(\omega)$ and $S_a(\omega)$, respectively, and evaluate the GHF using Equation (4.32).

Step 3: Evaluate $\xi(t)$ from $\eta(t)$ using Equation (8.5), and then $\mathrm{a}(t)$ and $\phi(t)$ are obtained according to their description.

Step 4: The envelope $a_s(t)$ is computed using

$$a_s(t) = \eta_a(t) + \tilde{a}_s, \tag{4.34}$$

where

$$\tilde{a}_s = (m_{\eta,0}\,\pi/2)^{1/2}. \tag{4.35}$$

Step 5: The wave group $\eta_1(t)$ is constructed with $a_s(t)$ as an envelope and $\phi(t)$ as a phase function:

$$\eta_1(t) = a_s(t)\cos\phi(t). \tag{4.36}$$

On the assumption of the gaussian distribution of sea waves, the simulated time series is expected to deviate from the target in terms of spectral characteristics. The remedial procedure to correct the deviation in the GHF is described in the following:

Step 1: Evaluate $S_{\eta 1}(\omega)$ from the time history of $\eta_1(t)$ using FFT and estimate its zeroth moment $(m_{\eta 1,o})$.

Step 2: Compute the envelope $a'_s(t)$:

$$a'_s(t) = \eta_a(t) + \tilde{a}'_s, \tag{4.37}$$

where

$$\tilde{a}'_s = ((m_{\eta 1,0} + \delta m_o)\pi/2)^{1/2}, \tag{4.38}$$

$$\delta m_o = m_{\eta,o} - m_{\eta 1,o}. \tag{4.39}$$

Step 3: The corrected wave group $\eta_2(t)$ is constructed with $a'_s(t)$ as an envelope and $\phi(t)$ as a phase function. The new signal is

$$\eta_2(t) = a'_s(t) \cos\phi(t). \tag{4.40}$$

In most practical situations, the time series is required to be corrected only once. In cases where accuracy is not obtained, the above-mentioned procedure can be conducted more than once.

As the GHF characterizes the height of wave groups, the relative height of wave groups is rather sensitive to the GHF; a larger GHF corresponds to a higher wave group. It is to be noted that the deviation in simulated significant wave height is observed to be more as the GHF increases. It is concluded thus that the simulation procedure fails for larger values of GHF as it could not maintain the target characteristics. On observing the variation of the GLF, for a constant value of GHF, for accurate simulation of wave groups, the GLF should be more: GLF $\ll 1$. The group length increases as the group height decreases. The simulation procedure is applicable to cases in which sea surface elevation is represented by narrow-banded spectra. The envelope spectral peak should be sufficiently low to keep the value of the GLF higher. The group height decreases with an increase in group length. A typical wave group profile is shown in Fig. 4.17.

4.6.5 *Shallow-water wave generation in the laboratory*

According to water depth conditions, wave theories are classified into deep-water and shallow-water wave theories. This is based on the water depth to wavelength ratio (d/L). The small-amplitude wave theory developed by Airy (1845) can be applied to all ranges of d/L. Finite-amplitude wave theories are complicated due to the relative importance of the additional parameters, namely H/d and H/L. Keulegan and Carpenter (1958) classified finite-amplitude wave theories depending on d/L as shown in Fig. 4.18.

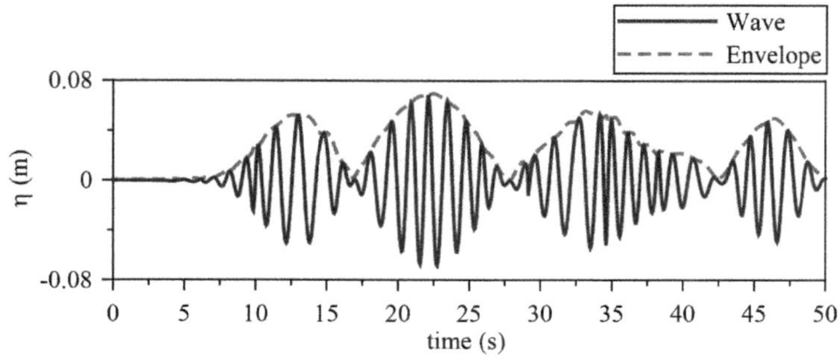

Fig. 4.17. Typical theoretically simulated wave groups.

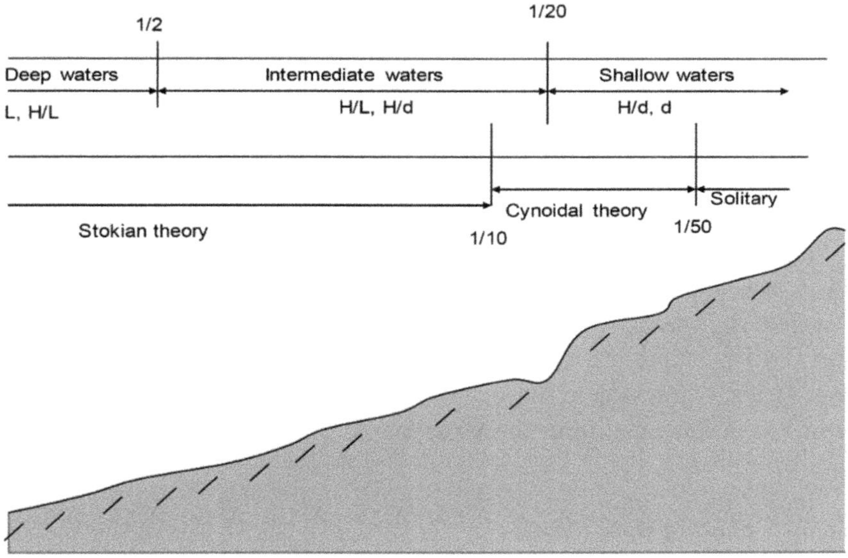

Fig. 4.18. Regions of validity of finite-amplitude wave theories.

Small-amplitude wave theory is based on the premise that motions are sufficiently small to allow free surface boundary conditions to be linearized, i.e. the terms involving wave amplitude to the second order and higher orders are considerably negligible. If the wave amplitude is large, it is a must to retain higher-order terms to obtain an accurate representation of wave motion.

The degree of wave nonlinearity escalates with an increase in the convective inertia term, a factor chiefly governed by three essential ratios: H/L,

H/d and L/d. Here, H represents wave height. In the context of deep-water conditions, typically characterized by small H/d and L/d ratios, the primary parameter dictating nonlinearity is H/L, also known as wave steepness. Transitioning from deep to shallow waters, the influence of wave steepness diminishes compared to the pivotal parameter H/d, termed relative height. This shift occurs as waves begin interacting with the seabed, thereby amplifying its influence on wave characteristics. In intermediate water depths, wave profile nonlinearity is influenced by both H/L and H/d ratios. Consequently, the relative significance of the convective inertia term can be elucidated using both ratios.

The Ursell parameter, derived from the ratio of wave elevation between first- and second-order wave theories, serves as a common metric for classifying linear and nonlinear wave theories in both deep and shallow waters. Alternatively, replacing H with the maximum wave surface elevation above the still water level (η_0), wherein it equals $H/2$ in linear theory, facilitates evaluation. If $\mathrm{Ur} \ll 1$, the linear small-amplitude theory applies. However, while the Ursell parameter provides a useful guideline, it may not sufficiently gauge the relative importance of nonlinear effects.

Figure 4.19 (Le Méhauté, 1976) illustrates a clear demarcation of various wave theories, along with limitations regarding wave height and period. Notably, the characteristics of Cnoidal waves become significant if $d/L < 1/8$ and $\mathrm{Ur} > 26.0$ (Le Méhauté, 1976; Chakrabarti, 1987). Applying the linear transfer function in these shallow-water regions results in spurious or error waves generated from the wave paddle.

4.6.5.1 *Solitary wave theory*

In very shallow waters, wave crests become peaked and trough becomes flattened. The entire crest is above the SWL, and the absence of the trough is a particular case of a shallow-water wave, called a solitary wave. Solitary waves have proved useful in engineering problems for studying very long waves like tsunamis, for determining wave properties near breaking in shallow water and for studying waves of maximum steepness in deep waters. Under such conditions, the wave characteristics are independent of L and T and depend only on H and d.

The generation of solitary waves by prescribing a "piston"-type wavemaker motion can be determined from the first-order solitary wave theory used by Goring (1978). The displacement of the wavemaker is given by the

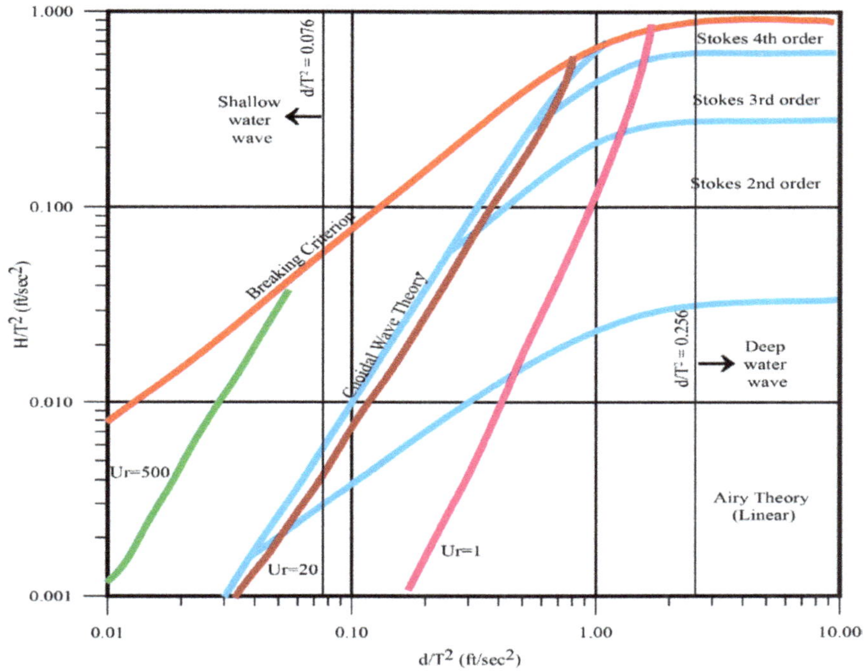

Fig. 4.19. Limits of validity of wave theories.
Source: Reworked from Le Méhauté (1976).

following:

$$x_p(t) = \frac{H}{\kappa}\left[\tanh\chi(t) + \tanh\frac{\kappa}{d}\lambda\right], \tag{4.41}$$

$$\chi(t) = \frac{\kappa}{d}(ct - x_p(t) - \lambda), \tag{4.42}$$

where H is the wave height, $\kappa = \sqrt{\frac{3H}{4d}}$ and celerity $c = \sqrt{g(d+H)}$.

The initial motion of the wave paddle is controlled by choosing a location where the theoretically infinitely long solitary wave is truncated. If at the place of truncation, the height is $\varepsilon_z H/d$, then it turns out that $\lambda/d = l/\kappa$, where $l = \cosh^{-1}\left[\frac{1}{\sqrt{\varepsilon_z}}\right]$.

It is clear from Equations (4.41) and (4.42) that they are implicit in x_p and χ, which can be solved by using the Newton–Raphson method. According to Grilli and Svendsen (1990), the value of ε_z can be taken as 0.002.

4.6.5.2 *KdV models*

An asymptotic approximation of the Euler equations in the limit of weak dispersion and weak nonlinearity is the well-known Korteweg–de Vries equation:

$$\eta_t + c_o\eta_x + \frac{3c_o}{2h}\eta\eta_x + \frac{1}{6}h^2 c_o\eta_{xxx} = 0. \tag{4.43}$$

Equation (4.43) is for right-running waves only; a similar KdV equation applies to left-running waves. The solitary wave solution of (4.43) is

$$\eta(x,t) = a_o \sec h^2\left[\sqrt{\frac{3a_o}{4h^3}}(x - ct)\right], \tag{4.44}$$

in which $a_o = H$, "x" is the origin at the crest and the speed of the wave is

$$c = c_o\left(1 + \frac{a_o}{2h}\right). \tag{4.45}$$

Wayne & Wright (2002) have formally demonstrated that, at the KdV order of weak nonlinearity and dispersion, left-running and right-running solitary waves interact linearly during their collision. Hence, linear superposition can be used as an approximate model of head-on collisions for comparison with the fully nonlinear, numerical solutions of the Euler equations.

The motion of the wavemaker was programmed to generate a solitary wave by forcing a (horizontal) velocity field in the water that is "close" to that occurring during the passage of a solitary wave. A procedure similar to that introduced by Goring and Raichlen (1980), which accounts in part for the finite displacement of the wavemaker paddle and the propagation of the wave during generation, can be adopted and is in practice. The generation was based on the KdV solitary wave whose horizontal velocity field is given by

$$u(x,t) = u_o \sec h^2\left[\sqrt{\frac{3u_o}{4h_o^2 c_o}}\left(x - c_o t - \frac{1}{2}u_o t\right)\right], \tag{4.46}$$

in which h_o is the quiescent water depth, $c_o = \sqrt{gh_o}$, g is gravitational acceleration, a_o is the wave amplitude and $u_o = a_o c_o/h_o$ is the maximum horizontal velocity. The displacement x_p of the wavemaker paddle from its initial position ($x = 0$) is then found numerically by solving the differential equation:

$$x_p'(t) = u(x_p, t), \tag{4.47}$$

which gives the Lagrangian path of a water particle.

4.6.5.3 *Generation of Cnoidal waves*

For the generation of Cnoidal waves, the displacement of the piston-type wave paddle is given as

$$x_p(t) = \frac{L}{2K_e d}\left[(y_t - d)\theta + \frac{H}{m}(E(\theta/m) - m'\theta)\right], \tag{4.48}$$

$$\theta = 2K_e\left(\frac{t}{T} - \frac{x_p(t)}{L}\right), \tag{4.49}$$

in which, $E(\theta/m)$ is the second incomplete elliptic integral and m' is the complementary parameter, $m' = 1 - m$. In these equations, m is the elliptic parameter, T is the wave period, L is the wavelength, y_t is the distance to the trough from the bottom and K_e is the first complete elliptic integral. Both the above-mentioned equations are also implicit and should be solved by the Newton–Raphson method. However, the most important problem in Cnoidal wave generation is to find out the m' for a given T and H. This can be estimated by using the trial-and-error method suggested by Goring (1978), Weigel (1964) or Sarpkaya and Isaacson (1981).

4.6.5.4 *Second-order wavemaker theory*

Daemrich and Götschenberg (1988) suggested that if wave-induced kinematics is to be reliably predicted for a shallower water depth, the second-order frequency difference terms should be included either using second-order wavemaker theory or using a self-correcting algorithm. It should be noted that while using first-order wavemaker theory, these sub- and superharmonics are not provided, and due to the mismatch of the water particle kinematics and the wave paddle velocity (piston type, most widely adopted), spurious-free harmonics will also be generated apart from the actual sub- and superharmonics. Thus, second-order wave paddle displacement is required as argued by Schäffer (1996) to remove the generated spurious-free components.

The complex form of first-order wave paddle displacement of each wave component is given by

$$X_0^{(1)} = \sum_{j=1}^{N} \frac{1}{2}\{-iX_{aj}\exp(i(\omega_j t) + c.c)\}, \tag{4.50}$$

where $c.c$ represents the complex conjugate of the previous term, j denotes the number of frequency components and X_{aj} is the constant first-order paddle amplitude at still water depth, evaluated using $X_{aj} = A_j/c_{0j}$, where c_{0j} is the Biesel transfer function and A_j is the complex amplitude.

In the case of second-order wave generation, surface elevation is obtained without any spurious-free wave generation and is given by

$$\eta_0^{(21)\pm} = \{G_{nm}^{\pm} A_n A_m^{-:*} \exp i(\theta_{0n} \pm \theta_{0m}) + c.c\}, \tag{4.51}$$

where the superscript $^{(21)\pm}$ denotes the second-order super- and subharmonics components of the first-order wave, $G_{nm}^{\pm}$ is the second-order wave elevation transfer function (Mansard *et al.*, 1988), A_n and A_m are complex amplitudes of two angular frequencies ω_n and ω_m, respectively, $^{-:*}$ denotes the complex conjugation for subharmonics and is ignored in the case of superharmonics, and $\theta_{0j} = k_{0j}x_t - \omega_{0j}t_t$, with j being n or m. Due to the interactions between two wave components of different angular frequencies ω_n and ω_m, there will be sum and difference terms in second-order wave theory. Therefore, a bi-chromatic analysis is necessary for the frequency domain.

The second-order wave generation (Sriram *et al.*, 2015) basically includes four components:

i. The first-order velocity potential,
ii. The wave–wave interaction potential-bound waves (super- and subharmonics),
iii. The spurious-free wave potential,
iv. The compensating free wave potential to remove the spurious waves.

The second-order paddle position in its final form is represented as

$$X_0^{(2)\pm} = \frac{1}{2}\left\{-iF^{\pm}\frac{A_n A_m^{-:*}}{d} \exp i\left(\omega_n \pm \omega_m\right)t + c.c\right\}, \tag{4.52}$$

where $F^{\pm}$ is the second-order transfer function that corresponds to sum and difference frequency terms and d is water depth. The major difference between $G_{nm}^{\pm}$ and $F^{\pm}$ is that $G_{nm}^{\pm}$ consists of only the bound waves generated due to wave interactions between two first-order components, whereas $F^{\pm}$ consists of compensating free waves besides, which is capable of removing spurious-free waves formed due to boundary condition mismatch at the wavemaker. A transfer function was derived by Schäffer (1996) using a double infinite series, which requires an enormous number of terms to achieve reasonable accuracy. Therefore, a special asymptotic summation method (Schäffer, 1993) is used to solve the double infinite series in second-order transfer functions to overcome the above-mentioned drawback. Moreover, this representation of the transfer function is simple, which considerably reduces computational complexity. This is advantageous for the computation of second-order control signals for the generation of irregular waves, as it avoids the requirement of a narrow-band assumption. The linear superposition of all wave components generates the irregular waves. The second-order long-wave correction is important during critical forcing, harbour seiching and mooring of floating structures. Further for wave run-up of the group waves, the spurious waves cancel the setdown, and hence the large waves in the group can uprush faster. Hence, second-order corrections are important under these circumstances.

4.7 Wave Absorbers

In order to investigate the effect of structure on long-term wave effects, in the physical tank in a facility, wave absorbers are placed normally. The absorbers are classified into two types: (a) *passive absorbers* that damp incident wave motion by a variety of techniques and (b) *active absorbers* that move in response to incident waves, thus absorbing them. In the case of passive absorbers, different types exist such as voided chambers, perforated screens, open metal cages with additional porous materials, sloping faced systems (permeable/impermeable), parabolic faced systems (permeable/impermeable), nests of open pipes (variable length) forming arrays, and combinations of the others. In the case of active absorbers, a control algorithm will be implemented in such a way that the reflected wave that reaches the wave paddle will be deducted and corrections will be given for the stroke in real time.

4.8 Summary

Physical Modelling is one of the important tools that has been applied by coastal engineers over several decades. Scale and model effects are of pivotal importance for the transfer of experimental results to the field. The method of dimensional analysis and principles of similitudes to facilitate the said task are discussed in this chapter. In particular, wave structure interaction problems need a detailed consideration of scale and model effects due to the processes inside the structures (such as porous breakwaters) and the difficulties of applying scale models (such as breaking waves). Hence, typical case studies for the geometric, kinematic and dynamic similitude considering wave–structure interaction problems are reported. The scale effects in particular for extreme waves and flows are critical, which are discussed in detail with typical experimental investigations. Unidirectional laboratory wave generation is well entrenched in the literature. However, the details on the generation of oblique as well as higher-order waves are rather scanty in the literature and are included for the benefit of the readers.

Physical Modelling stands as a cornerstone tool in the arsenal of coastal and river engineers, having been honed and refined over decades of practice. The intricacies of scale and model effects loom large, dictating the faithful translation of experimental findings to real-world applications. The methodical application of dimensional analysis and the guiding principles of similitude, essential for bridging the gap between laboratory experimentation and practical implementation, are discussed in this chapter.

In particular, the complex interplay between waves and structures demands meticulous attention to scale and model effects, especially considering the internal processes within structures like porous breakwaters and the challenges posed by phenomena like wave breaking. Thus, a series of illustrative case studies elucidating geometric, kinematic and dynamic similitude, tailored specifically to address wave–structure interaction challenges, are presented.

The ramifications of scale effects, particularly in the realm of extreme waves and flows, cannot be overstated. Through a comprehensive exploration of noteworthy experimental inquiries, the critical nuances surrounding scale effects, offering invaluable insights for practitioners and researchers alike, are unravelled.

While unidirectional laboratory wave generation techniques are well documented in the existing literature, the same cannot be said for the generation of oblique and higher-order waves. Recognizing this gap, the authors endeavoured to fill this void by providing detailed methodologies for the generation of such waves, enriching the knowledge base of our readers.

References

Airy, G.B. (1845). *On Tides and Waves: Encylopaedia Metropolitana*, Vol. 5 *(Mixed Sciences)*. London, pp. 241–396.

Ashlin, S.J., Sundar, V. and Sannasiraj, S.A. (2017). Pressures and forces on an oscillating water column type wave energy caisson breakwater. *Journal of Waterway, Port, Coastal, and Ocean Engineering*, 143, 1–18. DOI: 10.1061/(ASCE)WW.1943-5460.0000405.

Banner, M. L. and Phillips, O. M. (1974). On small scale breaking waves. *Journal of Fluid Mechanics*, 65, 647–657.

Battjes, J.A. and Sakai, T. (1981). Velocity field in a steady breaker. *Journal of Fluid Mechanics*, 11, 421–437.

Baldock, T.E., Swan, C. and Taylor, P. (1996). A laboratory study of nonlinear surface waves on water. *Philosophical Transactions of the Royal Society London, Series A*, 452, 649–676.

Bansal, R.K. (2010). *A Textbook of Fluid Mechanics and Hydraulic Machines*, 9th edn. Laxmi Publications, New Delhi.

Bonmarin, P. (1989). Geometric properties of deep-water breaking waves. *Journal of Fluid Mechanics*, 209, 405–433.

Bretschneider, H. (1980). Models of hydraulic structures: discharge conditions, energy dissipation, erosion. In *Hydraulic Modeling* (ed. H. Kobus), pp. 183–203. Germany: Paul Parey.

Bullock, G. N., C. Obhrai, D. H. Peregrine, and H. Bredmose. (2007). Violent breaking wave impacts. Part 1: Results from large-scale regular wave tests on vertical and sloping walls. *Coastal Engineering*, 54(8), 602–617.

Chakrabarti, S.K. (1987). *Hydrodynamics of Offshore Structures*. Computational Mechanics Publications, Southampton, Boston.

Chakrabarti, S.K. (1994). *Offshore Structure Modelling*. Advanced Series on Ocean Engineering. World Scientific, USA.

Chan, E. and Melville, W. (1988). Deep-water plunging wave pressures on a vertical plane wall. *Proceedings of the Royal Society London Series A: Mathematical, Physical and Engineering Sciences*, 417, 95–131.

Chanson, H., (2006). Tsunami surges on dry coastal plains: Application of dam break wave equations. *Coastal Engineering*, 48(04), 355–370.

Cuomo, G., Allsop, W., Bruce, T. and Pearson, J. (2010). Breaking wave loads at vertical seawalls and breakwaters. *Coastal Engineering*, 57(4), 424–439.

Daemrich, K.F. and Götschenberg, A. (1988). Wave generation and analysis in channels of the SFB 205. In *Proceedings of the 2nd International Symposium on Wave Research and Coastal Engineering*. SFB 205, Hannover.

Dean, R.G. (1990). Freak waves: A possible explanation. In Tørum, A. and Gudmestad, O.T. (eds.) *Water Wave Kinematics*. NATO ASI Series, Vol. 178. Springer, Dordrecht. https://doi.org/10.1007/978-94-009-0531-3_39.

Den Adel, H., Hendrikse, C.S.H. and Pilarczyk, K. (1996). Design and application of geo-tubes and geo-containers. In *Proceedings 1st European Conference on Geosynthetics: Applications, Design and Construction*, pp. 925–931.

Duncan, J.H. (1981). An experimental investigation of breaking waves produced by a towed hydrofoil. *Proceedings of the Royal Society London, Series A*, 377, 331–348.

Duncan, J.H. (1983). The breaking and non-breaking wave resistance of a two-dimensional hydrofoil. *Journal of Fluid Mechanics*, 126, 507–520.

Duncan, J.H., Philomin, V., Qiao, H. and Kimmel, J. (1994). The formation of Spilling breaker. *Physics of Fluids*, 6, 52.

Fritz, H.M., Borrero, J.C., Synolakis, C.E., and Yoo, J. (2006). 2004 Indian Ocean tsunami flow velocity measurements from survivor videos. *Geophysical Research Letters*, 33(24).

Funke, E.R. and Mansard, E.P.D. (1979). On the Synthesis of Realistic Sea States in a Laboratory Flume. Report LTR-HY-66, National Research Council, Canada.

Funke, E.R. and Mansard E.P.D. (1980). On the Synthesis of Realistic Sea States, *Coastal Engineering Proceedings: No. 17: Proceedings of 17th Conference on Coastal Engineering*, Sydney, Australia, 1980.

Goda, Y. (1970). Numerical Experiments on Wave Statistics with Spectral Simulation. Report of Port and Harbor Research Institute, Vol. 9, No. 3.

Goda, Y. (1976). On wave groups. In *Behaviour of Offshore Structures, BOSS'76.* The Norwegian Institute of Technology, pp. 115–128.

Goda, Y. (1983). Analysis of Wave Grouping and Spectra of Long-travelled Swell. Report of the Port and Harbour Research Institute, 22(1).

Golda Percy, V.P., Sriram, V. and Sundar, V. Holger Schüttrumpf (2023). Effect of the buffer blocks in attenuating a tsunami-like flow. *Ocean Engineering*, Volume 286, Part 1.

Golda Percy, V.P., Sriram, V., Sundar, V., and Holger Schüttrumpf (2024). Experimental investigation of the buffer blocks on the extreme flows generated using dam-break mechanism. In *Proceedings of 44th International Conference on Ocean, Offshore and Arctic Engineering – OMAE 2025 2023*, 115489.

Goring, D.G. (1978). Tsunamis- the Propagation of Long Waves onto a Shelf. Ph.D thesis of the California Institute of Technology, Pasadena, California.

Goring, D.G. and Raichlen, F. (1980). The generation of long waves in the laboratory. In Edge B.L. (ed.) *Proceedings of 17th Conference on Coastal Engineering*, Sydney, Australia. ASCE, p. 763783.

Goseberg, N., Wurpts, A., and Schlurmann, T. (2013). Laboratory-scale generation of tsunami and long waves. *Coastal Engineering*, 79, 57–74.

Greenhow, M. and Vinje, T. (1982). A theoretical and experimental study of the capsize of Salter's Duck in extreme waves. *Journal of Fluid Mechanics*, 118, 221–239.

Grilli, S.T. and Svendsen, I.A. (1990). Computation of nonlinear wave kinematics during propagation and runup on a slope. In Torum, A. and Gudmestad, O.T. (ed.) *Water Wave Kinematics.* NATO ASI Series E: Applied Sciences, Vol. 178, pp. 387–412. Klüwer Academic Publishers. http://dx.doi.org/10.1007/978-94-009-0531-3_24.

Harish, S. (2023). Tsunami-like flow induced forces on the shorefront structures. Ph.D-thesis, RWTH Aachen University. https://doi.org/10.18154/RWTH-2023-06471.

Harish, S., Sriram, V., Schüttrumpf, H. and Sannasiraj, S.A. (2021). Tsunami-like flow induced force on the structure: Prediction formulae for the horizontal force in quasi-steady flow phase. *Coastal Engineering*, 168(2021), 103938.

Harish, S., Sriram, V., Schüttrumpf, H. and Sannasiraj, S.A. (2022). Tsunami-like flow induced forces on the structure: Dependence of the hydrodynamic force coefficients on Froude number and flow channel width in quasi-steady flow phase. *Coastal Engineering*, 172(2022), 104078.

Heller, V. (2007). Landslide Generated Impulse Waves: Prediction of Near Field Characteristics. Doctoral dissertation, Dept. of Civil Engineering and the Environment, Eth Zurich.

Heller, V. (2011). Scale effects in physical hydraulic engineering models. *Journal of Hydraulic Research*, 49, 293–306.

Hudson, R.Y. (1958). Design of Quarry Stone Cover Layers for Rubble Mound Breakwaters. WES, Vicksburg, Research Report 2.2.

Hudson, R.Y., Herrmann, F.A. and Sager, R.A. (1979). Coastal Hydraulic Models (No. SR-5), US Army, Corps of Engineers Research Centre.

Hughes, S.A. (1993). *Physical Models and Laboratory Techniques in Coastal Engineering.* Advanced Series on Ocean Engineering, Vol. 7. https://doi.org/10.1142/2154.

Ivicsics, L. (1978). *Hydraulic Models.* Vizdok, Budapest.

Kamphuis, J.W. (1974). Practical scaling of coastal models. In *Proceedings of 14th Coastal Engineering Conference*, Copenhagen 3. ASCE, New York, pp. 2086–2101.

Kamikubo, Y., Murakami, K., Irie, I. and Hamasaki, Y. (2000). Study on practical application of a non-wave overtopping type seawall. In *Proceedings of 27th International Conference on Coastal Engineering*, Sydney, Australia, Vol. 126, pp. 2215–2228.

Kjeldsen, S.P. (1984). The experimental verification of numerical models of plunging breakers. In *Proceedings of Coastal Engineering.* https://doi.org/10.1061/9780872624382.002.

Kjeldsen, S.P. (1990). Breaking waves. In Torum, A. and Gudmestad, O.T. (eds.) *Water Wave Kinematics.* Birkhaeuser, Kluwer Academic Publishers, pp. 453–473.

Kobus, H. (ed.) (1980). Hydraulic modelling. German association for water resources and land improvement. Bulletin No. 7. Parey, Hamburg.

Kortenhaus, A., Pearson, J., Bruce, T., Allsop, N. and Van der Meer, J. (2004). Influence of parapets and recurves on wave overtopping and wave loading of complex vertical walls. *Coastal Structures 2004*, pp. 369–381.

Keulegan, G.H. and Carpenter, L. (1958). Forces on cylinders and plates in an oscillating fluid. *Journal of Research of the National Bureau of Standards.* DOI: 10.6028/JRES.060.043. Corpus ID: 124303029.

Kway, J.H.L., Loh, Y.S. and Chan, E.S. (1998). Laboratory study of deep-water breaking waves. *Ocean Engineering*, 25(8), 657–676. https://doi.org/10.1016/S0029-8018(97)00039-5.

Le Méhauté, B. (1976). *An Introduction to Hydrodynamics and Water Waves.* Springer Science Business Media, Ilc, New York.

Le Méhauté, B. (1990). "Similitude." *Sea Ocean Eng. Sci.* 9, 955–980.

Le Méhauté, B. and Hanes, D.M. (eds.). (2005). *Ocean Engineering Science*, Vol. 9. Harvard University Press.

Longuet-Higgins, M.S. (1974). Breaking waves in deep and shallow water. In *10th Conference in Naval Hydrodynamics*, Publ. ACR-204, pp. 597–605, Office of Naval Research, Arlington, VA.

Longuet-Higgins, M.S. and Stewart, R.W. (1964). Radiation stresses in water waves; a physical discussion, with applications. *Deep Sea Research and Oceanographic Abstracts*, 11(4), 529–562. https://doi.org/10.1016/0011-7471(64)90001-4.

Mansard, E.P.D., Sand, S.E., Klinting, P. (1988). Sub and super-harmonics in natural waves. *Journal of Offshore Mechanics and Arctic Engineering*, 110, 270.

Martin, H., and R. Pohl (eds.). (2000). Technische Hydromechanik 4 (Technical hydromechanics). Verlag für Bauwesen, Berlin [in German].

Mase, H., Kita, N. and Iwagaki, Y. (1983). Random wave simulation considering wave groups. *Coastal Engineering in Japan* (JSCE), 26, 61–75.

Melville, W.K. (1983). Wave modulation and breakdown. *Journal of Fluid Mechanics*, 128, 489–506.

Nepf, H.M., Wu, C.H. and Chan, E.S. (1998). A comparison of two- and three-dimensional wave breaking. *Journal of Physical Oceanography*, 28, 1496–1510.

Nolte, K.G. and Hsu, F.H. (1972). Statistics of ocean wave groups. In *Proceedings of OTC Conference*, Houston, Texas.

Novak, P. (1984). Scaling factors and scale effects in modelling hydraulic structures. In Rouse, H. (ed.) *Symposium on Scale Effects in Modelling Hydraulic Structures*, Vol. 3, pp. 1–5. Technische Akademie, Esslingen.

Ochi, M.K. and Tsai, C.H. (1983). Prediction of occurrence of breaking waves in deep water. *Journal of Physical Oceanography*, 2008–2019. https://doi.org/10.1175/1520-0485(1983)013.

Peterka, A.J. (1984). Hydraulic Design of Stilling Basins and Energy Dissipators Engineering Monograph No. 25. US Bureau of Reclamation, Denver, USA.

Ramberg, S.E. and Griffin, O.M. (1987). Laboratory study of steep and breaking deep water waves. *Journal of Waterway, Port, Coastal, and Ocean Engineering*, 113(5). https://doi.org/10.1061/(ASCE)0733-950X(1987)113:5(493).

Ramsden, J.D. (1993). Tsunamis: Forces on a Vertical Wall Caused by Long Waves, Bores, and Surges on a Dry Bed. Ph.D. thesis, California Institute of Technology, Pasadena, CA.

Ramsden, J.D. (1996). Forces on a vertical wall due to long waves, bores, and dry-bed surges. *Journal of Waterway, Port, Coastal, and Ocean Engineering*, 122(3), 134–141.

Rapp, R.J. and Melville, W.K. (1990). Laboratory measurements of deep-water breaking waves. *Philosophical Transactions of the Royal Society London, Series A*, 331, 735–780.

Ravindar, R. and Sriram, V. (2021). Impact pressure and forces on a vertical wall with different types of parapet. *Journal of Waterway, Port, Coastal, and Ocean Engineering* (ASCE), 147(3). https://doi.org/10.1061/(ASCE)WW.1943-5460.0000635.

Ravindar R., Sriram V., Schimmels S. and Stagonas D. (2021). Approaches in scaling small-scale experiments on the breaking wave interactions with a vertical wall attached with recurved parapets. *J. Waterw. Port Coast. Ocean Eng.*, 147(6), Article 04021034.

Sand, S.E., Hansen, N.E.O., Klinting, P., Gudmestad, O.T. and Sterndorf, M.J. (1990). Freak wave kinematics. In Torum, A. and Gudmestad, O.T. (eds.) *Water Wave Kinematics*. Birkhaeuser, Kluwer Academic Publishers, pp. 535–549.

Sarpkaya, T. and Isaacson, M. (1981). *Mechanics of Wave Forces on Offshore Structures*. Van Norstrand Reinhold Company.

Schäffer, H.A. (1993). Second order irregular wave generation in flumes — computation of transfer functions by an asymptotic summation method. In *Proceedings of Waves'93: 2nd International Symposium Ocean Wave Measurement and Analysis*, New Orleans, LA, USA.
Schäffer, H.A. (1996). Second order wavemaker theory for irregular waves. *Ocean Engineering*, 23(1), 47–88.
Schüttrumpf, H. and Oumeraci, H. (2005). Layer thicknesses and velocities of wave overtopping flow at Seadikes. *Coastal Engineering*, 52(6), 473–495. http://dx.doi.org/10.1016/j.coastaleng.2005.02.002.
Skyner, D.J. and Easson, W.J. (1998). Wave kinematics and surface parameters of steep waves travelling on sheared currents. *Journal of Waterway, Port, Coastal, and Ocean Engineering*, 124(1), 1–6.
Skyner, D.J., Gray, C. and Greated, C.A. (1990). A comparison of time-stepping numerical predictions with whole-field flow measurement in breaking waves. *Water Wave Mechanics*, Vol. 178. Part of the NATO ASI Series Book Series. NSSE.
Sriram, V., Schlurmann, T. and Schimmels, S. (2015). Focused wave evolution using linear and second order wavemaker theory. *Applied Ocean Research*, 53, 279–296. https://doi.org/10.1016/j.apor.2015.09.007.
Streicher, M., Kortenhaus, A., Altomare, C., Hughes, S., Marinov, K., Hofland, B., Chen, X., Suzuki, T. and Cappietti, L. (2019). Non-repeatability, scale- and model effects in laboratory measurement of impact loads induced by an overtopped bore on a dike mounted wall. In *International Conference on Offshore Mechanics and Arctic Engineering*, Vol. 58783. American Society of Mechanical Engineers, p. V003T02A032.
Svendsen, I.A. (1985). Physical modelling of water waves. In Dalrymple, R.A. (ed.), *Physical Modelling in Coastal Engineering*, Balkema, Rotterdam, The Netherlands, pp. 13–47.
Sukanya, R., Sundar, V. and Sannasiraj, S.A. (2022). Hydrodynamic performance assessment of a geo-synthetic seawall. *Journal of Ocean and Coastal Engineering*, 2250001-1–2250001-30. https://www.worldscientific.com/doi/epdf/10.1142/S2529807022500014.
Sundar, V. (2021). *Ocean Wave Dynamics for Coastal and Marine Structures.* Advanced Series on Ocean Engineering, Vol. 52. World Scientific. https://doi.org/10.1142/12268.
Sundar, V. and Sannasiraj, S.A. (2019). *Coastal Engineering- Theory and Practice.* Advanced Series on Ocean Engineering, Vol. 47. https://www.worldscientific.com/worldscibooks/10.1142/11148.
Van dorn, W.G. and Pazan, S. (1975). Laboratory investigation of wave breaking Part 2, deep water waves. Scripps, SIO 75~21.
Wayne, C.E. and Wright, J.D. (2002). Higher order modulation equations for a Boussinesq equation. *SIAM Journal on Applied Dynamical Systems*, 1, 271–302.
Weigel, R.L. (1964). *Oceanographical Engineering.* Prentice-Hall, Inc., Englewood Cliffs.

Xu, D., Hou, W., Zhao, M. and Jin Wu, J. (1993). Statistical simulation of wave groups. *Applied Ocean Research*, 15(4), 217–226. https://doi.org/10.1016/0141-1187(93)90010-U.

Yalin, M. S. (1971). *Theory of Hydraulic Models.* London: Macmillan.

Yalin, M.S. (1977). *Mechanics of Sediment Transport*, 2nd edn. Pergamon Press, Oxford, UK, 360 p.

Yasuda, T., Mori, N. and Ito, K. (1992). Freak waves in unidirectional wave trains and their properties. In *Proceedings of Coastal Engineering*, https://doi.org/10.1061/9780872629332.056.

Chapter 5

Physical Modelling of Nearshore Structures

5.1 Introduction

Physical and numerical models, field measurements and theoretical and analytical approaches are applied to solve a variety of coastal engineering and coastal zone management problems. Each method has its own significance, possibilities, strengths and weaknesses. The aim of coastal engineers should be to optimally apply, combine and couple the different methods in order to use synergies and come up with the best possible result. The application of only one method leads to hydraulic monoculture and often results in a waste of investment. Research in coastal engineering and practice should, therefore, apply and combine different techniques to reduce costs, increase the reliability of coastal measures and optimize their performance characteristics. Apart from the above, the application of these methods does not replace good expert judgement and should be considered more as decision support.

Physical modelling is one of the important tools that has been used by coastal engineers for decades. It is the replication of processes and structures in nature on a reduced scale. The choice of physical models depends first on a careful comparison with other tools in coastal engineering. The most suitable model has to be chosen to provide the best possible outcome. It also depends on the testing facilities that will decide how large a scale ratio could be adopted, which dictates the reproduction of physical phenomena in the field as closely as possible in the laboratory.

All physical models are influenced by scale and model effects, which might influence the final result. The scale and model effects are of pivotal importance for the transfer of experimental results to the field. In particular, wave–structure interaction problems need a detailed consideration of scale and model effects due to the processes inside the structures (such as porous breakwaters) and the difficulties in applying scale models (such as breaking waves).

This chapter provides a short introduction to physical modelling. More literature is available on this subject and should be considered if necessary.

Coastal and hydraulic structures are very complex and expensive, as they have to be adapted to individual boundary conditions. Local conditions, such as bathymetry, waves, currents and sediments, determine the size, dimensions and, finally, the cost of coastal structures. In order to optimally plan and realize these coastal structures, both economically and functionally, various tools, i.e. coastal engineering tools, can be used. Coastal engineering tools include numerical simulations, physical model tests and experiments, measurements and observations in nature, as well as theoretical, analytical and empirical approaches. These tools are used in practice and research to address coastal challenges. The choice of the suitable tool depends on numerous factors, which in turn depend on the advantages and disadvantages of the respective tool as well as other factors which vary over time. Coastal engineering – like many other sciences – has undergone a dramatic change in methods over the past decades. With the development and implementation of computing capacity and numerical models that are easily accessible, numerical modelling tools have been established as an additional coastal engineering tool in research and practice and, in some cases, have completely replaced the conventional methods.

Numerical models, however, may not be capable of solving several of the coastal engineering problems in practice due to their inherent disadvantages. Therefore, physical model tests and experiments in coastal labs or research facilities are still warranted.

Therefore, practice shows that, often, a combination of different tools is required, as indicated in Fig. 5.1. While numerical investigations have become widely accepted, especially for large-scale processes, much reliance is still placed on the advantages and possibilities of physical model tests for structural investigations and in the reproduction of physics involved in flow–structure interaction problems.

Experimental methods and field measurements thus continue to have a very high significance in coastal engineering research. Leonardo da Vinci's guiding principle is, therefore, still valid today:

Leonardo da Vinci (1452–1519)
In treatises on water, experimental experience must always come first.

For many of the practical tasks, extensive test facilities (including field tests) are available worldwide and can be used for adequate testing that

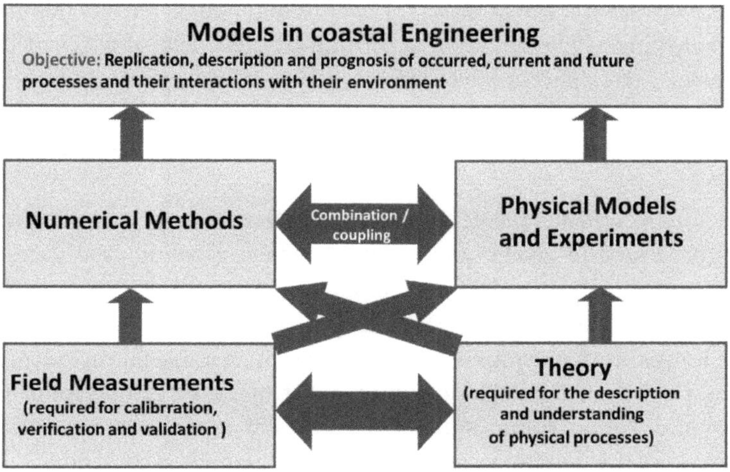

Fig. 5.1. Tools for coastal engineering.
Source: Schüttrumpf (2014).

can facilitate basic as well as applied research in the field of classical and maritime hydraulics.

The aim of this chapter is to present the significance and possibilities of coastal engineering tools as well as the challenges in the future. For this purpose, the terms surrounding the topic of coastal engineering tools are first defined. This is followed by a presentation of the advantages and disadvantages of the various methods, with some concluding statements on the choice of each method. Finally, the main challenges for setting up a physical model are presented.

Tools: Four hydraulic engineering investigation tools (or methods) are distinguished: (i) physical, (ii) numerical, (iii) theoretical and (iv) field measurements.

Model: Reproduction (Duden, 2000) of an object, state or process (a model can be of an intellectual, physical or mathematical/numerical nature).

Model experiment: Experiment to replicate an object, condition or process for practice-related questions.

Experiment: Scientific experiment (Duden, 2000) (also called laboratory experiment).

There are several possibilities for classifying models into model families. In the following, a classification according to their field of application into flow models, transport models and structure models is used. Other classifications (e.g. by application area, such as river, groundwater, estuary, or coast, or by task, such as flood protection or erosion control) are also possible.

The aim of flow models is to provide small- and large-scale descriptions of flow processes in rivers, lakes, dams, canals, pipelines, estuaries, seas, oceans and groundwater. Different modelling methods are required for the different areas of application. While, for example, the overflow of a sea dike or a breakwater requires the simulation of small-scale turbulence phenomena, the effect of the Coriolis force must be taken into account for the simulation of tidal dynamics in the ocean. Flow models represent the core of every coastal engineering and coastal zone management study.

The aim of transport models is to provide small- and large-scale descriptions of the various transport processes of bedload, suspension, salt, heat, oxygen and other water constituents caused by hydrodynamics. Transport models also reflect the requirements of the respective field of application and are directly coupled with a flow model.

Structural models aim at the constructive and functional design of structures such as dikes and dams, levees, revetments, groynes and breakwaters. Functionally, the different types of coastal structures influence the hydrodynamics with the aim of flow regulation, current control, improvement of maintenance, etc. Structurally, coastal structures must be designed for extreme hydraulic loads from high water levels, currents, waves, floods, etc.

For each model family (flow models, transport models and structural models), there are applications of the various coastal engineering methods.

5.2 Advantages and Disadvantages

5.2.1 *Introduction*

In the following, the advantages and disadvantages of the four basic hydraulic engineering methods are described, as illustrated in Fig. 5.2. The general validity of the statements is only given to a limited extent, as advantages and disadvantages have to be critically examined and discussed as individual cases. Therefore, exceptions to the following statements are always possible and are also presented in part if deemed necessary. The decision of which of the methods to choose in a specific case thus depends on the critical discussion and weighing of the respective advantages and disadvantages. Therefore, in many cases, a combination of several coastal

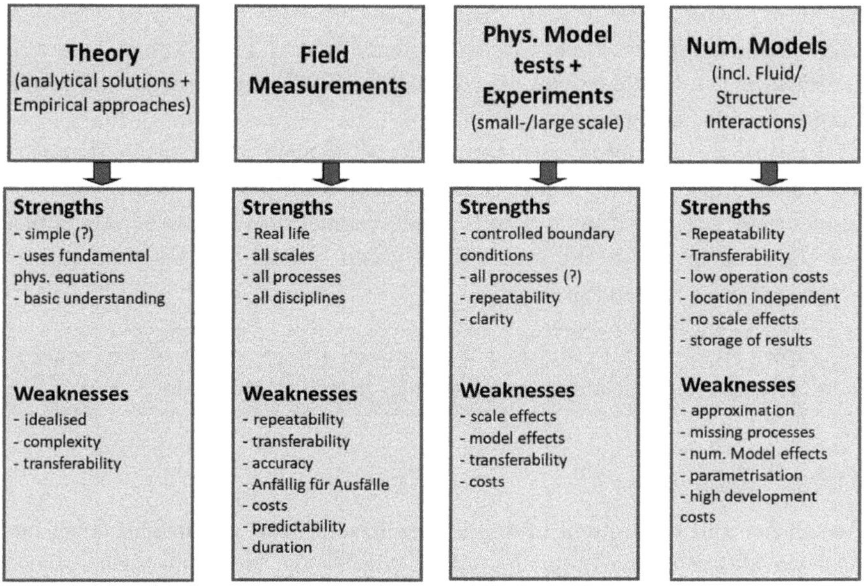

Fig. 5.2. Strengths and weaknesses of coastal engineering methods.

engineering tools, in the sense of composite or hybrid modelling, represents an ideal solution.

5.2.2 *Analytical and theoretical methods*

5.2.2.1 *Advantages*

Analytical and theoretical methods use equations to reproduce fundamental physics (conservation of mass, conservation of momentum, and conservation of energy). Analytical solutions can be simple and are often suitable for use in spreadsheets or calculators. Simple analytical solutions and theoretical procedures are also suitable for the verification of numerical model results or the plausibility checks of measurements in nature or in the laboratory. Examples of theoretical methods are the continuity equation, the Lagrangian equation for surge and sunk, and the various theoretical wave models (Airy–Laplace, Stokes, etc.).

Empirical or semi-empirical approaches represent a subgroup of theoretical methods. In a semi-empirical method, solutions to the basic physical equations are derived theoretically. Unconsidered processes, such as variable geometric, hydrodynamic, morphodynamic or other factors influencing the theoretical solution, are then taken into account on the basis of physical model experiments or measurements in nature with the help of empirical coefficients. Examples of this are the Poleni equation for determining a weir overflow or the Gaukler–Manning–Strickler (GMS) equation for determining the free level discharge.

Empirical approaches are determined on the basis of dimensional analytical considerations when a purely theoretical description of highly complex processes is not possible, and thus, parametrization is required. Examples are the description of the onset of sediment erosion or the wave run-up height for coastal structures.

> Empirical and semi-empirical approaches are particularly necessary when highly complex processes need to be parameterized.

5.2.2.2 *Disadvantages*

Analytical and theoretical methods idealize the field conditions. The natural variability of hydraulic boundary conditions, the complexity of bed topography and the geometry of structures, as well as their interactions, are often insufficiently taken into account. Thus, these methods are unsuitable for complex situations and cannot be transferred. In many cases, analytical solutions are mathematically highly complicated and inconvenient for practical use. Therefore, analytical approaches, such as continuity and motion equations, require the use of numerical methods for their solution.

Empirical and semi-empirical approaches are, in strict terms, only applicable under the boundary conditions for which they were determined. Complex situations are not taken into account, and natural conditions and processes are strongly idealized. Therefore, transferability to "similar conditions" should be approached with extreme caution.

> Analytical solutions and empirical and semi-empirical approaches idealize natural processes. Analytical equations can often only be solved by using numerical methods.

5.2.3 *Field measurements*

5.2.3.1 *Advantages*

Field measurements and investigations in nature represent "real-time" considerations, i.e. all processes, scales and disciplines are fully considered, along with their inhomogeneities, turbulence, scattering, etc. Thus, measurements in nature are independent of the scale and model effects of the physical model experiments as well as the numerical effects, parametrizations, empirical equations and coefficients of the numerical model procedures. Measurements in nature are also advantageous when, for example, initial and boundary conditions for physical model experiments and numerical models have to be determined, when the credibility of the description of processes has to be checked, or in the context of measurements for the preservation of evidence when the predicted effects of measures have to be verified.

Measurements in nature represent real-life observations and take into account all scales, processes, disciplines and their interactions.

Certain investigations are only possible in nature, as the corresponding initial and boundary conditions cannot be reproduced either in numerical models or in model tests. An example of this is wave overtopping tests on dikes with grass cover. Physical and numerical experiments fail due to the difficulty of taking into account the growth of the grass as well as the weathering (ageing) of the accumulating soil in the model. In this case, measurements in nature are recommended (van der Meer *et al.*, 2006).

5.2.3.2 *Disadvantages*

Measurements in nature have certain weaknesses, especially in forecasting, i.e. in describing the effect of measures of the watercourse, since the effect of a measure naturally occurs only after its implementation. The advantage of real-life observations is also often a major disadvantage in regard to the transferability and repeatability of achieved results. Inhomogeneities, natural scatter and the variability of processes, material properties and boundary conditions often do not allow identifying and describing a clear cause–effect relationship. This aspect is of particular importance in the case of measurements for the preservation of evidence when the signal caused by the measure is overlaid and possibly influenced by natural changes in a system. Measurements in nature can rarely be repeated precisely and transferred completely.

Real-time observations of measurements in nature often do not allow a clear identification of processes, interrelationships and interactions.

For example, the investigation of a dike or a dam failure as an effect of an extreme event (due to hurricanes, tsunamis, storm surges, floods, etc.) is not possible due to the rarity and unpredictability of the event in nature. Measurements in nature have to be designed for the natural stresses, i.e.

extreme weather conditions involving ice, storms or rain, frequently missing infrastructure (e.g. missing power supply) and undesirable human impacts (vandalism, theft, etc.). This results in higher maintenance and upkeep costs, a need for expensive measuring devices, susceptibility to failures and undesired interference signals.

5.2.4 *Physical model tests and experiments*

5.2.4.1 *Advantages*

Physical model tests and experiments reproduce the relevant physical processes, including their nonlinear interactions, on a suitable model scale. This also makes it possible to investigate the interactions between hydrodynamic processes, structure and bed as well as other interdisciplinary issues (e.g. interactions with flora and fauna, interactions between ship and waterway). Multiphase investigations (water, air, sediment, etc.) can also be ideally carried out in physical model tests and experiments. In particular, if the processes to be investigated are still unknown or inadequately described, results similar to those obtained from tests in nature can be expected if the model is set up appropriately. This is an advantage in the context of basic research in coastal engineering. Therefore, physical model tests and experiments can be used for descriptive, diagnostic and prognostic purposes.

In contrast to measurements in nature, physical model tests and experiments can be repeated at any time under the same boundary and initial conditions. Chaotic influences due to variable environmental conditions can be largely avoided. Innovative measuring systems allow measurements of hydrodynamic, morphodynamic and other parameters with high temporal and spatial resolutions.

Physical model experiments are highly illustrative and are therefore used in several procedures and projects to communicate project results to non-specialists. Due to their clarity, physical model tests and experiments are an important instrument for universities teaching the fields of coastal engineering and coastal management (teaching laboratories and experimental halls) (Naudascher, 1979).

> Basic physical processes (e.g. transport formulae) are usually derived from physical model tests and experiments. From this follows the necessity of physical model tests and experiments as an indispensable tool for basic research in coastal engineering.

Physical model tests and experiments continue to be an important instrument of university teaching due to their illustrative nature.

5.2.4.2 *Disadvantages*

The results of physical model experiments reflect the knowledge of the modeller and the techniques used by the modeller with regard to the preparation, implementation, analysis and interpretation of the modelled, observed and measured processes. Only a few physical model tests and experiments can be carried out on a natural scale (i.e. a scale of 1:1) and are therefore not influenced by scale effects that arise on a reduced scale due to the impossibility of replicating all force vectors to fulfil dynamic similarity in a model that is also intended to be geometrically and kinematically similar. The impossibility of fulfilling full dynamic similarity consists in the requirement of adjusting the physical properties of water (viscosity, surface tension, elasticity, etc.), which is only possible with certain special difficulties (Miller, 1972). For the morphodynamic models, in addition to the similarity of the hydrodynamics, the similarity of the solid transport must also be faithfully ensured (DWA, 2021).

Scale effects arise due to the insufficient fulfilment of the relevant similarity laws in the scaled-down model.

Model effects arise due to the approximate reproduction of the realities in the model and the decisive forces, as well as due to data acquisition and data analysis.

In addition to scale effects, model effects also influence the results of physical model tests. Model effects arise from the approximate replication of the prototype in the model and the governing forces as well as due to data collection and data analysis. Sources of model effects are, for example, the uncertainties of measuring the geometry of a coastal structure within a water body, the construction of a coastal structure in the model (e.g. surface roughness), the boundary conditions, measurement (the position and number of measurement positions and the type of measurement device), data acquisition (e.g. sampling rate), data analysis (e.g. analysis and the simulation period) as well as neglected processes (e.g. wind). Biogenic processes (soil consolidation, vegetation, etc.) cannot be simulated in the model either. Due to the numerous possibilities of model and scale effects, a high degree of experience is required in planning, executing, analyzing and interpreting the results obtained from the physical models and experiments. The relatively high costs are also a disadvantage for physical model tests since the model is rarely transferable to similar projects or applications.

5.2.5 *Numerical modelling techniques*

5.2.5.1 *Advantages*

The main advantage of numerical modelling is its flexibility, location-independence and consideration of all relevant two- and, possibly, three-dimensional processes for descriptive, diagnostic and prognostic studies. Numerical models are also particularly suitable for large-scale and long-term investigations that cannot be carried out in the physical model or in nature due to scale effects (or non-existent experimental facilities) and a lack of time. Therefore, with today's requirements, large river basins, estuaries and ocean models can only be studied with the help of numerical simulations. Small-scale turbulent processes are parametrized and play a subordinate role in informative value and accuracy. Due to their flexibility, different models can be built with the same numerical modelling method, and different questions can be solved. Due to this flexibility, extensions and additions to the model procedure (i.e. in the programme code) can be used simultaneously by several models. The repeatability of investigations in the numerical model is inherent. This also makes it easy to carry out the same analysis under different scenarios. A subsequent analysis of other parameters is also possible. Numerical model procedures can be used without any special infrastructure as long as they run on commercially available PCs or workstations. Exceptions to this are, for example, high-performance supercomputing applications. In a numerical model, any initial and boundary conditions, as well as a large number of variants, can be investigated. This makes numerical models a cost-effective and flexible coastal engineering method, in particular, for practical use. Graphical presentation of simulation results can also be seen as an advantage for numerical models (however, an excessive use of animated graphics can also raise doubts about the reliability of model results).

> Large-scale and long-term investigations are a classic field of application for numerical models. In this field, the numerical model is the dominant hydraulic engineering method.

5.2.5.2 *Disadvantages*

Numerical model procedures and models reflect the knowledge of the system developer and the modeller. However, only those processes that have been considered in the model procedure (i.e. in the programme code) can be modelled and simulated. Even the best model procedure applied by an inexperienced user can lead to incorrect results if settings such as grid spacing, simulation step, and bottom roughness have been (unknowingly) incorrectly selected. Modern systems try to give the user only little room for decision-making. Applying a numerical modelling method without reasonable questioning can cause risks since the computer can apparently deliver results even if the settings are totally wrong. The "interpretation" and responsibility for the correctness of the model results are often all too readily transferred to the computer.

The effects that occur in physical model experiments also occur in numerical model procedures and models, which result from the identification and selection of the decisive physical processes (e.g. vegetation and biogenic effects can only be simulated in a simplified way). In addition, the mathematical formulations of these processes and their numerical implementation (e.g. solutions to the nonlinear terms in the differential equations and the resulting numerical dissipation, different descriptions of vortex structures, such as DNS, LES, or RANS simulations) and the choice of the turbulence model (e.g. k–ε or k–ω model) can lead to slightly different results under different scenarios. Furthermore, numerical model effects arise in the horizontal, vertical and temporal discretizations of the model and the processes to be considered, as well as in the initial and boundary value controls of the model. A large number of empirical equations and constants also lead to numerical effects, as do the subsequent data storage and analysis. Different effects can produce different results for different (calibrated and verified) numerical models. Therefore, a multi-model approach, i.e. the construction of similar numerical models with different model procedures to solve the same problem, is important for assessing the quality and informative values of a simulation result (Plüß and Schüttrumpf, 2004).

Numerical model effects arise from (a) the identification and selection of physical processes, (b) the mathematical formulation of the governing physical processes, (c) the numerical implementation of the mathematical equations and model procedures, (d) the spatial and temporal discretisation, (e) the initial and boundary control of the model, and (f) the data analysis and evaluation.

Numerical models are often seen as a cost-effective alternative to physical models, which are quite labour intensive. This is certainly true in many cases, especially in practical applications. In applied research as well as in basic research, when the construction of a numerical model procedure is required, the associated development costs can sometimes exceed the costs for the construction and operation of a physical model. The construction of the computational grid should not be underestimated either, especially in the case of large domains and complex models.

It may not be possible to answer interdisciplinary questions through numerical models due to a lack of coupling. An example of this is the investigation of the failure of revetments due to wave loading.

5.2.6 *Choosing the right coastal engineering method*

The choice of a coastal engineering method depends on a number of factors. The advantages and disadvantages of the different methods influence the choice. Furthermore, the degree of complexity or nonlinearity of the processes to be investigated, the scales (time, space and disciplines), the size and importance of a project, the field of application (research or practice), the level of investigation (pre-feasibility, design or research) as well as the equipment of the institution (research institute, engineering office or authority) and the preference of the client or contractor determine the choice of hydraulic engineering method. In particular, the commissioning of physical model tests and experiments is limited to a few institutes in universities and research institutions, while simple numerical model procedures are also procured, set up and operated by engineering offices, construction companies and coastal authorities.

Due to the different advantages and disadvantages of the different hydraulic engineering methods, a combination of methods, in the sense of composite or hybrid modelling, should often be chosen in order to use synergy effects (Oumeraci, 1999; Schäffer, 1999). Hybrid models use methodologically different models to solve a hydraulic engineering task by compensating the respective weaknesses of one method with the strengths of another method. Depending on the degree of coupling, a distinction is made between uncoupled, loosely coupled and fully coupled hybrid models (Rutschmann, 2005). This approach makes sense in particular due to the uncertainties in the models, natural variability and human influence factors. Model validation typically shows deviations between the model (numerical or physical) and natural measurements. These deviations can be attributed to model uncertainties (e.g. numerical effects and physical model effects) as well as to the measurement of the comparison parameters in nature and their natural variability (Fig. 5.3).

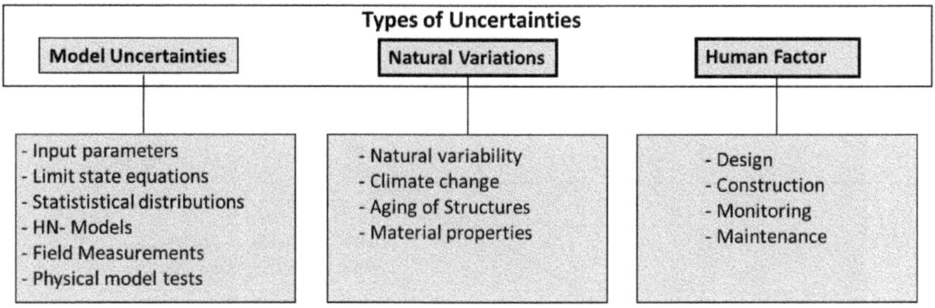

Fig. 5.3. Uncertainties.
Source: According to Kortenhaus (2003).

The necessity of sharing the different hydraulic engineering tools in the context of applied as well as basic research is therefore considered absolutely essential by numerous authors (e.g. Vollmers, 1998; Aufleger, 1998; Oumeraci, 1999; Plate and Köngeter, 2003; van Os *et al.*, 2004).

> The application of physical model tests is limited to a few coastal engineering laboratories at universities and research institutions within the framework of basic and applied research. Therefore, a broad application of numerical modelling methods in practice should not be seen as evidence of the lack of acceptance as well as the necessity of hydraulic engineering model tests. Restricting coastal engineering research to one coastal engineering method leads to scientific monoculture and hinders or prevents the use of research and development potentials.

5.3 Physical Modelling

The objective of most physical models is to provide either a good basis for any design of real structures or to improve our existing knowledge. Therefore, careful and excellent preparation of any physical model is necessary. The preparation depends on a number of factors.

5.3.1 *Selection of a suitable test facility*

As a first step, a suitable facility has to be selected. In general, coastal engineering distinguishes between small- and large-scale model tests and between tests in wave flumes and wave basins. Most of the model tests worldwide are performed on a small scale, which means that all geometrical parameters of the prototype are scaled down by a certain factor (called the

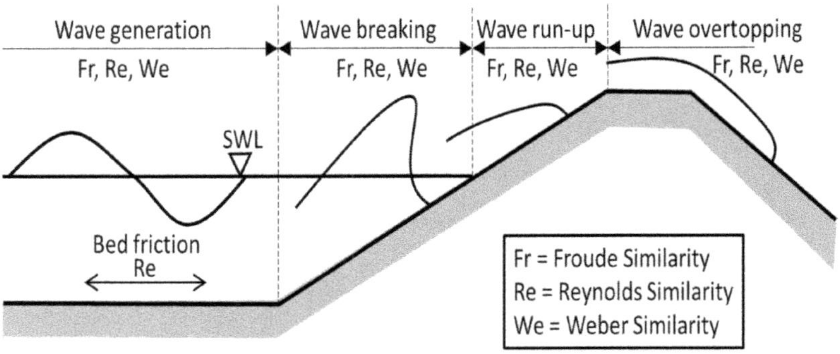

Fig. 5.4. Similarity laws associated with different processes.
Source: Schüttrumpf (2001).

scale factor). Small-scale model tests provide good information as far as Froude scaling is admissible. Froude scaling means that the most important forces are gravity and the inertia force. Other forces, such as capillarity, friction or elasticity forces, can be neglected (Fig. 5.4). Small-scale model tests are often performed on a scale between $1{:}\lambda_L = 1{:}10$ and $1{:}\lambda_L = 100$ (where λ_L is the scale factor). It should be noted that the final selection should always be directed towards larger scales to avoid scale effects.

Large-scale models try to test processes on a prototype scale or with a very large (between $1{:}\lambda_L = 1{:}1$ and $1{:}\lambda_L = 5$) scale factor. These tests are especially required if fluid–structure interactions are to be investigated. Large-scale model tests are more expensive than small-scale model tests due to construction and operation costs. As a first guess, the costs of large-scale model tests are in the same order of magnitude as the ratio of the scale factors. In addition, large-scale model tests are more difficult to perform since they need more time, space, instruments, etc. If it is required to understand the response of a structure, small-scale model tests are not suitable. The advantage of large-scale model tests is that the interactions between hydraulic loads and the sea bed and between hydraulic loads and the structure itself can be investigated on a prototype scale. It is impossible to scale the seabed or the material properties without considering scale effects.

Most of the tests are performed in wave flumes, which represent a 2D image of the real processes. This means that three-dimensional aspects (e.g. wave direction, shoreline bends, and the edges of a structure) are neglected. The most important factors in selecting a wave flume are cost and time. Any test in a wave flume is less expensive than a test in a wave basin. However, if three-dimensional aspects play a dominant role in the final result, tests should be carried out in a wave basin.

5.3.2 *Scale factor*

The second step is the determination of the scale factor. The scale factor is an indication of the similarity of the model. In general, it is defined as

$$\lambda_L = \frac{L_N}{L_M}$$

with λ_L = scale factor, L_N = dimension on the prototype scale, and L_M = dimension on the model scale.

The scale factor λ_L provides the geometrical similarity of the model. In addition, kinematic and dynamic similarity should be ensured as well. Kinematic similarity considers scaling of all time-dependent variables (time, velocity, discharge, etc.), whereas dynamic similarity considers the scaling of all force-dependent variables (pressure, force, etc.). Different forces influence the flow and wave processes in nature. The main forces are gravitational forces, inertia forces, elasticity forces, capillarity forces and friction forces. It has to be considered that scaling with a constant geometrical scale factor λ_L of all forces is not possible and that only tests on a prototype scale provide full dynamic similarity. Therefore, the most important law of similarity must be selected, which depends on the two governing forces. More than 99% of all models in coastal engineering are Froude models. This means that the two dominant forces are the gravity force and the inertia force. Other forces such as capillarity forces, friction forces and elasticity forces are neglected. This neglection results in scale effects, which can be reduced by a large scale factor and/or a careful interpretation of possible scale effects.

The scale factor λ_L influences not only all geometrical dimensions of the model but also the wave properties. Table 5.1 gives an overview of the scale factor for a number of selected parameters.

The scale factor λ_L depends on the possibilities of the selected facility as well and cannot be selected independently. In general, any facility is able to generate a maximum wave height and wave period or provide a maximum water depth which cannot be exceeded. The geometrical dimensions of the facility (the length and width of the wave flume or wave basin) represent another limiting factor for the determination of the scale factor. Therefore, the possibilities of the experimental facility are often the most common limiting factor for the choice of the scale factor.

Table 5.1. Overview of selected scale factors for a Froude model.

Parameter	Dimension	Scale factor
Length	[m]	λ_L
Height	[m]	λ_L
Width	[m]	λ_L
Slope	[—]	[—]
Area	[m^2]	λ_L^2
Volume	[m^3]	λ_L^3
Velocity	[m/s]	$\lambda_L^{1/2}$
Discharge	[m^3/s]	$\lambda_L^{5/2}$
Period	[s]	$\lambda_L^{1/2}$
Pressure	[N/m^2]	λ_L^3
Force	[N]	λ_L^3

5.3.3 *Measuring devices*

Careful selection of the measuring devices is the third step. Most of the tests in coastal engineering involve the generation of waves and their effects on shorelines and structures. This means that wave gauges should be selected that are able to measure the incoming waves as well as the reflected waves for later reflection analysis. Wave reflection analysis is an important aspect not only to avoid misinterpretation of data due to the overlay of incoming and reflected waves but also for wave absorption control.

Wave–structure interactions investigation require a variety of measuring devices (pressure cells, load cells, deformation measurements, etc.). Selection of the device depends on several criteria, the main being provided by the primary objectives and then possible secondary objectives of the tests. The objectives determine the parameters and thus the required instruments. As an example, for wave overtopping tests, an overtopping tank is required to collect the water to measure its volume. If it is necessary to distinguish between individual overtopping volumes and average overtopping rates, an overtopping tank must be equipped with more devices (either weighing cells or wave gauges) to distinguish between individual overtopping events. Other selection criteria for measuring devices are the number and type of instruments, sampling rate, measuring range and other parameters. The final number of instruments is always a compromise between the availability of instruments, the objective of the study, resolution, the available budget and other factors. In general, it is better to install more devices. The sampling rate depends on the speed of the processes dictated by the wave period to be investigated. While wave investigations can be performed with a sampling rate of 100 Hz, wave impact investigations need a sampling rate between 1000 and 2000 Hz. Finally, the measuring range should be determined at an early stage. If the measuring range is too high, the accuracy of the results is limited, and if the measuring range is not high enough, the peaks of the signal cannot be measured.

Experienced physical modellers know which instruments can be used. However, especially for new modellers like PhD students, the selection of instruments should be conducted through a preliminary study to avoid an expensive selection of the wrong instruments.

5.3.4 *Test programme*

The fourth step involves defining the test programme. The test programme depends on the objective of the tests. In fundamental research, it is recommended to set up the test programme on the basis of dimensional analysis. Dimensional analysis is a method to reduce the number of necessary tests by using dimensionless factors instead of real factors. Therefore, the Buckingham Π-theorem is often used in coastal engineering research. This theorem considers that an equation of n-dimensional variables can be reduced to a number of $(n - r)$-dimensionless variables, with r being the number of fundamental dimensions ([m], [s], [kg]).

A second advantage of dimensional analysis is the fact that results will be independent of the scale factor. This means that empirical formulae derived from experiments on smaller scales can be transferred to the prototype scale. A typical example are the wave overtopping formulae found in the EurOtop Manual (EurOtop, 2016). All these formulae were derived on the basis of small-scale model tests in wave flumes or wave basins. These formulae can be applied to any project in practice since they are independent of the model scale.

For practical investigations, it is recommended to set up the test programme on the basis of the design of a structure. It should be possible to test the design values with a safety margin in the selected facility. In addition, practical projects often require the testing of different geometrical variants instead of testing a wide range of wave parameters. This should be taken into account while setting up the programme.

5.3.5 *Testing*

The tests that need to be carried out come under the fifth step. Testing requires the availability of experienced staff to control experiments and model tests during the tests. The correct performance of all measuring devices, the stability of the model, and the correct generation of boundary conditions should be constantly observed and checked during the entire testing period. Protocols for the tests, including photos and videos, are required to be followed to record all observations for later interpretation of test results. In addition, it is important that tests are performed by the same person who will be involved in the later analysis and interpretation of the test results to avoid misinterpretation of the tests.

Observation of the tests and immediate control of the test results are important factors to avoid problems later. The functioning of instruments should be controlled and monitored on a permanent basis. This is also required to avoid problems later.

Tests should always be repeated to provide an indication of the scattering of the results. Therefore, it is recommended to rerun some tests at least three times as an indication of scattering.

5.4 Analysis and Interpretation of Test Results

The final step is related to the analysis of the results. A good and comprehensive presentation of the test results is required to ensure that interpretation is possible. Therefore, diagrams will have to be developed to present the measurement results, including the statistical analysis of the measured variables, the presentation of statistical factors and an error analysis. In addition, it is often necessary to provide all measured data in an annexe for later reanalysis of the results. All results are to be presented in the form of a report.

As already mentioned above, a chapter on possible scale and model effects is required for the correct interpretation of the results and observations (Fig. 5.5). Scale effects arise due to the violation or insufficient fulfilment of the relevant similarity laws in the scale model. Model effects occur due to the model and the governing forces not being faithfully reproduced, as well as due to data acquisition and data analysis.

Scale and model effects should be considered to avoid an expensive overdesign or a dangerous underdesign of coastal structures. These effects, if neglected, will result in additional costs, which might increase the construction costs of a coastal measure significantly. Therefore, the consideration of scale and model effects for the interpretation of model results is necessary not only for the economic design of coastal structures but also for the optimization of the physical model setup. Not taking into account scale

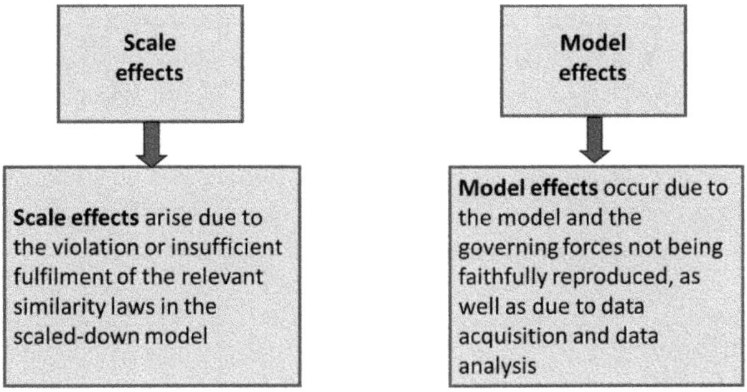

Fig. 5.5. Distinction between scale and model effects.

Table 5.2. Scale effects and critical limits.

Process	Relevant forces	Similitude law	Critical limits
Wave propagation	Gravity force	Fr_W,	$\mathrm{Re}_W >$
	Friction forces	Re_W,	$\mathrm{Re}_{W,\mathrm{crit}} = 10^4$
	Surface tension	We	$T > 0.35$s;
			$d > 2.0$cm
Wave breaking	Gravity force	Fr_W,	$\mathrm{Re}_W >$
	Friction forces	Re_W,	$\mathrm{Re}_{W,\mathrm{crit}} = 10^4$
	Surface tension	We	$T > 0.35\ s$; $d_b > 2.0$ cm
Wave run-up	Gravity force	Fr_A, Fr_q	$\mathrm{Re}_q > \mathrm{Re}_{q,\mathrm{crit}} = 10^3$
	Friction forces	Re_q,	$\mathrm{We} > \mathrm{We}_{\mathrm{crit}} = 10$
	Surface tension	We	
Wave overtopping	Gravity force	Fr_A, Fr_q,	$\mathrm{Re}_q > \mathrm{Re}_{q,\mathrm{crit}} = 10^3$
	Friction forces	Re_q,	$\mathrm{We} > \mathrm{We}_{\mathrm{crit}} = 10$
	Surface tension	We	

Notes: $\mathrm{Fr}_W = c/(g \cdot d)^{1/2}$; $\mathrm{Fr}_A = \mathrm{v}_A/(g \cdot h_A)^{1/2}$; $\mathrm{Fr}_q = \mathrm{v}_A/(2 \cdot g \cdot R_u)$; $\mathrm{Re}_W = c \cdot d/\nu$; $\mathrm{Re}_q = (R_U\text{-}R_C)^2/(\nu \cdot T)$; $\mathrm{We} = v_A \cdot h_A \cdot \rho_W/\sigma_W$ (Fr = Froude number, Re = Reynolds number, We = Weber number, T = wave period, d = water depth, g = gravity, v_A = run-up velocity, R_U = run-up, c = wave celerity, R_C = freeboard, ν = kinematic viscosity).
Source: According to Schüttrumpf (2001).

and model effects while setting up a physical model means that physical modelling becomes more of an art than a science, as stated by Oumeraci (1999).

Sources of model effects can be associated with the differences between the model and the real structure in the field (e.g. the development of underwater topography), the construction of the model (e.g. surface roughness), wave generation, measurement devices (e.g. position and type of measuring device), data acquisition (e.g. sampling rate), data analysis and neglected processes (e.g. wind, currents). Sources of scale effects can be associated with the dynamic forces being neglected by the application of the chosen law of similarity. Schüttrumpf (2001) derived certain limiting criteria for scale effects in wave run-up and wave overtopping investigations (Table 5.2). These criteria consider the influence of viscous and surface tension effects on wave run-up and wave overtopping investigations.

5.5 Summary

Each hydraulic engineering method has its own importance, possibilities, strengths and weaknesses. The aim must be to optimally use, combine or link the different methods in order to exploit synergy effects. An exclusive use of one method leads to coastal engineering monoculture. The further development of, for example, numerical models is then limited to numerical aspects, while our understanding of physical processes may not develop further. Therefore, coastal engineering model tests are indispensable research tools for the further development of transport formulae, design approaches of structures, failure models, etc.

Several present-day issues, such as the continuity of water bodies, the durability and functionality of hydraulic structures or the interaction between substances in the water and aquatic organisms, cannot currently be described with sufficient accuracy using numerical modelling methods. In many cases, the relevant issues cannot be modelled numerically due to the lack of a mathematical description of the relevant hydraulic, biological and chemical processes and their interactions. This has recently led to the development of hydromorphological, ecohydraulic, hydrotoxicological, hydrochemical or ethohydraulic methods that require an experimental approach (Schüttrumpf, 2013).

On the other hand, experimentally developed approaches have to be implemented in numerical modelling methods in order to find their way into coastal engineering practice. Measurements and observations in nature provide important boundary conditions for numerical and hydraulic models.

Consequence: Each coastal engineering method has its own significance, possibilities, advantages and disadvantages. The aim must be to optimally use, combine or link the different methods in order to utilize the synergy effects of hydraulic engineering in practice, research and teaching.

References

Aufleger, M. (1998). Flussmorphologische Modelle – Grundlagen und Anwendungsgrenzen. Berichte des Lehrstuhls und der Versuchsanstalt für Wasserbau und Wasserwirtschaft. Technische Universität München, Nr. 104, S. 198–207.

Duden. (2000). *Die deutsche Rechtschreibung*, 22 Auflage. Dudenverlag.

DWA. (2021). DWA-Merkblatt M-540 "Multidimensional morphodynamic-numerical modells for rivers". July 2021.

EurOtop. (2016). Manual on Wave Overtopping of Sea Defences and Related Structures. An Overtopping Manual Largely Based on European Research, But for Worldwide Application. Van der Meer, J.W., Allsop, N.W.H., Bruce, T., De Rouck, J., Kortenhaus, A., Pullen, T., Schüttrumpf, H., Troch, P., and Zanuttigh, B. www.overtopping-manual.com.

Kortenhaus, A. (2003). Probabilistische Methoden für Nordseedeiche. Dissertation. Technical University of Braunschweig.

Miller, R.L. (1972). The role of surface tension in breaking waves. In *Proceedings of the 13th Coastal Engineering Conference*, Vancouver, S. 433–449.

Naudascher, E. (1979). Einsatz des Hydrauliklabors in der Ausbildung von Bauingenieuren. *Bauingenieur*, 54, 121–124.

Oumeraci, H. (1999). Strenghts and limitations of physical modelling in coastal engineering – synergy effects with numerical modelling and field measurements. In *Proceedings of the Hydralab-Workshop in Hannover*, Germany, S. 7–38.

Plate, E. and Köngeter, J. (2003). Leitthema 5 "Wasser und Naturkatastrophen" in Wasserforschung im Spannungsfeld zwischen Gegenwartsbewältigung und Zukunftssicherung. *Deutsche Forschungsgemeinschaft. Denkschrift.* Wiley-VCH.

Plüß, A. and Schüttrumpf, H. (2004). Comparison of numerical tidal models for practical applications. In *29th International Conference on Coastal Engineering*, Lissabon.

Rutschmann, P. (2005). Hybride Modelle im Wasserbau. Berichte des Lehrstuhls und der Versuchsanstalt für Wasserbau und Wasserwirtschaft. H. 104. S87–S96.

Schäffer, H.A. (1999). On hybrid modelling in coastal and ocean engineering. In *Proceedings of the Hydralab-Workshop in Hannover*, Germany. S189–S197.

Schüttrumpf, H. (2001). Wellenüberlaufströmung an Seedeichen – Experimentelle und theoretische Untersuchungen. Dissertation. http://www.biblio.tu-bs.de/ediss/data/20010703a/20010703a.html.

Schüttrumpf, H.; Oumeraci, H. (2005) Scale and model effects in crest level design. In *2nd International Coastal Symposium on Iceland*, Höfn, the Town of Hornafjördur, Iceland.

Schüttrumpf, H. (2013). Brauchen wir wasserbauliche Versuchshallen? [Do we need water engineering Test facilities?]. In *Wasserwirtschaft: Hydrologie, Wasserbau, Hydromechanik*, Gewässer, Ökologie, Boden. – Wiesbaden: Vieweg + Teubner, GWV-Fachverl. [u.a.] - ISSN: 0043-0978, 2192–8762. -103 (2013) , 4, S./Art.: 3-3.

Schüttrumpf, H. (2014). Möglichkeiten und Perspektiven wasserbaulicher Modellversuche (Possibilities and perspectives of hydraulic engineering model tests). In Wasserbauliches Versuchswesen – Aus Anlass der Neuen Wasserbauhalle des IWW: 44. IWASA – Internationales Wasserbau-Symposium Aachen 2014; 9. und 10. Januar 2014, Aachen/Lehrstuhl und Institut für Wasserbau und Wasserwirtschaft, Rheinisch-Westfälische Technische Hochschule Aachen, RWTH. Editor: Holger Schüttrumpf.

Van der Meer, J.W., Bernardini, P., Snijders, W., and Regling, E. (2006). The wave overtopping simulator. In *Proceedings of the 30th International Conference on Coastal Engineering*, San Diego, California.

Van Os, A., Soulsby, R., and Kirkegaard, J. (2004). The future role of experimental methods in European hydraulic research. *PIANC Bulletin*, AIPCN No. 117. S41–46.

Vollmers, H.-J. (1998). Experimente und Modellversuche mit beweglicher Sohle. In *Dresdner wasserbauliche Mitteilungen*, Heft 13. S. 29–37.

Chapter 6

Wave Impact Response

6.1 Introduction

Coastal environments have posed a persistent challenge to engineers since the advent of sea trade. With the surge of the Industrial Revolution, sea routes expanded, necessitating the construction of coastal structures for berthing and the establishment of tranquil basins. Consequently, port development drove the acquisition of adjacent land for both commercial and residential purposes over time. The rise in human settlement near coastal zones underscores the need for the construction of wave buffers, emphasizing the importance of advancing our understanding of their interaction with ocean waves.

Vertical wall-type structures are commonly utilized for coastal protection works, breakwaters, and berthing structures. Over their operational life, these constructions endure constant exposure to wave action, with the impact of breaking waves posing a significant challenge. The force exerted by breaking wave impact is more than tenfold greater than the force of non-impact waves within a short range, leading to notable structural instability, damage and local fatigue concerns. Consequently, there has been growing emphasis on comprehending wave–structure interactions to facilitate optimal design, propelling research endeavours undertaken by engineers and scholars in this domain. Nonetheless, the breadth of investigation is limited by various factors, such as inadequate understanding of wave kinematics under breaking wave conditions, air entrainment phenomena and the intricate nature of flow dynamics.

In recent years, extensive research efforts have been underway to address the limitations of the physics associated with wave-breaking phenomena. While a few experimental studies have been conducted in the past, they have remained relatively scarce. Therefore, there is a pressing need for comprehensive investigations to delve deeper into the impact phenomenon associated with wave–structure interaction.

The advancement of impact studies significantly benefits from the concurrent development of theoretical and experimental research. While researchers have proposed a few mathematical models for wave impact problems, the complexity of breaking wave impact poses additional challenges. This complexity arises from the intricate nature of wave-breaking phenomena, which may result in a lack of sufficient knowledge regarding crucial parameters such as the kinematics of breaking waves and the duration of wave impact. Consequently, contemporary research is increasingly focused on conducting detailed studies to better understand the physics underlying these parameters. Therefore, numerical investigations into breaking wave impact problems are imperative to establish efficient design methods for sea-facing structures.

6.2 Types of Breakers

Depending on the mode of breaking, breaking waves may be classified as follows:

6.2.1 *Spilling breakers*

In this wave type, the crest experiences deformation and becomes unstable, leading to spillage over the front of the wave. This phenomenon often creates a frothy appearance. It typically occurs on gentle beaches. Waves with low steepness on mild slopes break by continuously spilling foam down the front face, a process sometimes referred to as white capping. This mode of breaking is gradual.

6.2.2 *Plunging breakers*

Plunging breakers are considered a dramatic form of wave breaking. The wave crest curls over and crashes into the base of the wave, generating a significant splash. This type of breaking typically occurs when medium- to large-steepness waves travel over medium-steepness beaches. The waves break instantaneously.

6.2.3 *Surging breakers*

On steeper beaches, a wave may progress without breaking, instead deforming and flattening from the bottom. The wave front advances towards the crest, leading to reflection. This phenomenon is observed with the steepest waves propagating over high-steepness bottom slopes. The base of the wave

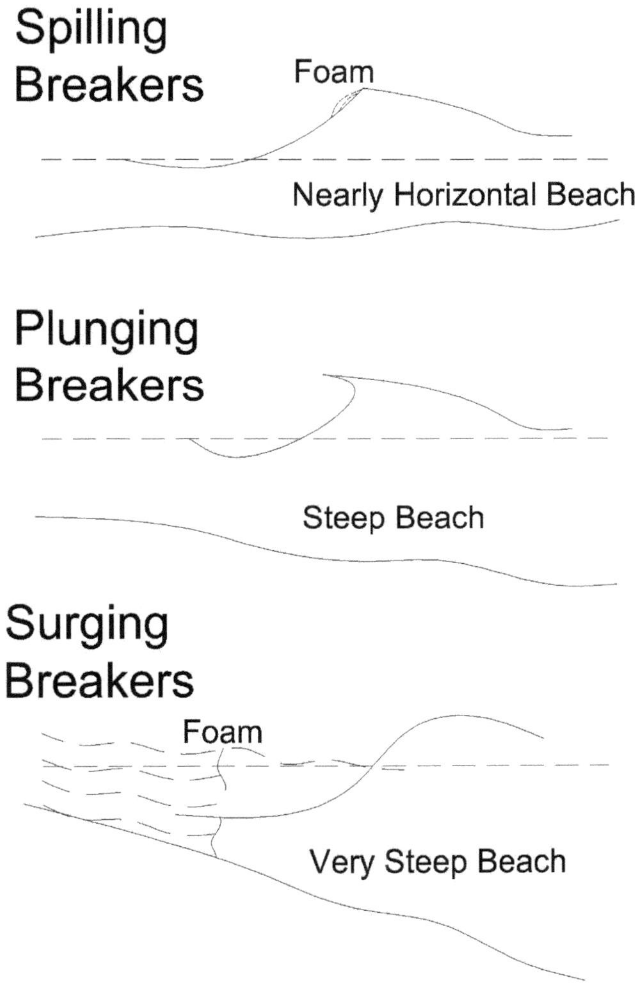

Fig. 6.1. Sketch of various types of wave breakers.
Source: Sundar (2021).

surges up the beach, producing significant foam accumulation. It appears as though it may transition into the plunging type of break.

6.2.4 *Collapsing breakers*

The classification of collapsing waves falls between plunging and surging, where the crest never fully breaks, but the bottom face of the wave steepens and collapses, leading to the formation of foam.

Illustrations of the different types of breakers are shown in Fig. 6.1.

The design of coastal and onshore structures is facing greater challenges due to the inclusion of tsunami-induced forces. The March 2011 Tsunami that hit the northeastern coast of Japan and the December 2004 Indian Ocean Tsunami have exposed the limitations in our engineering design of coastal and onshore structures. The conventional critical load combination and the factor of safety adopted for each type of load are not sufficient to withstand a tsunami impact.

The guidelines for the design of structures exposed to tsunamis have been detailed by the Federal Emergency Management Agency's Coastal Construction Manual (FEMA) (2005) and the City and County of Honolulu Building Code (CCH) (2000). Asakura *et al.* (2000) have carried out laboratory tests on the forces on structures exposed to tsunamis, the results of which were adopted by Okada *et al.* (2005) to establish design guidelines. Yeh (2007), through a critical review of the existing literature, proposed rational methodologies for the estimation of tsunami-induced loadings. In general, tsunami loads are considered in the following forms: hydrostatic, buoyant, hydrodynamic, surge, debris impact and wave-breaking forces. Among those, the impact force due to wave breaking has been defined inadequately. The plunging-type breaking occurs in a finite depth of water and, once the wave crest plunges, it becomes a broken wave in a shallower depth. A broken tsunami wave moves like a *bore* because of its long wavelength. The movement of tsunami on dry land is called a *surge*. In this chapter, the impact of tsunamis on coastal and onshore structures is examined by specifically considering the tsunami as a bore front. This approach considers the scenario where the wave breaks at a considerable distance from the shore.

The bore impact on a wall has been numerically predicted by Mohapatra *et al.* (2000) by solving 2D flow equations. The predictions were compared with the experimental data given by Ramsden (1993). In the present study, the bore impact force on a vertical-faced structure is explored using the pressure impulse theory proposed by Cooker and Peregrine (1995). The pressure impulse theory is useful for the prediction of high pressures that result from the sudden velocity changes at impact. The wave impact on slender cylindrical members needs a detailed computational fluid dynamics (CFD) simulation due to the influence of the wake on the impact on the structure.

Section 6.3 elaborates on the pressure impulse theory, which focuses on estimating impulse – an integrated quantity. When designing structures, predicting the maximum impact pressure and, consequently, the force poses a significant challenge, as these impacts occur over a very short duration, typically in the order of milliseconds. Therefore, the impulse, which represents the integrated energy form, serves as a convenient design parameter for estimating the impact of tsunami waves.

The following sections present the pressure impulse theory in the context of bore impact and the subsequent validation of the theory for tsunami impact on a nearshore elevated structure. Section 6.8 presents the CFD simulation of wave impact on cylindrical members.

6.3 Pressure Impulse Theory

The concept of pressure impulse was proposed by Cooker and Peregrine (1995). Pressure impulse is the time integral of the pressure during a violet wave impact of short time scale. The model assumes an inviscid, incompressible and flat free surface at the moment of impact. Air entrainment is neglected.

It has been reported (Wood and Peregrine, 2000) that the pressure impulse concept, together with simplifications of the wave geometry, gives easily calculated values of pressure impulse for application and provides insight into various aspects of wave impact. The original application of the pressure impulse method was for two-dimensional wave impacts on a vertical wall where there is a negligible amount of air trapped by the wave. Preliminary studies showed that pressure impulse distributions and magnitudes were surprisingly insensitive to details of the shape of the wave.

The pressure impulse approach to wave impact modelling can thus be useful for estimating the most severe loading on a coastal or offshore structure. The following forcing and structural parameters are needed such as wave height, impact velocity and the zone of impact over the wall. With considerable simplification of the geometrical shape, Laplace's equation gives relatively simple solutions for pressure impulse in either two or three dimensions.

The proper estimation of pressure impulse needs the forcing parameters to be known accurately, i.e. breaking wave height and water velocity during impact. In particular, the estimation of water velocity at impact is difficult to quantify due to complex phenomena near breaking. In recent years, wave kinematics under breaking waves has been analyzed by various researchers (Chanson and Fang, 1997; She *et al.*, 1997; Naik, 2005; Lynett, 2006). But the velocity estimates differ from one study to another to a great extent. However, the typical range tends to align closely with the wave velocity or group celerity. Hence, in this chapter, the water velocity is approximated to wave group celerity.

6.4 Mathematical Formulation of Pressure Impulse

Pressure impulse is defined as

$$P = \int_{t_b}^{t_a} p(x,t)dt, \tag{6.1}$$

where t_a and t_b are the times after and just before impact, respectively.

Euler's equation of motion can be used to the model wave impact phenomenon on the assumption of two-dimensional (2D) motion, inviscid fluid and incompressible flow. Assuming an impulsive event occurs over such a short period of time, then the nonlinear convective terms in the equation of motion can be ignored:

$$\frac{\partial u}{\partial t} = -\frac{1}{\rho}\nabla p. \tag{6.2}$$

Here, u is the velocity of water at impact, $p(x,t)$ is the impact pressure and ρ is the density of water.

Integrating Equation (6.2) with respect to time, $t = t_b - t = t_a$, gives

$$u_a \mathrm{u}_b = -\frac{1}{\rho}\nabla p, \tag{6.3}$$

where u_a and u_b are water velocity after and before impact, respectively.

Taking the divergence of Equation (6.3) leads to

$$\nabla^2 P = 0. \tag{6.4}$$

Equation (6.4) shows that the pressure impulse satisfies the Laplace equation. This equation can be solved numerically and in the present study, the Finite element method is used as a numerical tool.

It is possible to find maximum impact pressure if the impact pressure rise–fall time is known. The rise–fall time can be derived from experimental measurements. The shape of the impact pressure time curve can be well approximated by a triangle with peak pressure P and a base width of Δt (Fig. 6.2). Hence, from pressure impulse (P) and impact pressure rise–fall time (Δt), the maximum impact pressure can be estimated as follows:

$$p_{\text{peak}} = \frac{2P}{\Delta t}. \tag{6.5}$$

6.5 Model of an Idealized Wave Impact on a Vertical Wall

6.5.1 *General*

A two-dimensional idealized model depicting wave impact on a vertical wall is illustrated in Fig. 6.3. The Cartesian coordinate system (x, z) is utilized,

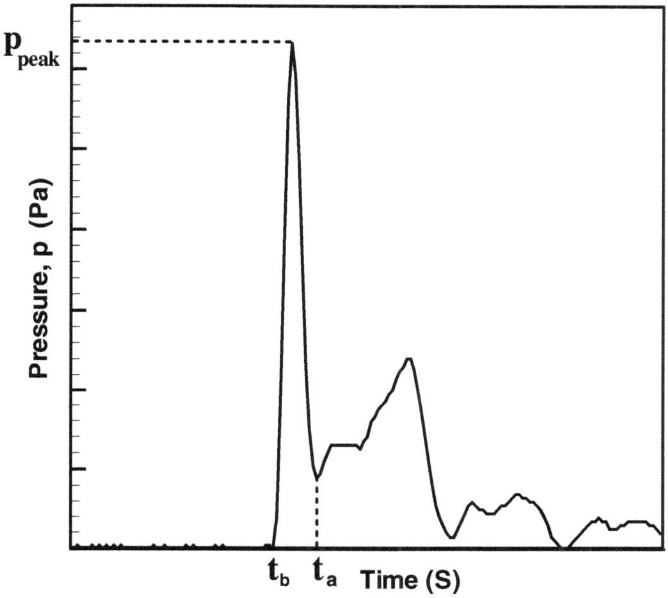

Fig. 6.2. Sketch of typical impact pressure time history.

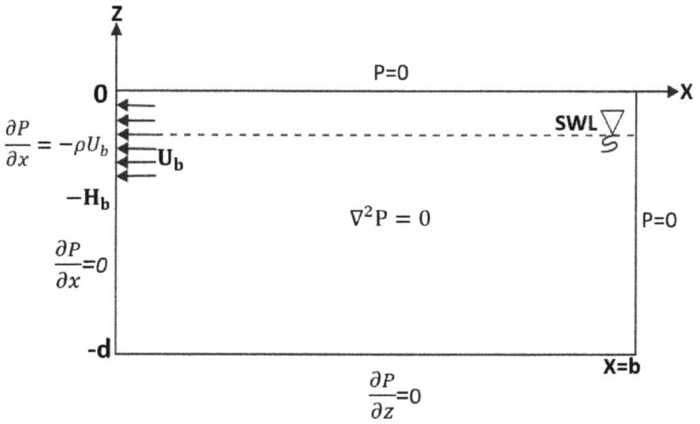

Fig. 6.3. Sketch of idealized model of wave impact on a vertical wall.

with the origin set at the intersection of the free surface and the vertical wall. The fluid domain is simplified to a rectangle, with the free surface corresponding to the elevation of the breaking wave crest, and the wall positioned at $z = 0$. The domain's dimensions are restricted to $x = b$. The impact zone of the breaking wave on the wall spans from the crest elevation to the trough, denoted by the breaking wave height, H_b. Below the impact region, the fluid encounters the wall up to the seabed ($z = -d$).

6.5.2 *Boundary conditions*

The general boundary conditions of the impulse model are given in the following:

(a) Free surface boundary:

$$(P)_{z=0} = 0. \tag{6.6}$$

(b) At the vertical wall:

Over the solid boundary at the impact region ($z = 0$, $z = -H_b$), where the breaking wave hits, the change in normal velocity gives the normal derivative of the pressure impulse:

$$\left(\frac{\partial P}{\partial x}\right)_{\mathrm{x=0}} = -\rho U_b. \tag{6.7}$$

Below the impact region, where the fluid is in contact with wall before and after impact,

$$\left(\frac{\partial P}{\partial x}\right)_{x=0} = 0. \tag{6.8}$$

(c) At the bottom boundary (x=0, $x = b$):

$$\frac{\partial P}{\partial n} = \frac{\partial P}{\partial z} = 0. \tag{6.9}$$

(d) Far-field boundary:

$$(P)_{x=b} = 0. \tag{6.10}$$

From Equation (6.7), it should be noted that the normal velocity of water at impact has to be known for computation. As we have discussed earlier, U_b can be well approximated to wave group celerity, C_G.

The pressure impulse (P) is defined in terms of instantaneous impact pressure $p(x,t)$ within the short duration $t_a < t < t_b$:

$$P = \int_{t_a}^{t_b} p(x,t)\,dt. \tag{6.11}$$

In a two-dimensional flow field, with the assumption of inviscid and incompressible fluids, Euler's equation of motion can be used to model the wave impact phenomenon. The nonlinear convective terms in the equation of

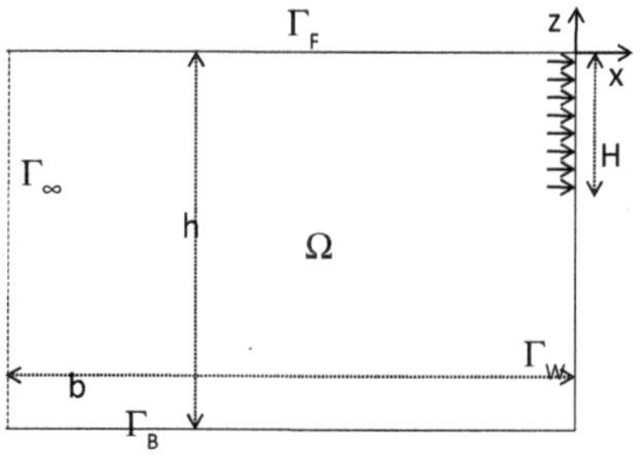

Fig. 6.4. Idealized fluid domain.

motion have been neglected under the assumption that large local pressures last only for a shorter duration. Therefore, Euler's equation can be written as

$$\frac{\partial u}{\partial t} = -\frac{1}{\rho}\nabla p, \tag{6.12}$$

where ρ is the density of water and u is the flow velocity at impact. Integrating Equation (6.12) over the impact duration ($t_a \leq t \leq t_b$) and taking the divergence lead to the Laplace equation:

$$\nabla^2 P = 0. \tag{6.13}$$

Figure 6.4 shows a two-dimensional idealized model of the wave impact on a vertical wall. The fluid domain (Ω) has been idealized with a horizontal free surface (Γ_F) at the top of a bore ($z = 0$) and a horizontal bottom surface ($z = -h$). The rigid vertical wall forms the right boundary (Γ_W) at $x = 0$. The total flow depth (h) is taken equal to the bore height (d_1). The impact on the wall occurs over a depth of ($d_1 - d_2$), where d_2 is the flow depth in front of the bore movement.

The boundary conditions imposed on the fluid domain are defined as follows:

(a) On the free surface Γ_F,

$$P = 0. \tag{6.14}$$

(b) The impulse on the far field (left boundary, Γ_∞) vanishes:

$$P = 0. \tag{6.15}$$

(c) Due to no velocity flux along the bottom boundary, the pressure impulse does not dissipate through the bottom boundary, $\Gamma_B(z = -h)$. Hence,

$$\frac{\partial P}{\partial z} = 0. \tag{6.16}$$

(d) On the rigid vertical wall, the variation in normal flow velocity leads to pressure impulse flux on the impact region. Below this region, where the fluid is initially in contact with the wall prior to and after the impact ($z < d_2$), the pressure impulse flux vanishes. This assumption can be justified by considering the flow field of a turbulent bore:

$$\frac{\partial P}{\partial n} = \begin{cases} -\rho U_b & -(d_1 - d_2) \le z \le 0 \\ 0 & -d_1 < z < -d_2 \end{cases}. \tag{6.17}$$

6.5.3 *Numerical procedure*

The pressure impulse is computed during the flow impact. The impulse is taken as a field variable in the entire domain, and the governing Laplace equation (Equation (6.13)) can be solved using the finite element method. The governing equation along with the four boundary conditions may be expressed in weighted residual form to obtain the finite element system of equations as follows:

$$\sum_e \int_{\Omega^e} \nabla N_i \nabla N_j d\Omega^e \{P_j\} = \sum_e - \int_{\Gamma_w} N_i \frac{\partial P}{\partial n} d\Gamma^e. \tag{6.18}$$

In the finite element approach, the domain is discretized into sub-domains (Ω^e) represented by discrete elements, which exchange information through the joining nodes. The pressure impulse at any point inside the sub-domain can be represented in the form of a prescribed shape function $N\ (x, y)$ and is presented as

$$P = \sum_{j=1}^{n_e} P_j N_j(x, z), \tag{6.19}$$

where P_j denotes the nodal variables and n_e is the number of element degrees of freedom. The shape functions are explicitly written as the functions of quadratic iso-parametric coordinates. The summation implies the assembly of the element property matrices over the entire domain Ω (Zienkiewicz and Taylor, 1989), and the resulting simultaneous equations are solved for the pressure impulse at every node, P_i, $i = 1, 2, \ldots, M_n$,

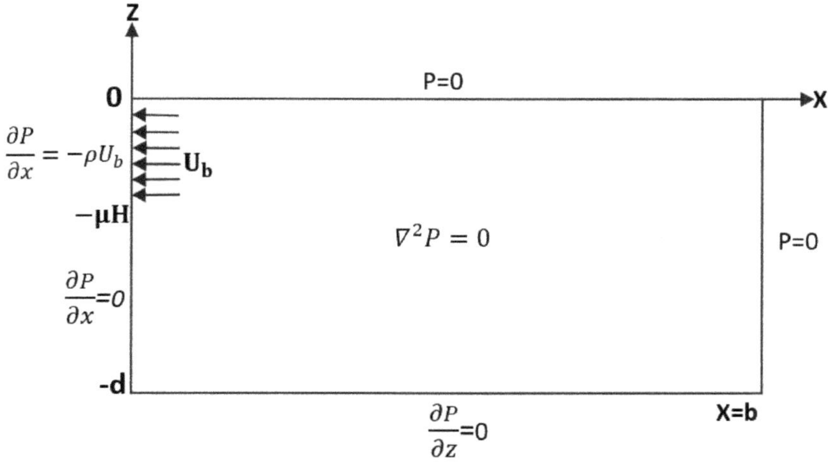

Fig. 6.5. Impact of an idealized wave on a vertical wall.
Source: Cooker and Peregrine (1995).

taking advantage of the symmetric banded nature of the matrix. The method is discussed in detail by Muni Reddy *et al.* (2007).

Herein, the domain length (b) is a crucial parameter for computation. This influences the accuracy of pressure impulse to a greater extent. So, determining the length of the domain is an important task for proper computation of pressure impulse as well as for reducing computational complexity. The effect of the domain length on the pressure impulse has been discussed by Cooker and Peregrine (1995), and they recommended that the length of the domain (b) be equal to or greater than the height of the domain (h) for proper computation of pressure impulse. In the present study, the length of the domain is set equal to the total fluid depth, i.e. $b = h$.

The finite element solution of the governing Laplace equation is compared with the series solution. The governing equation was solved analytically by Cooker and Peregrine (1995) using the separation of variables in Laplace's equation and Fourier analysis. The proposed idealized domain solved by Cooker and Peregrine (1995) is shown in Fig. 6.5. Herein, H is the distance from the sea bed to the wave crest and μ is the fraction of this height.

The test conditions used for computations (Cooker and Peregrine, 1995) are defined by $H = \rho = U_b = 1$ and $\mu = 0.5$. The series and numerical

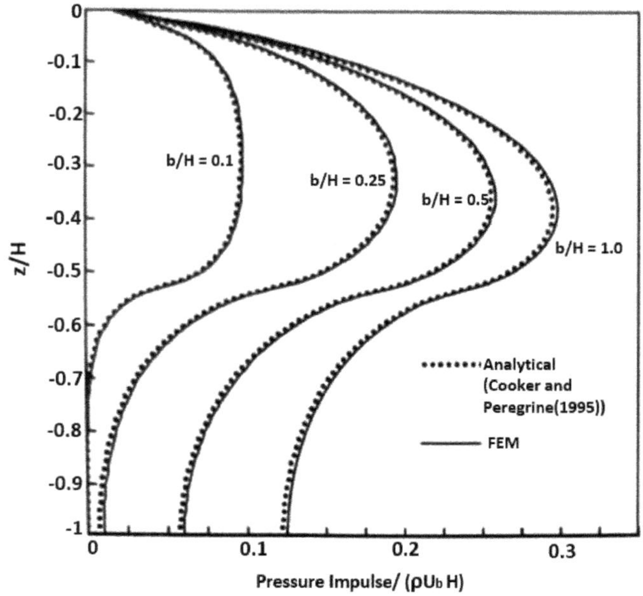

Fig. 6.6. Comparison of FEM with analytical solutions.

solutions for different b/H values are given in Fig. 6.6. The maximum variation in the results between analytical and FEM solutions is observed to be about 6%.

6.6 Discussion on Salient Results

6.6.1 *Impulse on a vertical wall*

To validate the present finite element model with the analytical solution of Cooker and Peregrine (1995), the pressure impulse on a vertical wall induced by a plunging wave was first simulated. The fluid impact depth (H) is taken as half the fluid depth. Various domain lengths (b/h) are considered in the range from 0.1 to 100. Pressure impulse is normalized with the fluid velocity at impact (U_b), the depth of fluid impact (H) and the fluid density (ρ). The vertical elevation (z) is normalized with the total fluid depth (h). The domain width of the order of fluid depth ($b/h = 1.0$) is sufficient for the estimation of pressure impulse: The deviation at the peak impulse from the case of $b/h = 100$ is 5%. It is emphasized that the said numerical prediction as detailed by Senthil Kumar and Sannasiraj (2011) and Manjula *et al.* (2013) exactly matches the analytical solutions of Cooker and Peregrine (1995).

The domain used for computation is shown in Fig. 6.5. It can be noted that the height of the domain can be calculated by using the following formula:

$$\text{Height of the domain } h = d + \frac{H_b}{2},$$

where d is the water depth and H_b is the breaking wave height.

However, the length of the domain (b) is also needed for computation. This influences the pressure impulse to a greater extent. So, fixing the length of the domain is an important task for proper computation of pressure impulse and as well as reducing computational complexity. The effect of the length of the domain on the pressure impulse has been reported by Cooker and Peregrine (1995), and shows that $\frac{b}{h} \geq 1$ is probably more appropriate for the proper computation of pressure impulse.

Hence, in this present study, the length of the domain has been fixed equal to the height of the domain, $\frac{b}{h} = 1$. The fluid velocity U_b is taken as the group celerity C_G.

6.6.2 *Comparison with experiments*

Naik (2005) conducted an experimental study of breaking wave impact pressures on vertical walls for regular, random and breaking waves at the Department of Ocean Engineering, Indian Institute of Technology Madras. Deep-water wave plunging was simulated by generating a frequency-modulated wave packet programmed to yield constructive interference of the different components at the desired location. The description of experimental data is given as follows:

- Input breaking distance = 11.40 m (from wavemaker),
- Mean frequency = 0.76 Hz,
- Mean time period, $T_c = 1.31$ s,
- Characteristics wavelength, $L_c = 2.57$ m,
- Span of impact location = 0.2 $L_c = 0.51$ m.

The time series of induced pressures from pressure transducers and the incident wave time history from the wave probes were measured.

Table 6.1. Experimental data from Naik (2005) (Plunging).

Run No	Water Depth, d (m)	Breaking Wave Height, $H_b(m)$	Group Celerity, C_G (m/s)
231	0.8	0.44	1.15
232	0.8	0.44	1.20
222	0.8	0.44	1.09
223	0.8	0.44	1.12
224	0.8	0.44	1.09
213	0.8	0.44	1.12
214	0.8	0.44	1.13

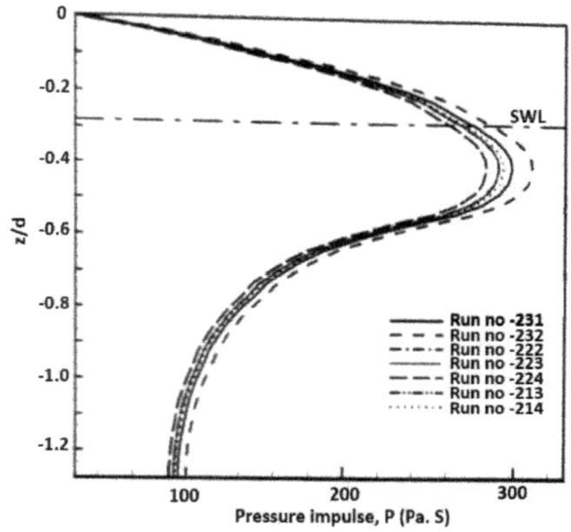

Fig. 6.7. Measured pressure impulse.

6.6.3 *Plunging breakers*

Numerical results have been compared with experimental results by Naik (2005). The experimental data given by Naik (2005), which are used for computations, are given in Table 6.1. The run number is used as an experimental data reference to distinguish different input parameters. The mean wave period is 1.31 s.

The vertical distribution of the computed pressure impulse (P) on the vertical wall for different runs is shown in Fig. 6.7, in which it can be noted that the pressure impulse is a maximum slightly below the still water level.

As observed, determining the rise and fall time of impact pressure is crucial for assessing the impact pressures accurately. However, estimating Δt poses challenges due to the intricate nature of the breaking phenomenon. The research is still in progress to theoretically evaluate the Δt. Results

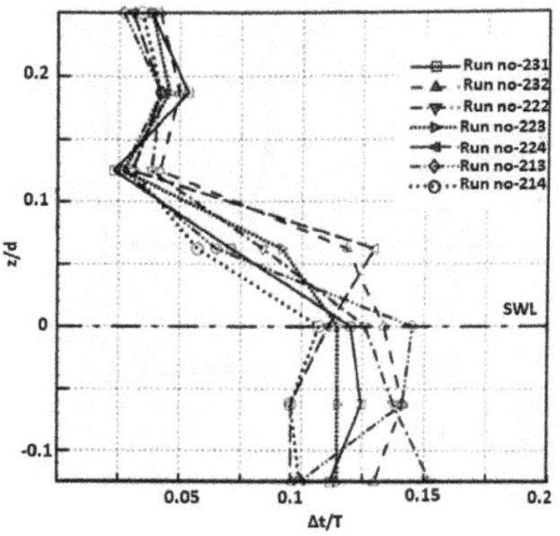

Fig. 6.8. Experimental impact pressure rise.

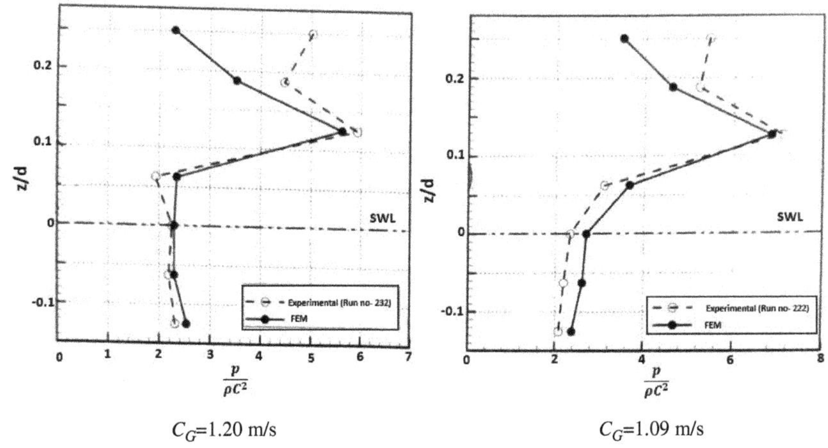

Fig. 6.9. Impact pressure on the vertical wall.

may be achieved from further theoretical and experimental studies. In this study, the pressure rise time (Δt) has been taken from the experiments by Naik (2005) for a better prediction of impact pressures. The experimental impact pressure rise–fall time (Δt) for different runs is depicted in Fig. 6.8. Numerical impact pressures can be calculated from Figs. 6.7 and 6.8. by using Equation (6.5). Typical comparisons of the results from numerical and experimental studies are shown in Fig. 6.9.

From the results, it is observed that maximum impact pressure always occurs slightly above the still water level. Numerical maximum impact peak values are well compared with experiments. So, the pressure impulse model is suitable for predicting wave impact pressures if the pressure rise–fall time is known. Complete theoretical prediction of impact pressures is possible only if theoretical pressure rise–fall time is known. Hence, it should be achieved through further experimental and theoretical analyses.

6.6.4 *Spilling breakers*

In this type of wave, the crest undergoes deformation and destabilizes, resulting in it spilling over the front of the wave. This wave tends to create a frothy appearance. This foam is made up of air pockets. So, the effect of the density of air has to be considered in wall-sided boundary conditions for proper computation of pressure impulse. Hence, from the top, 10% of the air effect has been included in the wall-sided boundary condition, where the water hits the wall. Therefore, the proposed new idealized model of wave impact on vertical walls is shown in Fig. 6.10.

Numerical results have been compared with experimental results by Naik (2005). The experimental data given by Naik (2005), which are used for computations, are given in Table 6.2. The run number is used as an experimental data reference to distinguish different input parameters. The mean wave period is 1.31 s. The vertical distribution of computational pressure impulses (P) on a vertical wall for different runs is given in Fig. 6.11, in which it can be noted that the pressure impulse gets a maximum slightly below the still water level.

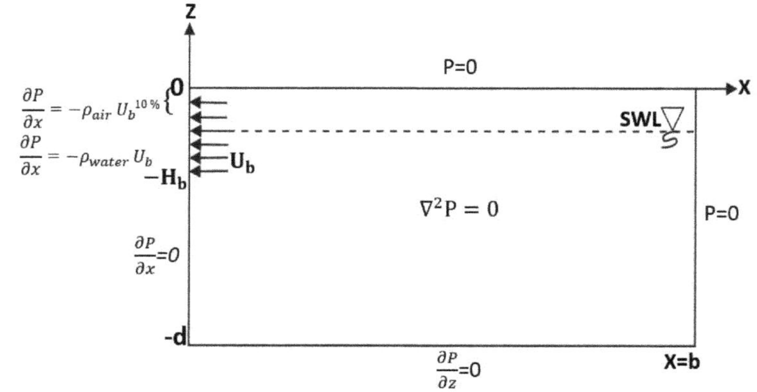

Fig. 6.10. Sketch of idealized model of wave impact on a vertical wall for spilling case.

Table 6.2. Experimental data from Naik, 2005 (Spilling).

Run No	Water Depth, d (m)	Breaking Wave Height, $H_b(m)$	Group Celerity, $C_G(m/s)$
243	0.8	0.41	1.11
244	0.8	0.42	0.98
235	0.8	0.42	1.05
236	0.8	0.42	1.01
227	0.8	0.44	1.10
219	0.8	0.44	1.02

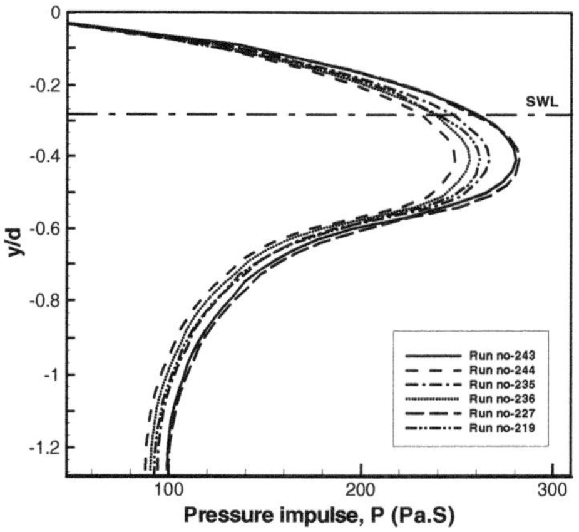

Fig. 6.11. Computed pressure impulse on the wall.

The numerical impact pressures can be calculated from the results in Figs. 6.11 and 6.12 through Equation (6.5). These numerical results for different runs were compared with the experimental results of typical breaking wave cases and are shown in Fig. 6.13.

Based on the results, it is evident that the numerical impact peak values closely align with experimental measurements. Therefore, the proposed boundary condition satisfactorily meets the requirements for analyzing the problem, and the pressure impulse model proves suitable for predicting wave impact pressures, even for spilling breaking waves, provided that the pressure rise–fall time is known. Full theoretical predictions of impact pressures are achievable only when the theoretical pressure rise–fall time is determined.

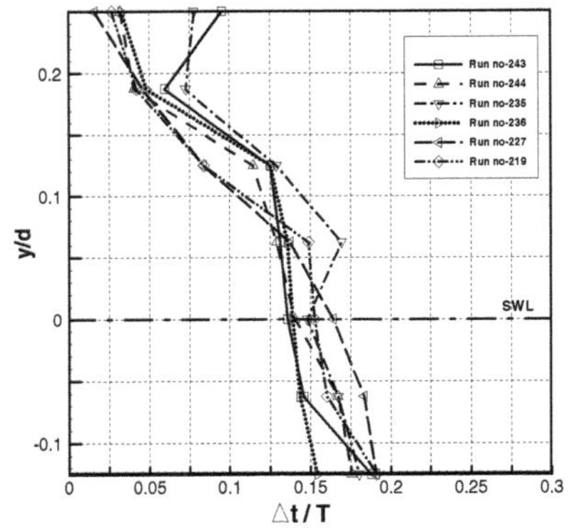

Fig. 6.12. Measured impact pressure rise–fall time on the wall (Spilling).

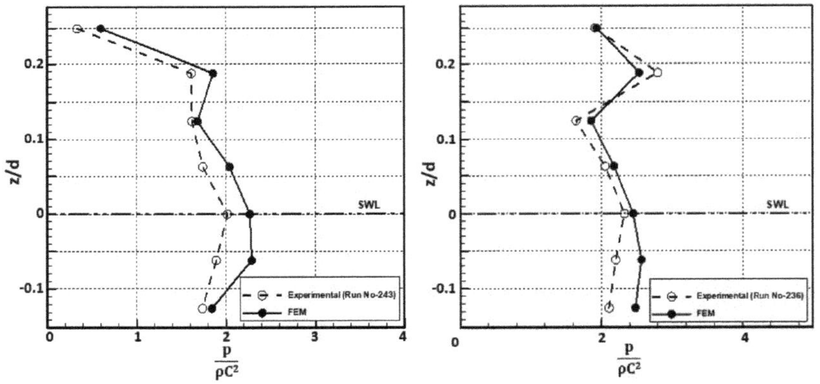

Fig. 6.13. Typical comparisons of measurements (Yeh, 2007) and computed impact pressure on a vertical wall.

6.6.5 *Parametric study*

The following parameters influence the pressure impulse and were used for the parametric study.

- Fluid velocity, U_b,
- Breaking wave height, H_b,
- Water depth, d.

The effect of fluid velocity on the pressure impulse has been investigated, maintaining the other two parameters of breaking wave height and water depth. The results for a breaking wave height $H_b = 0.44\,\text{m}$ and water depth

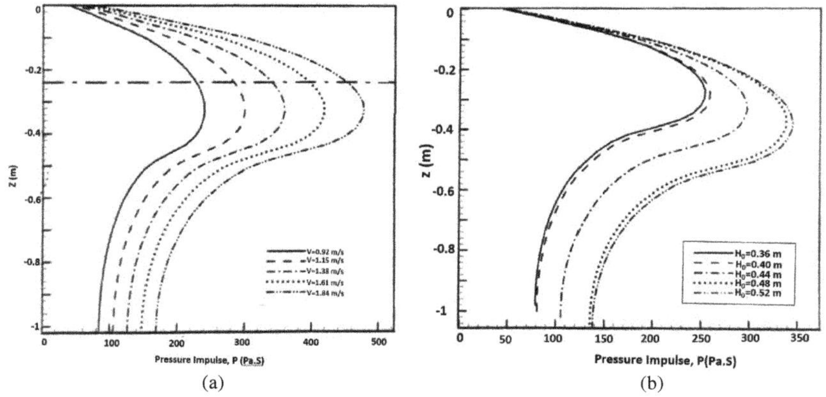

Fig. 6.14. (a) Pressure impulse for different velocities and (b) pressure impulse for different H_b.

$d = 0.8\,\mathrm{m}$ (Fig. 6.14(a)) show that the fluid velocity has a linear effect on the pressure impulse.

The effect of breaking wave height on the pressure impulse has been investigated with the other two parameters of fluid velocity and water depth remaining fixed. The results for a fluid velocity $U_b = 1.15\,\mathrm{m/s}$ and water depth $d = 0.8\,\mathrm{m}$, (Fig. 6.14(b)) show that the breaking wave height increases the span of wave impact such that the pressure impulse also increases.

The effect of water depth on the pressure impulse has been investigated maintaining the other two parameters of fluid velocity, $U_b = 1.15\,\mathrm{m/s}$, and breaking wave height, $H_b = 0.44\,\mathrm{m}$, constant. The results reveal (Fig. 6.15) that the effect of water depth on the pressure impulse is not significant.

Tsunami-induced forces on nearshore structures can also be modelled as turbulent bore-front impacts as detailed in the following.

6.7 CFD Modelling of Wave Impact on Cylinders

6.7.1 *General*

Modelling of fluid dynamics typically begins by adhering to the fundamental conservation equations of fluid mechanics and elasticity theory, particularly if fluid–structure interaction significantly influences the hydrodynamic response. In the case of thin cylindrical members, computational fluid dynamics modelling commonly employs the finite volume technique. This approach entails dividing the total region of interest into smaller subregions known as control volumes. The global equations are then discretized

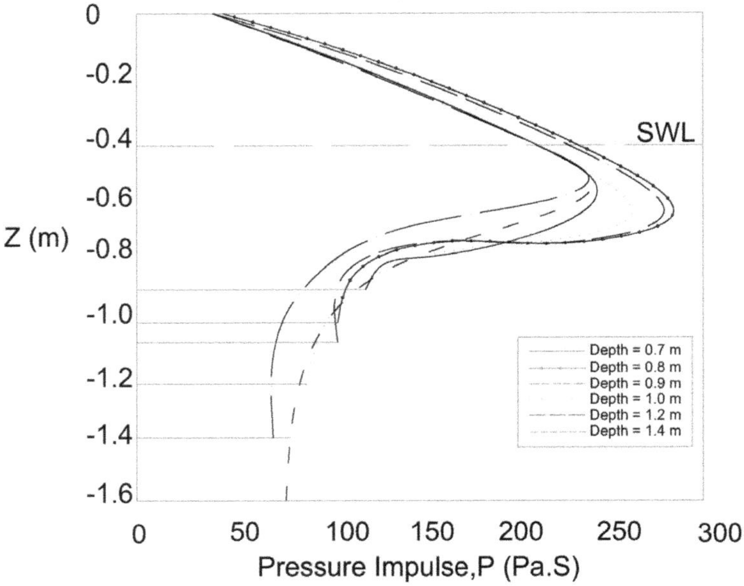

Fig. 6.15. Computational pressure impulse on the vertical wall for different water depths.

and solved iteratively within each control volume. The size of the control volumes is determined based on the mesh. With known variable values in each control volume, visualizing the flow behaviour across the entire region becomes straightforward.

6.7.2 *Volume of a fluid model*

The model is provided for systems containing two or more immiscible fluid phases, where each phase constitutes a large structure within the system (such as typical free surface flows). This approach captures the movement of the interface between the fluid phases and is often used for marine applications. Both the multi-phase segregated model and the volume of fluid model utilize distinct Eulerian phases, in contrast to a Lagrangian phase and a dispersed phase. Consequently, a unified framework has been established to designate phases for both models, known as the Eulerian multiphase model in Star-CCM+. It facilitates the creation and management of multiple Eulerian phases, requiring the specification of phase materials, viscous regime and the phase equation of state.

For the current problem, the volume of fluid model was selected due to its capability for interface tracking and handling interface reconnection without necessitating interface reconstruction. This model is also highly numerically efficient, particularly in simulations where each phase constitutes a substantial structure with a relatively minor contact area between phases. Within this approach, the spatial distribution of each phase at a given time is defined using a variable known as Volume Fraction.

6.7.3 *Modelling solid deformation*

The solid stress analysis in Star-CCM+ is carried out by adopting the Finite Volume Method (FVM). Unlike the fluid equations, solid stress is solved in a Lagrangian framework. That is, the mesh follows the material points in the material. This means that the mass in a particular cell will remain unchanged as it deforms because it physically represents the same material. This cell mass is determined to be the product of the original density of the material and the original volume. Therefore, the density in the deformed state will be the original mass divided by the current volume of the cell. In the small displacement Finite Volume (FV) stress solver in Star-CCM+, the solid mesh is not moved. It is assumed that the deformations are so small that any changes in mass are negligible. However, with large displacement capability, Star-CCM+ continually adjusts the solid vertices of the mesh according to the displacements.

In many aspects, the requirements for finite volume meshes to accurately resolve displacement and stresses parallel those of finite element meshes. Regions experiencing stress concentrations necessitate precise mesh resolution. Features such as fillets, the outer diameter of holes, parts undergoing necking, re-entrant corners, cracks or welded surfaces where two surfaces are incompletely joined require local refinement of the grid. There are some distinctions between the requirements for finite volume and finite element meshes. For instance, finite volume meshes require at least two cells spanning between surfaces, whereas a single finite element needs at least two cells to establish a gradient across the span. In practice, at least three cells are often employed to ensure accuracy.

The cell quality is the most important factor influencing the convergence and, to a lesser extent, the accuracy of the finite volume method. The finite volume stress analysis is more sensitive to the quality of the mesh than the typical CFD. Star-CCM+ offers many measures of the quality of the grid including skewness, cell-to-cell volume change and least square gradient. The quality of the grid over all the cells is registered in one field function, which is an indicator of bad cells. If less than 0.1% of the cells in the solid stress model are considered poor, then this mesh can be considered of high quality. If the number of cells of poor quality is greater than 1%, then this grid is generally poor and more attention is required for meshing.

In FEA models, the free surfaces are those faces that have no load applied to them (zero traction). A traction-free loading is the natural boundary condition and no special considerations are needed to maintain the zero-traction condition. In contrast, the FVA has no counterpart for the "natural" boundary condition, so there is no particular distinction between zero traction and non-zero traction in terms of the formulation. Both are traction conditions with one simply having a traction of zero. For this reason, the FVA model must identify the boundary regions of the traction boundary condition whether it is zero or not. In Star-CCM+, the default condition is the free boundary condition.

The traction boundary condition is particularly difficult for the FVA because an estimate of the displacement on the boundary face is still needed to compute the gradient (strain tensor) at the centre of the adjacent cell. This gradient is needed to properly define the complete stress tensor on the face. Remember, a zero-traction boundary condition does not imply a zero-stress tensor on the face, much less a zero-strain tensor. The boundary face displacement must be extrapolated using the gradient at the cell centre, but the boundary face centre is used to estimate the gradient. This iterative process to define the displacement may not converge – especially when the cell in question has few neighbours, is not orthogonal or is poorly shaped. The quality of a cell depends not only on its geometry but also on the number of neighbours and orientation of its neighbours, and the boundary condition that may be applied to a particular cell.

Linear tetrahedral cells tend to behave stiffer unless the grid is extremely fine. Similarly, linear brick elements with full integration exhibit the same behaviour. Usually, under these conditions, an under-integrated linear brick element is a reasonably good choice with at least two elements through the thickness of the thin member. Under-integrated quadratic bricks also perform reasonably well. Under the same conditions, the FVA brick will require a denser grid than the under-integrated linear brick element to give comparable stiffness behaviour. In general, for thin members where the distance between supports relative to the thickness of the plates is greater than $\sim$10, the FVA cell stiff behaviour will be more pronounced. To give acceptable stiffness behaviour, the FVA cell aspect ratio needs to be near a value of one for the cell. Furthermore, the iterative FVA solver will generally require more iteration to converge with thin members under bending. Under such loads, convergence is greatly accelerated by the successive over-relaxation (SOR) technique, but it is particularly important for the FVA mesh to be free of poor cells if the SOR technique is to be effective.

The FVA constraints are in two forms – either all displacement components are constrained on a face or the displacement is constrained only in the normal direction while allowing the tangential components to be unrestrained. There is also no FVA counterpart to the FEA multi-point constraints. Multi-point constraints allow a constraint of a degree of freedom (DOF) based on the DOF on other nodes that may be remote. Unless relatively stiff continuum cells are linking the nodes, the FVA solver cannot treat this restraint. In this respect, the FVA is not unlike an explicit FEA solver.

The FEA methodology is formulated to accept almost any type of material model. This is accomplished by separating the functions of the solver from the material model. The function of the material model is twofold:

- Compute the stresses given the current increment of the strain, temperature and other state variables which describe the internal degrees of freedom such as plastic strain.
- Return a tangent modulus (the change in the stress relative to the change in the strain) relative to the current state of stress of the model.

The FVA model must evaluate the material model at the centre of internal faces as well as the centres of faces where the displacement is applied. In addition, the FVA evaluates the material model at bonded material interfaces at the centre of the opposite face. Star-CCM+ has updated the Lagrangian formulation used in the FV stress model and it allows for large solid mesh motion. This is especially applicable to problems dealing with fluid–structure interaction.

A force acting on a solid body results in the formation of stress. Stress is a measure of the force, divided by the area over which it is applied, as the area tends to zero. When an object under a load deforms, strain energy develops. Strain energy is potential energy created within an object because of a load deforming an object. Strain energy may also be viewed as the area under the load–deflection curve, $U = \int P \times \partial d$. Here, U is the strain in the structure developed because of loading on the structure, P is the load acting on the structure and ∂d is the deflection in the structure due to the load.

6.8 Modelling Fluid–Structure Interaction

Fluid–structure interaction is the thermo-mechanical interaction of a fluid and a solid structure. FSI implies the two-way coupling of a fluid and a deformable structure, such that the deformation and rigid body DOF are coupled with the fluid (Johannessen, 2012). Even under this definition, there are classifications of the degrees of coupling from weak coupling to strong coupling.

6.8.1 *Weak coupling*

Consider a structure that bends in steady-state flow. This bending can significantly alter the flow patterns around the structure, which, in turn, affects the bending. When searching for a steady-state condition, one is interested in the shape the structure takes under the steady condition. As the structure and fluid reach the steady state, the structure material point velocities approach zero. In this situation, the coupling is weak since in the steady state the structure imparts no motion on the fluid. In fact, it is almost one way since the primary influence of the structure on the fluid is from its velocity, which in steady state is zero.

6.8.2 *Strong coupling*

In strong coupling, physical coupling is two way and the coupling between the codes is pronounced. Often, strong coupling is associated with dynamic problems where the hydrodynamic loads and the structural velocities change dramatically. In strong coupling algorithms, the fluid and structure solvers can reside in processor memory simultaneously. Data are passed at regular intervals (often referred to as coupling time or coupling step) from the memory that the structure solver uses to/from the memory that the fluid solver uses. Depending on the actual strength of the coupling, it can be necessary to exchange data between the solvers more than once per time step for the simulation to remain stable.

The resilient structure reacts to both pressure and viscous shear forces within the fluid medium. Conversely, the fluid interacts with the structure through diverse mechanisms. While the geometry of the structure influences the flow dynamics, it is the velocity of the structure, or the rate of displacement alteration, that exerts the most significant impact. In fact, it can be shown that the acceleration of the structure causes proportional pressure responses much like how the acceleration of the fluid leads to a hydrostatic pressure gradient, such as striking a hammer on a constrained pipe. In the fluid, the response would be a pressure wave that could be heard but not seen, and the small deformations in the pipe would be visually imperceptible. In this case, the response is determined by how fast the structure moves or, more accurately, how the structure accelerates. It is not determined by how much the structure moves.

The FVA stress analysis is implicit and has no restriction on the time step size to remain stable. The FVA employs an iterative solver, which is different from the iterative solvers used in the FEA but more like the segregated solvers found in the CFD. The FVA solver uses a very rough estimate of the stiffness matrix for the preconditioner or linearization. The preconditioner is based solely on the nearest neighbours of each cell even though the actual stencil of cells, which directly influences a cell, can be much larger. In addition, since the FVA uses a segregated solver, the matrix (generally referred to as coefficients in the CFD) only needs to store terms of the stiffness matrix which involves one component.

Since the interaction between a wave and the structure can be classified as strong coupling, implicit coupling is necessary to ensure stability. Such cases can be dealt with in Star-CCM+ by taking advantage of the implicit coupling between the CFD and FV stress solvers. This coupling algorithm is strong and deep, and the exchange of data occurs at each iteration within a time step. This coupling also considers the load balancing and the partitioning, and it minimizes the cross-talk among different processors between the fluid and the solid.

The FVA iterative solver updates any material nonlinearity on the fly at each iteration. The FEA linearizes the solution and then iterates to obtain the solution of this linear solution. After obtaining the linear solution, the FEA takes another Newton's step, computes a new linearization and then uses the iterative solver to converge to the next step. But the FVA iterative solver updates the linearization matrix (or more precisely the precondition matrix) on each iteration so that when it converges all the nonlinearities are considered. In this sense, the linearized iterative solver and Newton's iteration are rolled into one outer iteration (Davison *et al.*, 2008).

During the two-way FSI simulation, Star-CCM+ solves for the pressures caused by the fluid flow and applies these forces on the structure. The solid physics model subsequently calculates the resulting displacements at each iteration. The deformed geometry of the free span is transferred back to the fluid physics to solve the flow again, resulting in pressure on the new geometry.

The FVA precondition matrix is not easily inverted. Hence, like the CFD segregated solvers, the FVA employs an inner iterative solver (the AMG solver). This inner solver need not be converged to machine accuracy but only sufficiently converged to provide a reasonably smooth solution. The outer iteration then updates the solution based on the inner iteration solution and updates the residual, including any changes in the material nonlinearity. The inner iteration solver's default tolerance is smaller than the CFD momentum tolerance (0.03 as opposed to 0.1).

The FVA iterative stress analysis solver has so much in common with the CFD segregated solvers; the stress analysis memory requirements will be comparable to the CFD models of a similar number of cells. In addition, since the solver technology is common to both the CFD solver and the stress analysis, the scalability of parallel computing is comparable. Thus, there is a trade-off. The FVA elements likely require more elements to describe properly bending stress in thin members, but the cost per cell is less per typical cost per element in the FEA. The FVA offers the potential for analyzing much larger models than in the FEA, so the overall computer costs relative to the accuracy can be much lower than the FEA.

The FVA solver's stability and convergence will be sensitive to the quality of the cells. However, there is some leniency in how the FVA solver treats bad cells to avoid polluting the solution and causing the iterative solver to diverge. This is due to the principle in solid mechanics that stresses at points are independent of the actual load distribution provided the load is applied at some distance. Only the total force and moment applied have an effect on the stress distribution at some distant point removed from the load. Therefore, there is reason to support the fact that even though some poor cells may produce locally incorrect results, these poor cells will not affect the overall behaviour in the part and will not affect the local stress distribution at points distant from the poor-quality cells. Therefore, the FVA solver includes a unique strategy to identify poor-quality cells and then acts on these cells to minimize their impact on the surrounding cells and on the convergence of the iterative solver.

The mechanics of continua (both solid and fluid) are described by transport equations derived from the assumption that mass, momentum, energy and entropy are conserved in the continuum. In the finite volume method, the solution domain is subdivided into a finite number of small control volumes, corresponding to the cells of a computational grid. Discrete versions of the integral form of the continuum transport equations are applied to each control volume. The objective is to obtain a set of linear algebraic equations, with the total number of unknowns in each equation system corresponding to the number of cells in the grid. For nonlinear equations, iterative techniques that rely on suitable linearization strategies are employed. The resulting linear equations are then solved with an algebraic multi-grid solver.

6.8.3 *Breaking wave impact*

The results obtained from the numerical study performed in the CFD are discussed in the following. In the case of fluid–structure interaction problems, the major results are the total force acting on the cylinder, the strain induced in the cylinder due to the impact of breaking waves, and the reaction force and reaction moments at the fixed ends of the cylinder. In this particular problem, strains are found at various locations along the length of the cylinder, thereby making it easy for us to find the location of the maximum strain.

6.8.3.1 *Comparison of wave height for various breaking wave types*

Wave breaking is created by the linear superposition of linear first-order waves using the energy-focusing principle. The energy focusing method generates a plunging wave with maximum wave intensity ($P1$). Then, by reducing the input wave amplitude using the wave steepness parameter given below plunging with a lower order of intensity can be generated, such as $P2$, $P3$ and $P4$.

The wave elevation measured at the point of impact of breaking waves on the cylinder as shown in Fig. 6.16 and Table 6.3 reveal that the wave elevation decreases from $P1$ to $P4$. Figures 6.17(a)–(d) show snapshots of

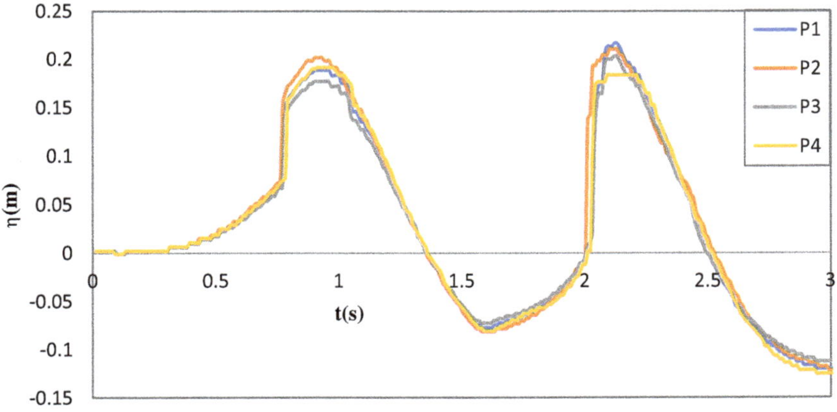

Fig. 6.16. Wave elevation at a point just before the cylinder.

Table 6.3. Types of breaking waves and their parameters.

Types of breaking	γ	H_b (m)
Plunging type $P1$	0.5773	0.2944
Plunging type $P2$	0.5702	0.2859
Plunging type $P3$	0.5541	0.2756
Plunging type $P4$	0.54513	0.2644

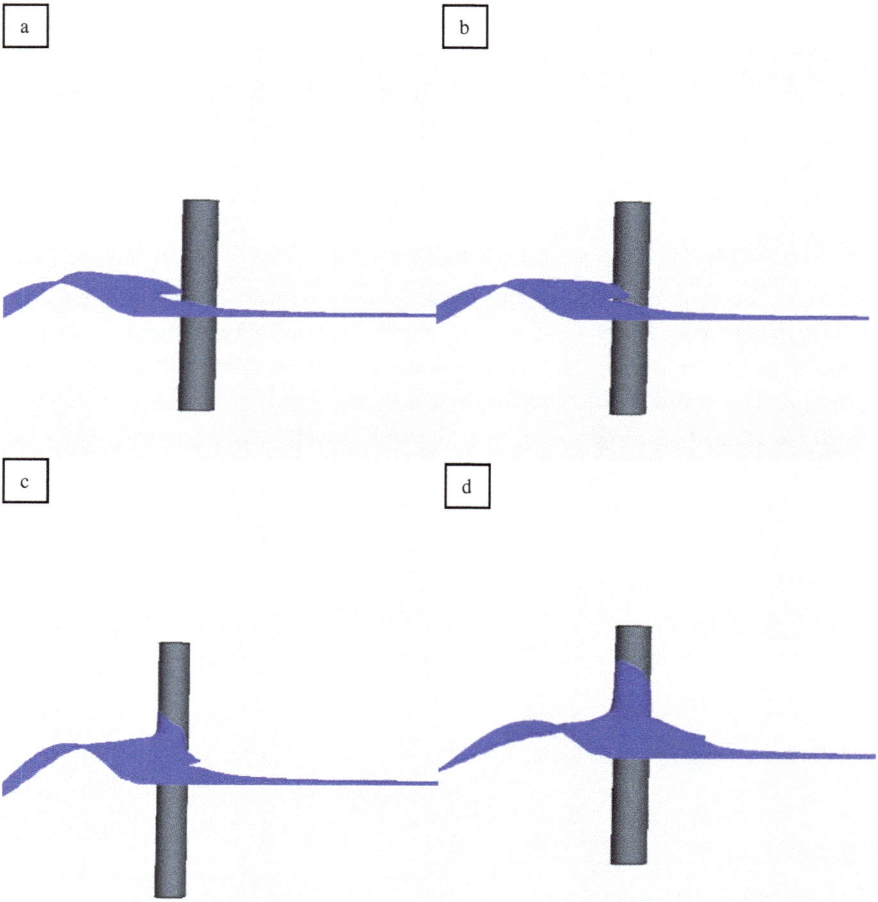

Fig. 6.17. Breaking wave impact on the cylinder at a time step of 0.05 s. The relative free surface profile with respect to the vertical thin cylinder is shaded in blue.

breaking wave incidence at 0.05-second intervals, capturing the maximum impact during the interaction between the steep wave front and the cylinder.

The abrupt increase in wave elevation suggests a high wave steepness parameter, indicating an imminent break. Consequently, around 2.05 s, the wave elevation peaks, followed by its inevitable break. For details, refer to Manjula *et al.* (2014).

6.8.3.2 *Force acting on the cylinder for various types of breaking waves*

Four types of plunging waves are generated and allowed to have an impact on the slender cylinder. The force variation with respect to time is plotted for all the cases, and a reduction in force when plunging wave intensities are reduced from *P*1 to *P*4 can be seen in Fig. 6.18.

Then, the variation of peak impact force with respect to the wave steepness parameter γ is plotted in Fig. 6.19, and the force variation is found to be linear with respect to the wave steepness parameter.

The strain obtained from the numerical study in the CFD is compared with the strain obtained from the experimental investigation of Saravanavel

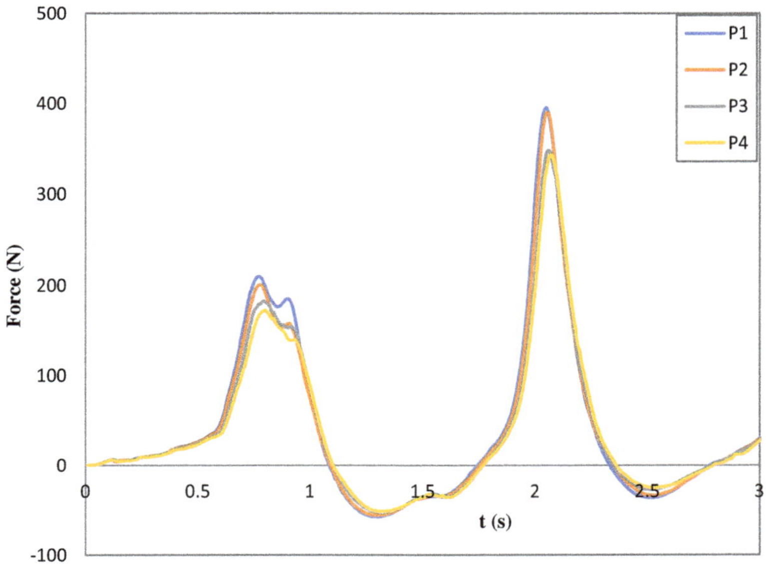

Fig. 6.18. Total force measured in the cylinder due to various types of plunging waves.

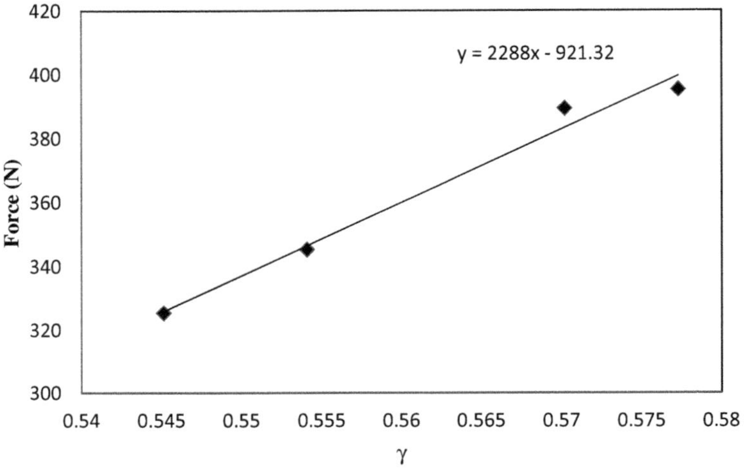

Fig. 6.19. Impact force variation with respect to the wave steepness parameter, γ.

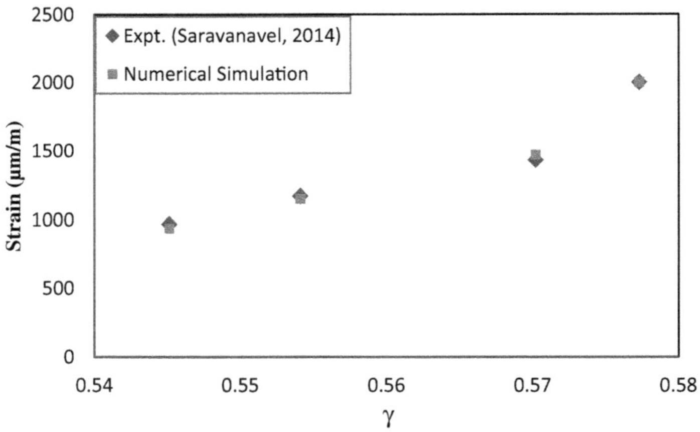

Fig. 6.20. Comparison of experimentally measured strain (Saravanavel, 2014) with the present numerical simulation.

(2014) in Fig. 6.20, in particular when the flexural rigidity of the slender cylinder is the same. Since the strain measured is the bending strain, it depends on the flexural rigidity of the slender cylinder. The values along with the percentage variations are given in Table 6.4.

From these results, it is seen that the numerical results and experimental results are almost matching and hence the results are validated. Only

Table 6.4. Comparison of strain.

Γ	Measured strain	Computed strain	Percentage difference
0.5773	2000	2000	0
0.5702	1430	1470	2.72
0.5541	1170	1150	1.74
0.54513	970	940	3.19

the maximum impact case, $P1$ is considered for the strain analysis across various materials of identical dimensions. The analysis aims to explore how strain varies with the flexural rigidity of the material and to examine the variation of reaction force in relation to the material's flexural rigidity.

References

Asakura, R., Iwase, K., Ikeya, T., Takao, M., Fujii, N. and Omori, M. (2000). An experimental study on wave force acting on on-shore structures due to overflowing tsunamis. *Proceedings of Coastal Engineering* (Japan Society of Civil Engineering), 47, 911–915.

Chanson, H. and Fang, L. (1997). Plunging jet characteristics of plunging breakers. *Coastal Engineering*, 31, 125–141.

City and County of Honolulu Building Code (CCH). (2000). Department of Planning and Permitting of Honolulu Hawaii. Chapter 16, Article 11, Honolulu, Hawaii.

Cooker, M.J. and Peregrine, D.H. (1995). Pressure-impulse theory for liquid impact problems. *Journal of Fluid Mechanics*, 297, 193–214.

Davison, J.B., Ferguson, S.W., Mendonçą, F.G., Peck, A.F. and Thompson, A. (2008). Towards an automated simulation process in combined thermal, flow and stress in turbine blade cooling analyses. In *Turbo Expo: Power for Land, Sea, and Air*, Vol. 43161, pp. 2641–2648.

FEMA Coastal Construction Manual. (2005). FEMA 55 Report. Edition 3, Federal Emergency Management Agency, Washington, DC.

Johannessen, S.R. (2012). Use of CFD to study hydrodynamic loads on free-fall lifeboats in the impact phase: A verification and validation study. Master's thesis, Institutt for marin teknikk.

Kwon, Y.W. and Bang, H. (2000). The Finite Element Method Using MATLAB, 2nd edn. CRC Press, Boca Raton.

Lynett, P.J. (2006). Wave breaking velocity effects in depth-integrated models. *Coastal Engineering*, 53(4), 325–333. https://doi.org/10.1016/j.coastaleng.2005.10.020.

Manjula, R., Sannasiraj, S.A. and Palanichamy, K. (2013). Laboratory measurements of breaking wave impact pressures on a slender cylindrical member. *The International Journal of Ocean and Climate Systems*. https://doi.org/10.1260/1759-3131.4.3.151.

Manjula, R., Sannasiraj, S.A. and Palanichamy, K. (2014). Experimental investigations of acceleration on slender cylindrical member under breaking waves. *The International Journal of Ocean and Climate Systems*. https://doi.org/10.1260/1759-3131.4.3.151.

Mohapatra, P.K., Bhallamudi, S.M. and Eswaran, V. (2000). Numerical simulation of impact of bores against inclined walls. *Journal of Hydraulic Engineering*, 126(12), 942–945.

Muni Reddy, M.G., Sannasiraj, S.A. and Natarajan, R. (2007). Numerical investigation on the dynamics of a vertical wall defenced by an offshore breakwater. *Ocean Engineering*, 34(5–6), 790–798. https://doi.org/10.1016/j.oceaneng.2006.04.012.

Naik, M.S. (2005). Breaking wave forces on vertical wall. M. Tech thesis, Indian Institute of Technology Madras.

Okada, T., Sugano, T., Ishikawa, T., Ohgi, T., Takai, S. and Kamabe, C. (2005). Structure design method of buildings for tsunami resistance (proposed). English translation of The Building Letter of 2004.

Ramsden, J.D. (1993). Tsunamis: Forces on a Vertical Wall Caused by Long Waves, Bores, and Surges on a Dry Bed. Report No. KH-R-54, W.M. Keck Laboratory, California Institute of Technology, Pasadena, California, 251.

Saravanavel, A. (2014). Breaking wave impact load on rigid and deformable vertical cylinder. M. Tech. thesis, Department of Ocean Engineering, IIT Madras.

Senthil Kumar, N. and Sannasiraj, S.A. (2011). Numerical modeling of breaking wave impact pressure on a vertical wall. In *Asian and Pacific Coasts 2011*, December 14–16, 2011, Hong Kong. DOI: 10.1142/9789814366489_0197.

She, K., Created, C.A. and Easson, W.J. (1997). Experimental study of three-dimensional breaking wave kinematics. *Applied Ocean Research*, 19(5–6), 329–343. https://doi.org/10.1016/S0141-1187(97)00031-X.

Sundar, V. (2021). *Ocean Wave Dynamics for Coastal and Marine Structures*. Advanced Series on Ocean Engineering, Vol. 52. World Scientific, Singapore. https://doi.org/10.1142/12268.

Wood, D.J. and Peregrine, D.H. (2000). Study of wave impact against a wall with pressure-impulse theory: Part 2: A porous berm. *Journal of Waterway, Port, Coastal, and Ocean Engineering* (ASCE), 126, 191–195.

Yeh, H. (2007). Design tsunami forces for onshore structures. *Journal of Disaster Research*, 2(6), 531–536.

Zienkiewicz, O.C. and Taylor, R.L. (1989). *The Finite Element Method*, Vol. 1. McGraw Hill Book Company, London.

Chapter 7

Multi-Functional Coastal Structures

7.1 Introduction

Many coastal areas, especially those that are urbanized or semi-urbanized, face vulnerabilities to natural coastal hazards. Added to this is the sea level rise due to the melting of ice over the past century, which is anticipated to continue to increase and is a big threat to the stability of the coast, especially the low-lying zones of coastal stretches around the globe. The increased frequency of coastal flooding due to sea level rise (SLR) has been discussed by several researchers (Brody *et al.*, 2011; Bilskie *et al.*, 2014; Sukop *et al.*, 2018).

Development along the coast due to the migration of communities towards the coast for livelihood requires immediate attention due to the need for adaptation to sea level rise. The factors that dictate the adaptation strategy are technical, financial, planning and decision-making as well as socio-economic aspects. To combat this problem, the conventional method is for the affected stretch of the coast to be armoured with hard structures like seawalls, groyne fields, submerged reefs or detached offshore breakwaters. The main purpose of these structures is as a mitigation measure against perennial erosion or flooding. Such structures are made of natural rocks or cast as artificial concrete blocks, each of which would be of weight dictated by the design wave height. In addition, two breakwaters are usually preferred for the formation of artificial harbours to provide the desired tranquillity to facilitate smooth loading and unloading operations.

Coastal development, in particular involving coastal structures, henceforth has to incorporate new methods of planning, design and monitoring in order to handle the current as well as future coastal hazards dictated by the changing climate. To fulfil imperatives such as averting ecological decline, navigating climate shifts and attaining broad societal advantages, we must consider a range of solutions, encompassing nature-based strategies, robust physical interventions like structural construction and more subtle interventions such as vegetation management and sand bypassing. Multi-functional infrastructures must not only bolster resilience against extreme events but also mitigate various other disruptions that may emerge. Moreover, they should offer supplementary benefits, such as providing opportunities for beachfront relaxation in areas experiencing distress.

Recently, Morris *et al.* (2023) provided compelling evidence that rock fillets, as a nature-based approach for coastal protection, have proven effective in sediment accumulation and fostering mangrove colonization. Compared to traditional rock revetments, rock fillets are anticipated to offer a more environmentally and economically sustainable solution. Schoonees *et al.* (2019) discussed nature-based adaptations for coastal protection, advocating for (1) integrating soft solutions alongside hard solutions, (2) managing morphological and hydrodynamic changes, and (3) ecologically enhancing hard coastal structures. The success of these adaptations hinges significantly on local conditions, including hydrodynamic forces, spatial constraints, and socioeconomic factors. The overview offers general guidance on selecting appropriate solutions tailored to site-specific conditions, encompassing all critical aspects essential for environmentally sensitive coastal designs. Overall, the study aims to guide interdisciplinary design decisions in implementing nature-based coastal defense structures.

This chapter explores the methods for planning various kinds of marine structures as multi-functional structures.

7.2 Multipurpose Seawalls (Boardwalks)

Andiroglu and Suraneni (2021) delved into the multifaceted capabilities of coastal structures, as evidenced by their case study in South Florida. Their findings underscore the importance of integrating various elements into the future design of coastal infrastructure, including enhancing offshore marine habitats and submerged aquatic vegetation and extending coastal buffer zones landward. Additionally, they advocate for implementing berm and swale designs to manage runoff effectively, alongside the incorporation of tiered wall structures. For effective coastal resilience through multi-functionality, coastal boardwalk applications are found to repetition of effective provide economic and/or social benefits and mitigate some climate change impacts. A successful design can also facilitate access to the beach for tourism and recreation purposes (Panin and Nicolaev, 2002; Brand and Jax, 2007).

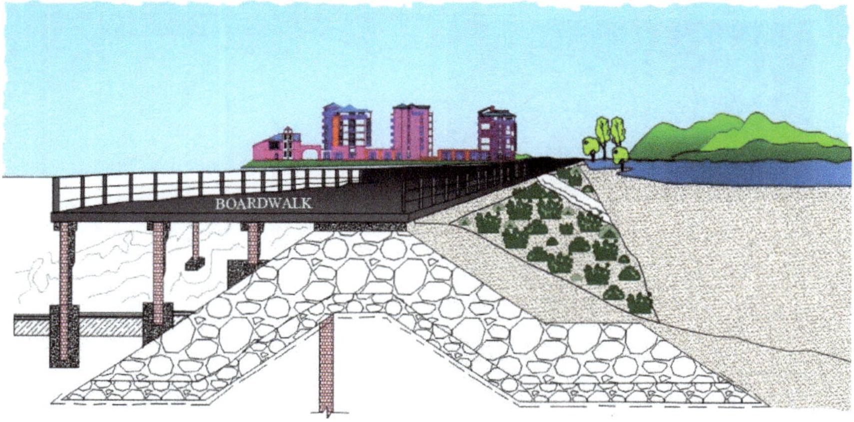

Fig. 7.1. Typical multipurpose seawall.

A walkway, roadway or a footpath constructed along the shore can be termed as a boardwalk. The first 9 km long boardwalk was opened in 1870 in Atlantic City, New Jersey and it is the busiest boardwalk. At its widest point, it is 18 m wide and stands 3.6 m above sea level (Britannica, 2023). A typical cross-section of a multipurpose seawall and boardwalk is shown in Fig. 7.1.

Cengiz *et al.* (2022) described the benefits of the Güzelcehisar Beach Boardwalk (GBB) in Turkey's Western Black Sea Region as it enhanced socio-economic aspects, aesthetics and accessibility, improved tourism and provided environmental protection. This multi-functional structure has an observation terrace, a beach boardwalk and connectivity from the boardwalk to a rocky island and the terrace. The location map and views of the boardwalk are shown in Fig. 7.2.

Figure 7.3 illustrates a unique seawall design implemented in Büsum, Germany, which combines coastal protection with recreational aspects. The walkway or the boardwalk is suitable in locations of high tidal range. Further, it is not just for protecting a coast exposed to moderate wave activity but also serves as a relaxing place for beach lovers.

Kamikubo *et al.* (2000) introduced a multipurpose flaring shaped seawall, FSS, featuring a top area for both traffic and leisurely walks along the beach. Through numerical and experimental analyses, they compared its hydrodynamic properties with those of conventional vertical seawalls, ultimately finding that the critical crest elevation is significantly lower in their

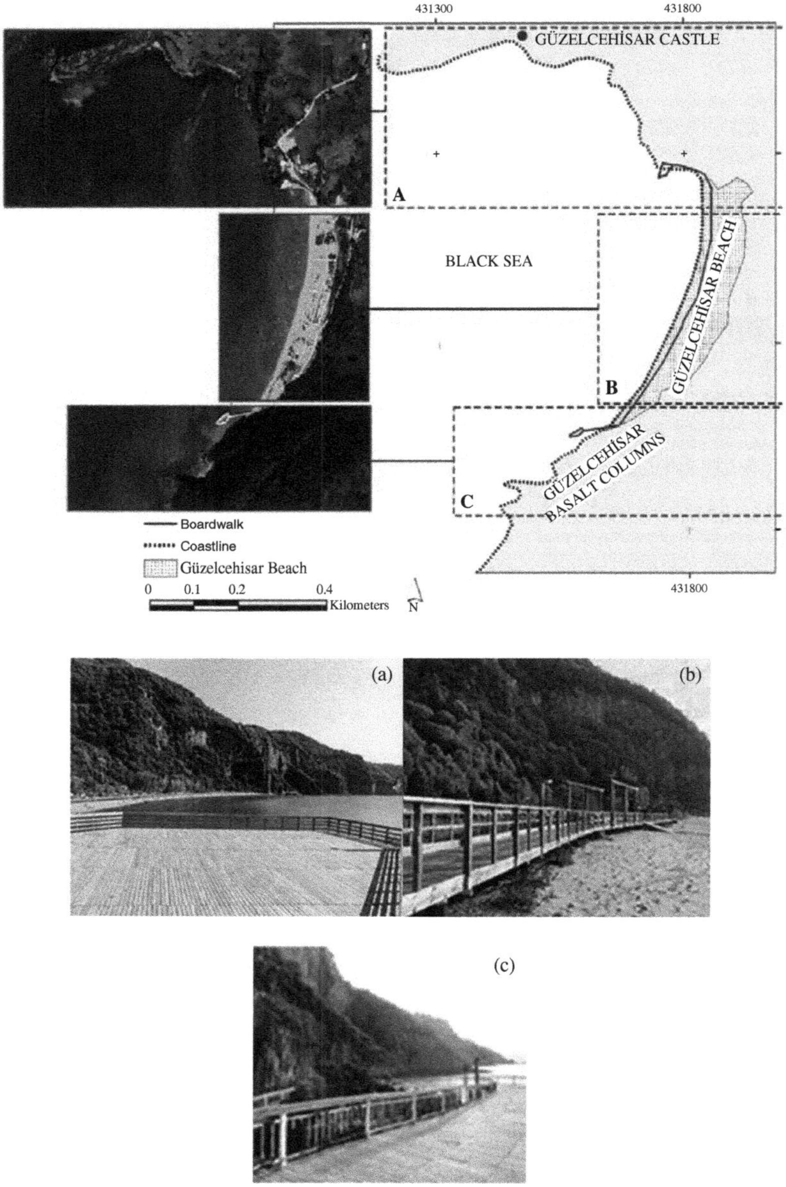

Fig. 7.2. Güzelcehisar Beach Boardwalk: (A): observation terrace, (B): beach boardwalk and (C): boardwalk connection to a rocky island and observation terrace.

Source: Cengiz *et al.* (2022).

Fig. 7.3. View of a seawall that integrates coastal protection and recreational features in Busum, Germany.
Source: Schüttrumpf.

structure. De Chowdhury *et al.* (2017) have studied the hydrodynamic performance of three prevalent curved front seawalls along with a vertical seawall for Ursell numbers between 8 and 16. The study employed Smoothed Particle Hydrodynamics (SPH) models and ANSYS-FLUENT to predict dynamic pressures and overtopping. These predictions were compared with experimental results, demonstrating that SPH models are effective in accurately predicting the behavior of run-up and potential over-topping.

At Muzhapilangad beach, India, for tourism development, a promenade is proposed to be constructed. The study area is located along the coast of Kerala, in the southwest of India at the coordinates of 11.778935 N, 75.454911 E and 11.80675 N, 75. 434659 E. The average erosional rate is found to be 0.91 m/yr. The proposed promenade structure will have the dimensions of 56 m × 7 m (Unit-I). The promenade structure will comprise

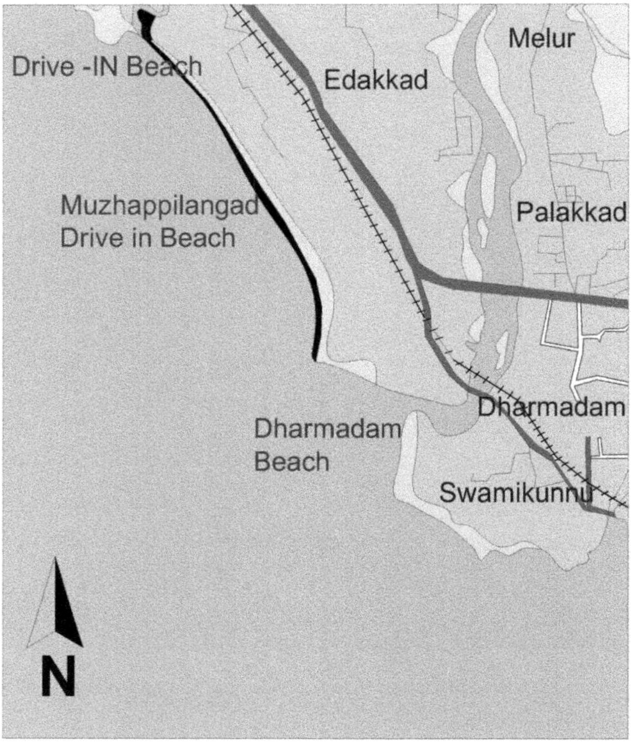

Fig. 7.4. Proposed master plan for Muzhapilangad beach, Kerala.

several panels of 56 m length along the coast and two rows of reinforced cement concrete (RCC) piles of 700 mm. The pile centre-to-centre spacing will be 5 m in the transverse direction. The longitudinal pile spacing will be 6 m. The deck level of the promenade is fixed at (+) 4.1 m with its founding level at (−) 23 m (CD). The slabs have been designed as two-way slabs of 30 mm thickness. The Proposed Master Plan for the conservation of beaches by protecting the coast of Muzhapilangad is shown in Fig. 7.4. The proposed promenade structure will be constructed such that the seabed disturbance is minimal. Further, revetment is proposed at the shorefront using gabions such that the porous gabions would dissipate the wave energy and, hence, the energy spent on sediment movement is minimized. A typical cross-sectional view of the general arrangement of one unit of the proposed promenade structure is shown in Fig. 7.5(a) and a view of a multipurpose seawall is shown in Fig. 7.5(b).

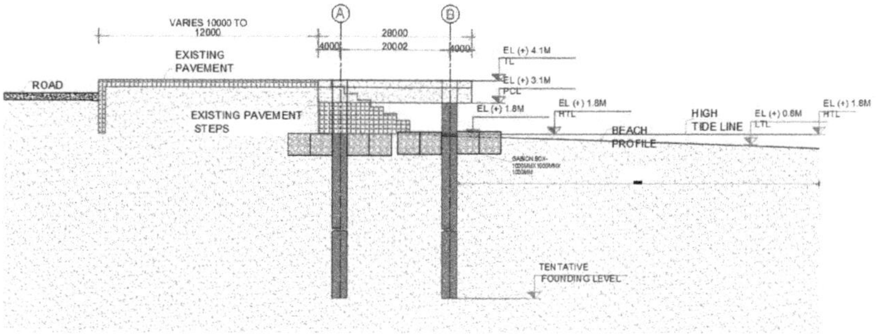

(a)

(b)

Fig. 7.5. (a) Cross-section of proposed promenade for Muzhapilangad beach, Kerala. (b) A view of a multipurpose seawall on the island of Norderney, Germany. *Source*: Schüttrumpf.

7.3 Multipurpose Groyne Fields

The coastal villages of Simon Colony, Vaniyakudi, Kurumbanai and Enayam near Colachel in Kanyakumari District, along the southwest coast of India [8° 11.5′N, 77° 13.5′E and 8° 10.3′N, 77° 14.8′E], had been

Fig. 7.6. Status of the coast prior to protection measures.

experiencing the problem of erosion at a rate of about 0.4 m/year (Fig. 7.6). Over a decade, a landmass of about 440 hectares along the coast was affected, hampering the livelihood of the coastal community. Based on site conditions, six groynes were proposed for this study area, with two groynes at Kurumbanai, one groyne at Vaniyakudi, two groynes at Kodimunai and one groyne at Simon Colony, the layout of which is shown in Fig. 7.7(a). The proposed groynes were oriented at about 15° inclination to the shore, normal from the shore towards the offshore direction, with the end of the groyne curved to face the waves coming during the Southwest monsoon. It was thought that this proposal would provide shore protection and also help the fisherman anchor or tow the boats during the months of June to September directly from the location of berthing. The groyne field had been functioning well in trapping the longshore sediment since 2003. Due to this measure, a beach of about 200 m width has formed and stabilized in between the groynes. The groyne field serves not only as a coastal protection measure for the coast of the villages but also acts as mini fishing harbours. Furthermore, the outcrops present near Simon Colony were to be connected, as seen in Fig. 7.7(b), which is now a relaxing place for beach lovers. Thus, the groyne field provides coastal protection, is a beach landing facility for the fishing boats and a place for relaxation (see Fig. 7.7(c)). The complete details of this study are discussed in Sundar and Sundaravadivelu (2007).

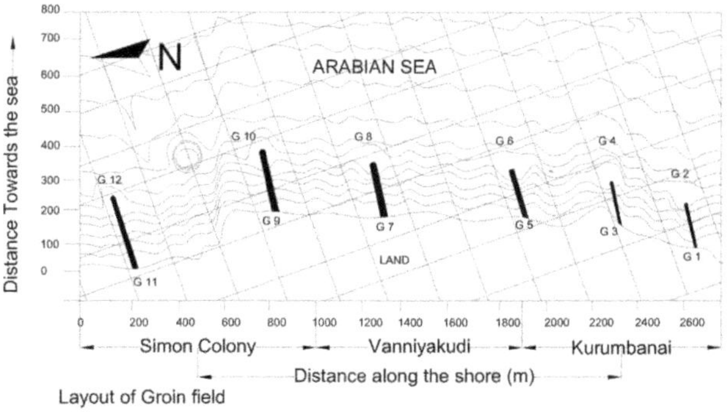

(a)

(b)

(c)

Fig. 7.7. (a) Layout of groyne field, (b) connecting G9 and G10 groynes with the outcro and (c) view of the multipurpose groyne field.

7.4 Submerged Structures/Reefs

7.4.1 *General*

The wave attenuation characteristics of a submerged breakwater have been investigated by several researchers in the past. Building on previous studies, Van der Meer and d'Angremond (1991) used physical modelling to derive an empirical relationship for the transmission coefficient K_t. This coefficient, defined as the ratio between the transmission wave height H_t and the incident wave height H_i, was found to be a function of wave characteristics, the mean diameter of the armor D_{50}, the crest width of the structure b, and the distance between the crest elevation and the water elevation R_c.

$$K_t = \left(0.031\frac{H_s}{D_{50}} - 0.24\right)\frac{R_c}{D_{50}} + b. \tag{7.1}$$

Submerged structures facilitate the premature breaking of waves to enhance the dissipation of energy in them. They are considered an ecofriendly solution to coastal protection. They also help avoid or reduce coastal flooding. Figure 7.8a illustrates an eroding beach and the function of a submerged breakwater or reef to rebuild the beach. Figure 7.8b shows a view of typical reefs.

As submerged structures were not found effective in locations of higher tidal range or locations frequented by storms, broad-crested structures were used as stated by Yoshioka *et al.* (1993). Womera (2011) has studied the advantages and disadvantages of submerged structures as presented in Table 7.1. Yoshioka *et al.* (1993) have discussed wave breaking, the dissipation of incident wave energy and the interaction of waves with submerged reefs based on a number of projects carried out in Japan. In addition, discussions on the design and the analysis to be conducted prior to the installation of submerged breakwaters are provided by Pilarczyk (2003) and CIRIA (2007). A comprehensive review of the functioning of about 150 artificial reefs, most of which are along the coast of New South Wales, Australia, by Blacka *et al.* (2013) serves as a useful source for detailed information. A similar detailed review has also been conducted by Lokesha *et al.* (2013).

Submerged reefs need not be only trapezoidal. Any submerged structure has the potential to function as an artificial reef. One such structure, semicircular breakwaters, was categorized by Sasajima *et al.* (1994) into several types: solid, front-wave dissipating, permeable and rear-wave dissipating.

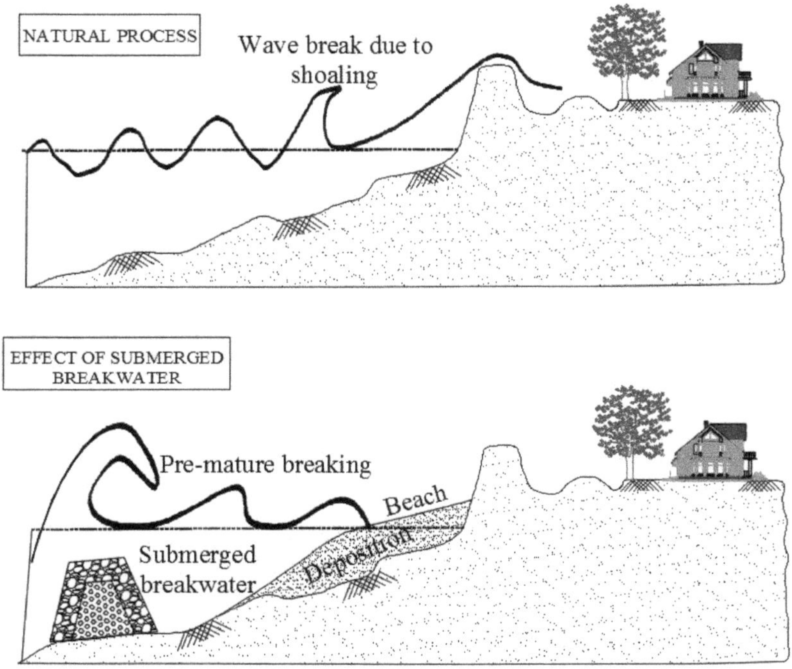

(a)

(b)

Fig. 7.8. (a) Effect of submerged breakwaters as a coastal protection measure. (b) Elements of permeable reefs.

Source: Schüttrumpf.

Table 7.1. Advantages and disadvantages of submerged reefs according to Womera (2011).

Advantages	Disadvantages
Aesthetically appealing.	The generation of currents in the nearshore due to overtopping and run-down would pose a danger to swimmers.
Relatively less expensive than structures piercing the free surface.	Dissipation of incident wave energy is partial.
The rate of build-up of beach between the shoreline and the structure is gradual.	The rate of protection of a coast depends on the design freeboard of the structure, which dictates the effectiveness of the protection measure and the rate at which the beach build-up takes place.
Circulation of water behind the structure is good.	A host of parameters are involved in the design, requiring greater attention from design and construction engineers.
Serves as a nest for marine life and promotes the ecosystem.	In the event the submerged reefs are discontinuous, the possibility of generation of rip currents is high.

In the solid type, both the front and rear walls are impermeable. The front-wave-dissipating type features only a perforated front wall, while the permeable type has perforations in both front and rear walls. Lastly, the rear-wave-dissipating type includes only a perforated rear wall. The impermeable type is particularly effective at dissipating wave energy when waves pass over the seaside quadrant face of a semi-circular breakwater. The front-wave-dissipating submerged breakwater (SBW) causes energy dissipation by creating turbulence inside the wave chamber in addition to the run-up on the structure. In the case of the permeable type, improved energy dissipation is achieved. Wave overtopping is significantly dissipated by the perforated rear wall in the rear-wave-dissipating SBW. These types are shown in Fig. 7.9, and the perforated semi-circular breakwater has been constructed at the Miyazaki Port, Japan.

Dhinakaran *et al.* (2010, 2012) conducted laboratory studies to investigate the hydrodynamic performance characteristics of semi-circular breakwaters across various water depths. They explored the relative depths of submergence (ratio of height of the breakwater to the water depth) at 1, 1.2 and 1.4, along with porosities ranging from 7% to 17%. Their

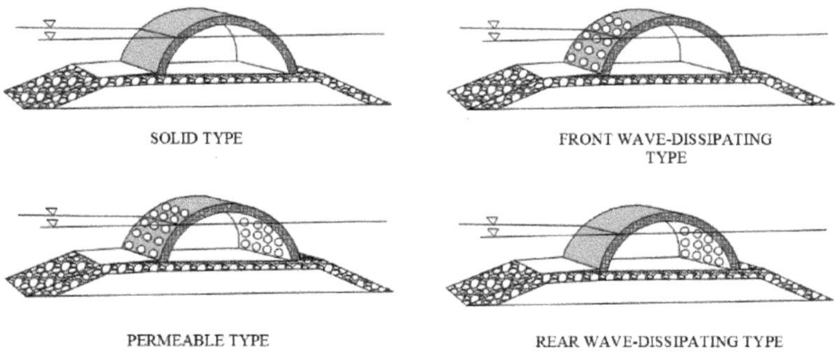

Fig. 7.9. Types of semi-circular breakwaters.

findings revealed that the optimal percentage of perforation for both seaward-side-perforated and fully perforated breakwaters is 11%. Additionally, they determined that the height of the submerged structure should be approximately 0.8 times the water depth to minimize transmission past the structure.

The effect of porosity on wave transmission past submerged and emerged breakwaters was investigated by Rahman and Aysha (2014); it was reported that minimum transmission was achieved for emerged breakwaters with the lowest porosity and minimum reflection was observed for submerged breakwaters with the highest porosity.

Gold Coast Beach, Australia, has experienced shoreline instability due to littoral drift to an extent of about 500000 m^3/year (Lenze *et al.*, 2002) with net erosion, which was mitigated by deploying geo-synthetic sand-filled containers in 1997. This solution stabilized the shoreline and resulted in an increased beach width, promoting regeneration of the ecosystem and increased tourism, including surfing. The reef survived the extreme events and the erosion (Jackson *et al.*, 2012). Similarly, Young-Jin Beach along the eastern Korean shoreline was stabilized by the installation of submerged two-layer geo-tubes with a scour apron mat (Shin and Oh, 2007).

Submerged structures can be broadly classified based on their utility either as breakwaters or as artificial reefs. The classification of submerged structures based on the shape, application and structural characteristics is shown in Fig. 7.10.

Artificial reefs or man-made barriers are submerged and rest on the ocean floor. A variety of materials like steel, concrete, fibreglass or composite materials are adopted for construction. Old wrecked cars, airplanes,

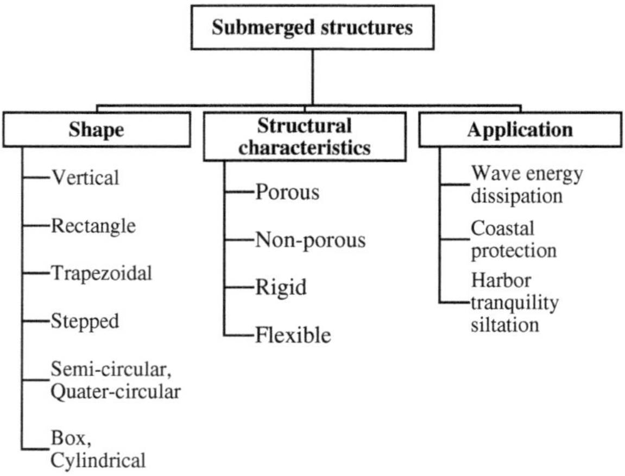

Fig. 7.10. Classification of submerged breakwaters.

military tanks, used truck or car tyres, scrapped appliances, docks, aged boats, ballistic missiles, withdrawn ships and outdated oil rigs have been sunk and designated as artificial reefs. In India, in order to increase the fish catch, three reef types, i.e. lobster reef, triangular modules and rubber tyres, were deployed on the coast of Kerala. There are several other types of reefs that have been adopted along the Indian coastline (Lokesha, 2019).

7.4.2 *Case study of submerged offshore reef*

7.4.2.1 *Valappu beach, Kerala*

This stretch (10° 1′4.17″N; 76°12′35.37″E) of beach extends for 1.5 km with a width of about 25 m and is located immediately to the north of the facilities of Indian Oil Corporation Ltd. (IOCL). The beach slope is very steep and the beach width oscillates depending on the wave climate, which in turn is seasonal. Cross-shore sediment transport is found to be significant during the Southwest Monsoon Season. Valappu beach is an active tourist attraction and thus the aesthetics need to be preserved. A submerged reef structure with geo-tubes as a continuous 800-m-long section is proposed. The tranquillity of waves behind the structure can be established through the well-known numerical Conjugate-Gradient wave (CG-Wave) model by Demirbilek and Panchang (1998) to arrive at a modified magnitude and phase variation of waves. The nonlinear wave propagation associated with most common wave deformations (e.g. wave reflection,

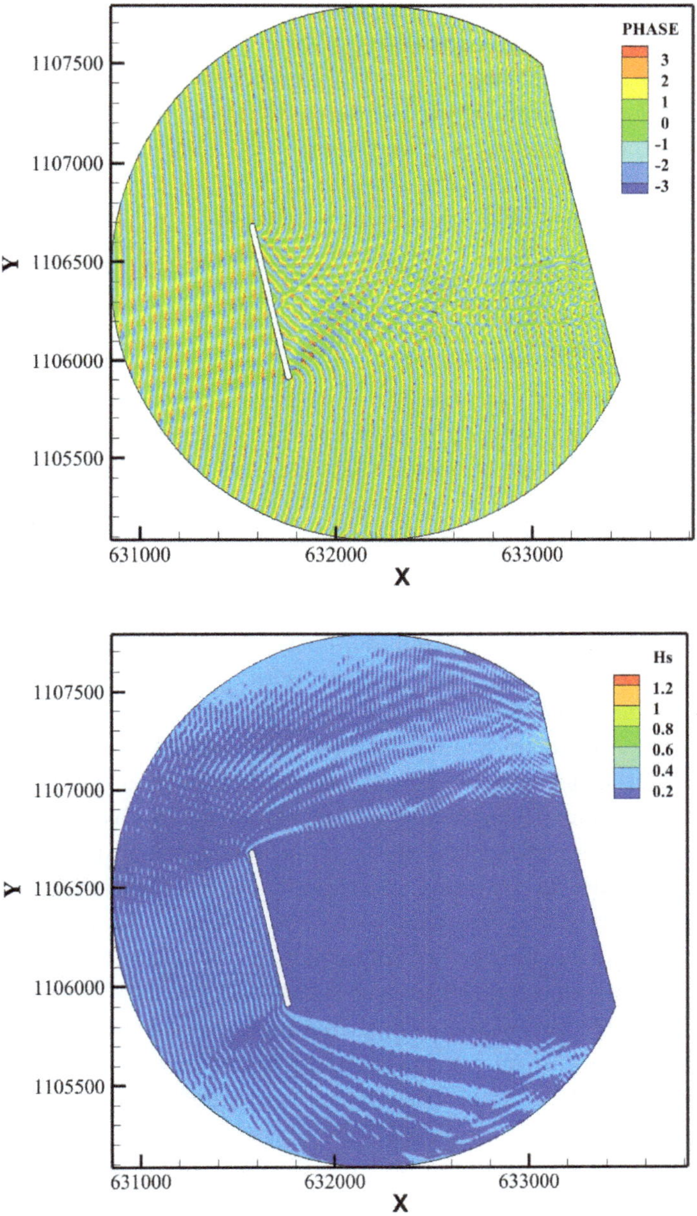

Fig. 7.11. Typical phase distributions and wave height (in meters) for a wave approach angle of 270°.

refraction and diffraction) is generally represented by the shallow-water mild-slope equation. The mild-slope equation proposed by Berkhoff (1972) is applicable for computations of refraction and diffraction of linear waves. The finite element model is used to solve the shallow-water equations and to analyze the wave phase and wave height distribution in the vicinity of the proposed structure (see Fig. 7.11). Based on the numerical study, it is concluded that the submerged reef can reduce the wave impact along the coast.

7.4.2.2 *Sagar island*

A wealth of literature exists aiming to optimize energy dissipation through submerged reefs, focusing on factors such as their geometry and positioning relative to wave characteristics and local geomorphology. Traditionally, artificial reefs consist of hollow concrete blocks crafted in diverse sizes and shapes. The optimized sections are selected from the literature or through physical model studies for site-specific conditions. Typically, these reefs are considered for island protection where the transport of huge quantities of concrete pre-cast or cast-in-situ materials is difficult. Hence, the novel idea of deploying geo-containers and geo-bags with a toe section is considered.

The temple of Kapil Muni (21°38′15.35″N, 88°4′30.56″E) is located on the southwestern side of Sagar Island, off the coast of West Bengal, India, and is famous for an annual religious festival that involves the participation of about a million pilgrims. The pertinent erosion problem at a rate of about 5m/year is prevalent at the site and has considerably reduced the beach width, thereby resulting in reduced space for religious as well as recreational activities along the coast. The average tidal range is of the order of 3 m and 4 m and the horizontal distance between the high tide level (HTL) and low tide level (LTL) over each tidal cycle is about 200–250 m. Thus, the construction of shore-connected protection structures would pose great difficulty in terms of execution and maintenance.

Geo-textile tubes are ecofriendly and are widely adopted for coastal protection and harbour formation. The likely problem being encountered in employing them for the site conditions at Sagar Island is the preparation of the seabed; if not done properly, this may lead to unequal settlements, thereby leading to the failure of the tubes and improper anchoring of the tubes that may lead to their rolling down towards the offshore side. Hence, geo-textile containers are preferred over geo-textile tubes, as they possess similar advantages. Sundar *et al.* (2022) have described the stability criteria for sand-filled geo-synthetic containers under various loading conditions. An extract of the existing formulae to aid in the design of sand-filled containers is presented in Table 7.2. It was recommended to construct a continuous 2.3-km-long submerged reef to protect the stretch of the Kapil Muni coast. The geo-textile containers are dimensionally smaller and are prone to disturbances due to wave and tidal currents; to enhance the stability of the containers in the submerged state, weighted geo-textile mattresses are to be positioned over the containers. These mattresses are to be extended over a distance of about 3 m on either (lee or sea) side to provide sufficient anchorage and to combat scour. A combination of geo-textile containers and

Table 7.2. Literature on hydraulic stability of geo-containers.

Authors/ Structures	Formulae/Nomograms	Remarks
Hudson (1956)/ Rubble mound structures	$N_s = \frac{H}{\Delta D_n} = (K_D \cot\alpha)^{1/3}$ $W_{50} = \frac{\rho_s g H^3}{K_D \left(\frac{\rho_s}{\rho_w} - 1\right)^3 \cot\alpha}$	Formula based on geometrical considerations of the balance of wave-generated flow forces acting on an armour stone in a slope of a breakwater.
Wouters (1998)/ GSC revetments	$N_s = \frac{H_s}{\left(\frac{\rho_E}{\rho_w} - 1\right) D} = \frac{C_w}{\sqrt{\xi_o}}$ $D = l \sin\alpha$ $l_c = \frac{H_s \sqrt{\xi_o}}{\left(\frac{\rho_E}{\rho_w} - 1\right) \sin\alpha \, C_w}$	Stability number expressed in terms of surf similarity parameter based on experimental data. Instead of the required weight of the armour unit, thickness D of the cover layer is defined by $D = l_c \sin\alpha$.
Oumeraci *et al.* (2002a, 2002b, 2003) GSC revetments and low-crested structures	$N_{s,\text{slope}} = \frac{H_s}{\left(\frac{\rho_E}{\rho_w} - 1\right) D} < \frac{C_w}{\sqrt{\xi_o}}$ $l_c = \frac{H_s^{3/4} \sqrt{T}}{C_w \left(\frac{2\pi}{g}\right) \left(\frac{\rho_{GSC}}{\rho_w} - 1\right) \sqrt{\frac{\sin(2\alpha)}{2}}}$ $N_{s,\text{crest}} = \frac{H_s}{\left(\frac{\rho_{GSC}}{\rho_w} - 1\right) D} <$ $0.79 + 0.09 \frac{R_c}{H_s}$ $l_c = \frac{H_s}{\left(\frac{\rho_{GSC}}{\rho_w} - 1\right)\left(0.79 + 0.09 \frac{R_c}{H_s}\right) \sin\alpha}$	The stability number is formulated and postulated to be a function of the surf similarity parameter. Two different formulae are presented for slope and crest GSCs for high-overtopping revetments and low-crested structures based on small- and large-scale experiments.

(Continued)

Table 7.2. (*Continued*)

Authors/ Structures	Formulae/Nomograms	Remarks
Recio (2007)/ GSC revetments and submerged GSC structures	$l_{c(\text{sliding})} \geq u^2 \dfrac{[0.5KS_{CD}C_D + 2.5KS_{CL}C_L\mu]}{[\mu KS_R\Delta g - KS_{CM}C_M \frac{\partial u}{\partial t}]}$ $l_{c(\text{overturning})} \geq u^2 \dfrac{[0.05KO_{CD}C_D + 1.25KO_{CL}C_L]}{[0.5\Delta KO_R g - 0.1KO_{CM}C_M \frac{\partial u}{\partial t}]}$ Stability nomograms to be used for feasibility studies based on Recio (2007)	Process-based stability formulae presented for different types of GSC structures considering two key hydraulic failure modes: overturning and sliding. Force coefficients were found by conducting several small-scale experiments.
Mori *et al.* (2008)/ Submerged GSC breakwater	$N_s = \dfrac{H_s}{\left(\frac{\rho_E}{\rho_w} - 1\right) D} < \dfrac{C_w}{\sqrt{\xi_o}}$	Showcased the applicability of Oumeraci *et al.*'s formula (2003) to estimate the stability of the crest of the submerged GSC structure studied through experiments.
Recio *et al.* (2010)/Sub-merged GSC	Stability nomograms to be used for feasibility studies based on Recio (2007)	Discussed the extension of Recio's (2007) formulae to design prototype structures and provide nomograms for different GSC applications.
Coghlan *et al.* (2009) and Hornsey *et al.* (2011)/Non-overtopping GSC revetments	Stability nomograms were developed for two specific GSC geometries by using the data from tests.	Nomograms indicate the design wave height at the structure, which should be used as an initial damage criterion for the given structure slope.

Note: GSC – geo-synthetic sand-filled container; C_D, C_L and C_M – drag, lift and inertia coefficient, respectively; C_w– empirical parameter derived from N_s; g – acceleration due to gravity; H_s – incident significant wave height; K_D – stability coefficient; KO – the coefficient for deformation due to overturning; KS – the coefficient for deformation due to sliding; l_c– length of the container; N_s – stability number; R_c – crest freeboard; $L_o = gT^2/2\pi$ – deep-water wavelength; u – horizontal velocity; α – slope angle of structure slope; $\xi_o = \tan\alpha/(H_s/L_o)^{1/2}$ – surf similarity parameter; $\Delta = (\rho_s/\rho_w - 1)$, $\frac{\partial u}{\partial t}$ – horizontal acceleration; ρ_s – density of armour unit; ρ_w – density of water; and ρ_E – density of GSC.

Source: Dassanayake and Oumeraci (2012).

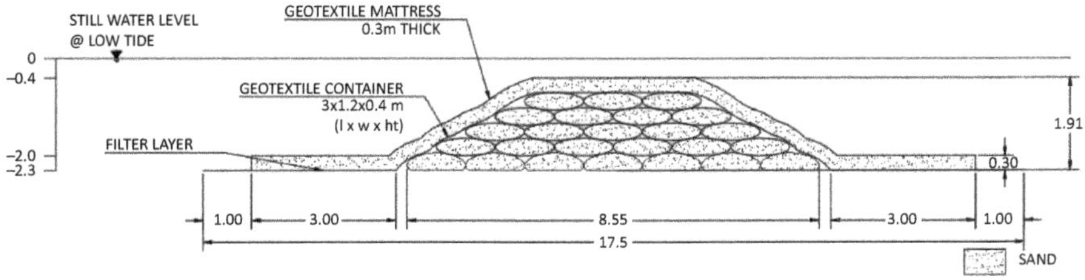

(a)

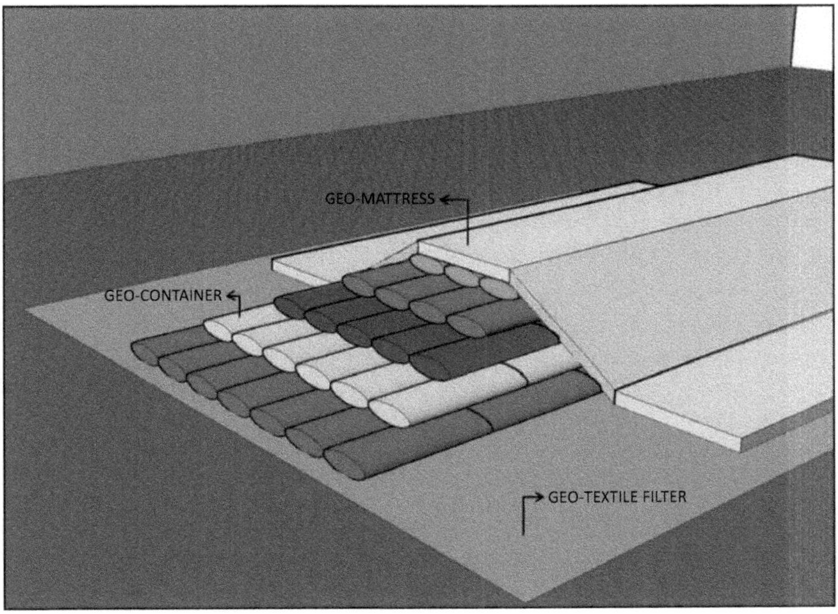

(b)

Fig. 7.12. Proposed submerged reef. (a) cross-section of submerged reef and (b) a 3D view of the submerged breakwater.

mattresses is to be employed to create a stable cross-section of submerged reef as shown in Fig. 7.12.

Hydrodynamic modeling was carried out by solving the shallow water equations using the finite element method to calculate current velocities and directions at each time step for the given bathymetry and tidal boundary condition. In the sediment transport modelling, Ackers and White's (1973) formula was discovered to yield a close estimate of the real-field conditions at Sagar Island. Based on the aforesaid study, the total quantity of sediment deposition was computed to be $8.2 \times 10^{-3}\,\mathrm{m^3/s}$ per metre length of the structure; since the structure's span is about 2.3 km along the coast, the quantity is $24.6\,\mathrm{m^3/s}$ along the coast. On extending the annual estimate with spatial distribution growth and a factor of safety, the expected sedimentation after construction of the submerged geo-synthetic reef at Sagar Island is estimated to be 13 meters per year.

7.4.3 *Effectiveness as coastal protection (a few case studies)*

7.4.3.1 *Vaan island*

Vaan Island (8° 50′16″ N 78° 12′42″ E) situated in the Gulf of Mannar, close to southeast coast of India, had decreased in area from 0.21 sq. km in 1969 to 0.023 sq. km in 2015 (Jayanthi *et al.*, 2020). A total of 10600 perforated trapezoidal artificial reef units were deployed in two rows in 2 m and 3 m water depths along the eastern side of the island to protect against waves. This resulted in an increase in the area of the island. Figure 7.13 shows a precast module before installation and its effect under the sea, proving the growth of marine habitat. The complete details of this study can be found in Sundar and Sannasiraj (2019).

7.4.4 *Multipurpose artificial reefs (MFARs)*

To promote surfing or to serve as a coastal protection measure, multipurpose artificial reefs (MFARs) have been implemented at a few locations around the globe. The first such reef named Cable Station was installed close to Perth, Australia, for the purpose of surfing, and the work was completed in December 1999, the details of which are discussed in Ranasinghe *et al.* (2001). MFARs, apart from being planned for surfing, can be implemented to protect the coast against erosion and can also serve as wave regulators through controlled refraction and diffraction of waves (Voorde *et al.*, 2009).

In pursuit of coastal protection, artificial reef ball units were deployed off the coast of the Dominican Republic. Harris (2009) contends that this intervention has enriched marine life, flora and fauna. An array of reef ball units, each of a height of 1.3 m and diameters of 1.5 m and 1.6 m, resulted in beach stability along the shore of the Dominican Republic. Twenty-eight reef ball units and 108 overlapping concrete block units were deployed

Fig. 7.13. Triangular and trapezoidal reef units.
Source: Patterson *et al.* (2016).

by Pradjoko *et al.* (2015) at different locations off the coast of Indonesia as a coastal protection measure that has also resulted in the growth and reproduction of marine organisms, such as fish, corals, and other sea life, which can thrive due to the habitat and resources provided by the reefs.

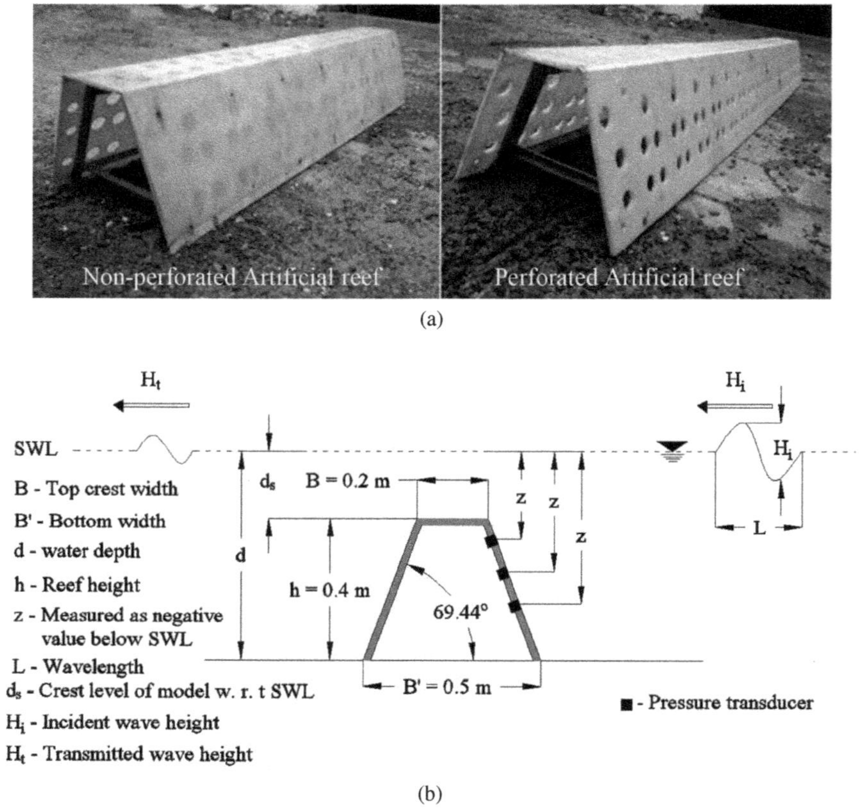

Fig. 7.14. (a) View of non-perforated and perforated trapezoidal artificial reef units, STARUs. (b) Cross-section of artificial reef model fitted with pressure transducers and its distance measured with respect to the still-water level for a particular water depth.

Artificial reefs replace natural rock formations as submerged breakwaters due to their minimal impact on the marine environment and their flexibility to mould into any desired shape and size is the greatest advantage. Lokesha *et al.* (2019) carried out an experimental investigation to examine the effect of perforations on a single submerged trapezoidal artificial reef unit (STARU) with perforations of 11% (Fig. 7.14(a)) exposed to regular and random waves. This artificial reef unit was part of a scaled-down model that was employed to protect Vaan Island (8° 50′16″ N 78° 12′42″ E), Tuticorin, on the southeast coast of India, as discussed in detail by Sundar and Sannasiraj (2019).

(a)

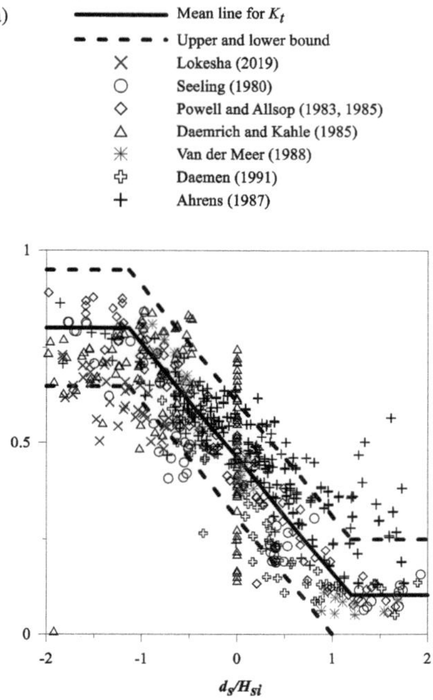

(b)

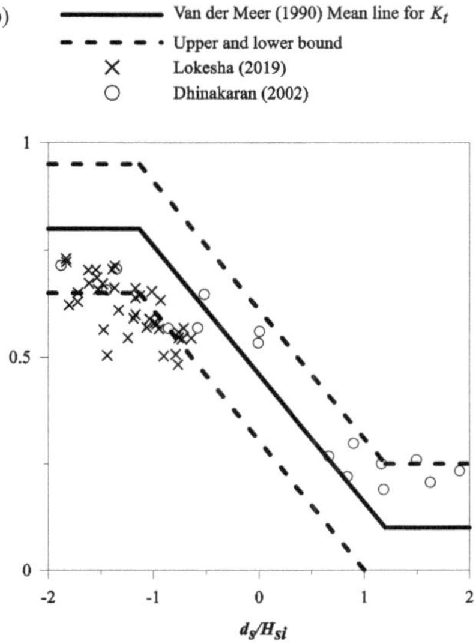

Fig. 7.15. (a) Comparison of K_t for a single unit of the STARU-11 model ($B_t/B = 0$) and a rubble mound breakwater. (b) Comparison of K_t for a single unit of the STARU model (B_t) and a semi-circular submerged breakwater. *Source*: Dhinakaran (2002).

The cross-section of the STARU, the locations of the pressure ports and the relevant parameters used in the experimental setup to arrive at the predictive equations for the transmission coefficient (transmitted wave height/incident wave height) K_t are shown in Fig. 7.14(b).

Although the tests on both impermeable and permeable single and multiple STARU models were exposed to the action of both regular and random waves (Lokesha, 2019), there was a variation of K_t for a single STARU model with 11% perforations, with the results of past investigations presented in Fig. 7.15(a). Herein, B_t is the trench width and $B_t/\mathrm{B} = 0$ for a single reef unit. The results demonstrate that K_t for the STARU model mostly falls between the mean line and the lower bound for conventional rubble mound breakwaters. Dhinakaran (2002) reported the results of experiments on the hydrodynamic characteristics of semi-circular breakwaters with 7–11% perforations. The prediction of K_t from the STARU model is compared with Dhinakaran (2002) for semi-circular submerged breakwater with 11% perforation. The wave dissipation for a trapezoidal reef unit (Van der Meer, 1990) is more than the semi-circular submerged reef.

The regression equation for a single STARU model with 11% perforations for the K_t as proposed by Lokesha *et al.* (2019) is

$$K_t = -0.02 - 1.92\frac{B}{L_p} + 0.52\frac{d}{h} - 0.01\frac{d_s}{H_{\mathrm{si}}}. \tag{7.2}$$

The test ranges for the equations subjected to random waves are

$$0.007 \leq \frac{H_{\mathrm{si}}}{L_p} \leq 0.055; \quad 1.25 \leq \frac{d}{h} \leq 1.5;$$

$$0 \leq \frac{B_t}{B} \leq 0.57; \quad -3.5 \leq \frac{d_s}{H_{\mathrm{si}}} \leq -0.5.$$

7.5 Breakwaters (Multipurpose)

7.5.1 *Piled breakwaters*

7.5.1.1 *General*

The demand for handling deep-draft vessels has been ever-increasing, forcing a search for alternative options for berthing vessels in deeper waters through which maintenance dredging, a persisting problem involving huge cost and risk, can be avoided. A rubble mound consisting of layers of graded stone with large concrete blocks forming an armour layer is uneconomical for deeper waters. Further, in offshore locations with a severe wave climate, the construction of rubble mound breakwaters is not practical as the stones are easily carried away by the waves. Concrete caisson-type breakwaters have inherent problems like scour and instability because of the excessive reflection of waves from their vertical front faces. The basic concept of a pile-supported breakwater is to serve as an impervious barrier near the free surface, where the water particle movements are greater, while leaving the lower region unobstructed. Construction of a pile-supported breakwater will allow the free passage of currents and allow the wave-induced littoral drift to pass through without leading to the instability of the shoreline that occurs in the case of shore-connected breakwaters. Further, at locations with soft soil and no hard stratum within a reasonable depth, a pile-supported breakwater can be an alternative.

The efficiency of such structures depends on the depth of immersion, its type, whether the skirt is permeable or impermeable and the incident wave characteristics. Closely spaced piles and submerged skirts as alternatives for breakwaters have been tried in the past. The hydrodynamic characteristics of pile breakwaters consisting of just closely spaced piles have been discussed in detail by Karthik Ramnarayan *et al.* (2020b). Sundar and Subba Rao (2003) and Suh *et al.* (2006) reported that the relative spacing between the piles (ratio of the space between piles to pile diameter) does not affect the reflection coefficient of a quadrant front-face, pile supported breakwater. The performance of a pile breakwater consisting of closely spaced piles alone is not very encouraging and driving the piles in very close proximity is practically difficult. However, if we resort to a pile-supported breakwater with a skirt, deep-draft vessels can approach it closer and berth on its leeward side for loading and unloading operations. In fact, the sea side of the skirt can also be converted to an oscillating water column device for the extraction of energy from ocean waves, the details of which will be discussed later in this chapter.

Different configurations for the seaward side face of pile-supported breakwaters have been investigated in the past as discussed by Takahashi (1994). A curved-front or alternatively a porous barrier on the seaward side can reduce reflection. Interestingly, the curved-front breakwater models investigated to date are of a convex shape on the seaward side, while concave profiles have also received attention in the recent past. The different curved-front-type breakwaters that have been reported to experience less reflection are presented in Fig. 7.16.

The hydrodynamic coefficients of different types of pile-supported breakwaters are summarized in Table 7.3.

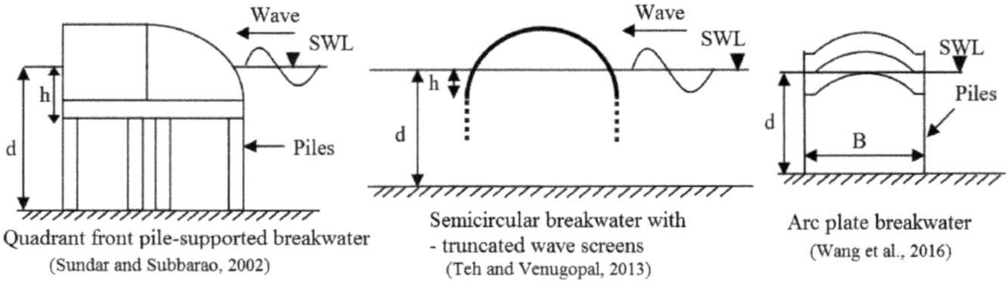

Fig. 7.16. Curved-front-type breakwaters.

Based on the merits of the curved seaward side face as discussed by Anand *et al.* (2011), Karthik Ramnarayan *et al.* (2020b) adopted a model scale of 1:25 and carried out tests in a laboratory on a pile-supported

Table 7.3. Hydrodynamic performance of pile-supported breakwaters.

Literature	Breakwater type	Optimal water depth (m)	Hydrodynamic Coefficients	
			K_t	K_r
		Vertical front type		
Neelamani and Vedagiri (2002)	Twin vertical barrier	10–25	0.05–0.80	0.40–0.90
Neelamani and Rajendran (2002)	T-type breakwater	10–25	0.15–0.67	0.65–0.95
Rageh *et al.* (2009)	Caisson supported on piles	5–10	0.03–0.75	0.45–0.99
Rageh and Koraim (2010)	Vertical wall with horizontal slots	10–15	0.20–0.81	0.15–0.75
Koraim *et al.* (2011)	Double row of vertical wall with horizontal slots	10–15	0.05–0.70	0.20–0.99
		Porous type		
Nakamura *et al.* (2001)	Inclined plate array on seaward side	6.5–9.5	0.05–0.65	0.05–0.70
Brossard *et al.* (2003)	Semi-immersed breakwater with absorbing caisson	—	0.05–0.45	0.05–0.65
Wang et al (2006)	Multiple layers of horizontal plates	12	0.10–0.95	0.10–0.60
Koraim *et al.* (2014)	C-shaped bars supported on piles	2.5–5	0.40–0.99	0.03–0.70

Table 7.3. (*Continued*)

Literature	Breakwater type	Optimal water depth (m)	Hydrodynamic Coefficients	
			K_t	K_r
	Curved front type			
Sundar and Subba Rao (2003)	Quadrant front-face breakwater	20–25	0.10–0.50	0.25–0.85
Teh and Venugopal (2013)	Semi-circular breakwater with truncated wave screens	20	0.05–0.90	0.12–0.48
Wang *et al.* (2016)	Arc plate breakwater	12	0.05–0.90	0.05–0.40

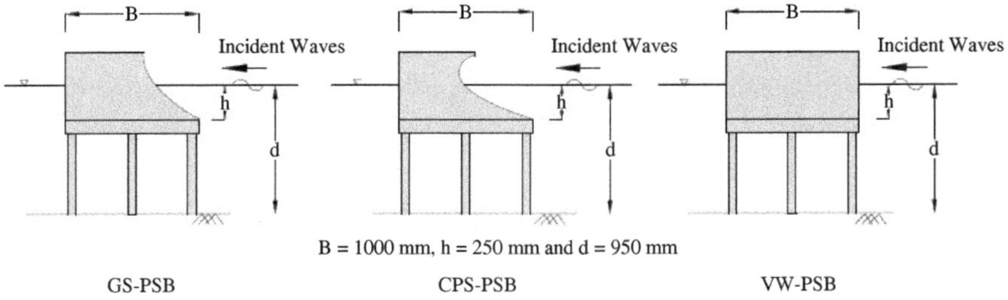

Fig. 7.17. Sectional view of the GS-PSB and CPS-PSB models.
Source: Karthik Ramnarayan *et al.* (2020b).

superstructure with a GS (Galveston seawall), termed as GS–PSB and CPS (circular cum parabola shape), termed as CPS-PSB. Since the piles were used to merely support the topside structure and not for energy dissipation, a large pile spacing was maintained. Schematic views of the GS-PSB and CPS-PSB models are shown in Fig. 7.17.

Two stainless steel wires of 1 mm thickness each and placed at a spacing of 10mm were placed along the concave profile surface to measure the wave run-up. Five pressure sensors were fixed on the concave profiles of both models to measure the wave-induced dynamic pressures. The elevation of the pressure sensors was the same for both models as shown in Fig. 7.18.

The non-dimensional parameters obtained from the measurements of the rigorous physical modelling were subjected to multiple regression analysis

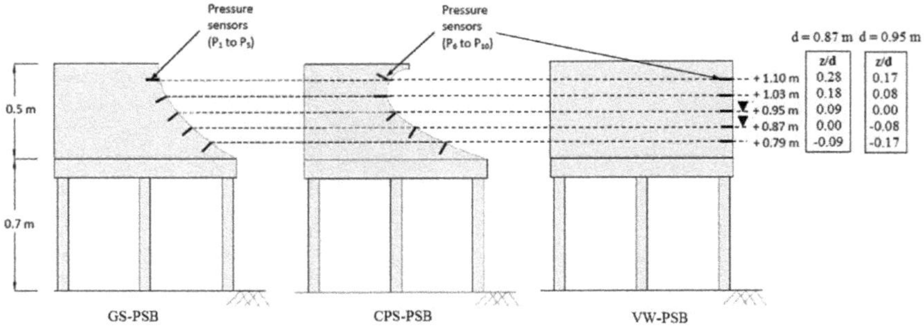

Fig. 7.18. Locations of the pressure sensors.
Note: The elevation of pressure sensors was maintained the same for both models.

based on the method of least squares. K_t and K_r are expressed as a function of dimensionless parameters, with relative breakwater width B/L and relative depth of submergence h/d. The non-dimensional pressure $P_{\text{sh}}/\gamma H_i$ is expressed as a function of d/L and z/d. The relative run-up R_u/H_i is expressed as a function of d/L and h/d. Herein, P_{sh} is shoreward peak pressure, γ is specific weight of water, H_i is wave height, d is water depth, R_u wave run-up, L is wavelength, z is pressure port location negative below the still-water level, h is the depth of submergence of the curved-face superstructure and B is the width of the breakwater. The equations using the aforementioned different parameters for GS-PSB and CPS-PSB as presented through Equations (7.3) to (7.6) display a better performance than VW-PSB as discussed by Karthik Ramnarayan *et al.* (2020b):

For transmission

$$\textbf{GS-PSB:}\, K_t = 0.003 \times (B/L)^{-1.68} \times (h/d)^{-1.24}. \tag{7.3a}$$

$$\textbf{CPS-PSB:}\, K_t = 0.005 \times (B/L)^{-1.58} \times (h/d)^{-1.11}. \tag{7.3b}$$

For reflection

$$\textbf{GS-PSB:}\, K_r = 0.91(B/L) + 2.01(h/d) - 0.11. \tag{7.4a}$$

$$\textbf{CPS-PSB:}\, K_r = 0.98(B/L) + 1.69(h/d) - 0.29. \tag{7.4b}$$

For wave-induced pressures

$$\textbf{GS-PSB:}\, P_{\text{sh}}/\gamma H_i = -1.61(d/L) + 2.31(z/d) + 1.43; \quad (\text{For } z \leq 0). \tag{7.5a}$$

$$\textbf{CPS-PSB:}\, P_{\text{sh}}/\gamma H_i = -1.65(d/L) + 3.20(z/d) + 1.51; \quad (\text{For } z \leq 0). \tag{7.5b}$$

For run-up

$$\textbf{GS-PSB:}\, R_u/H_i = 1.54(d/L) + 0.67(h/d) + 0.75. \quad (7.6a)$$

$$\textbf{CPS-PSB:}\, R_u/H_i = 1.01(d/L) - 3.71(h/d) + 2.01. \quad (7.6b)$$

The above-mentioned equations are valid for a d/L range of 0.13–0.61 (B/L of 0.14–0.64). The energy dissipation offered by the curved profiles does not affect the wave transmission past the structure in deep-water conditions but results in lesser transmission in intermediate-water depths. The energy dissipation affects the wave reflection more than its transmission. The wave reflection is the least for CPS-PSB which is more receptive to the incoming waves due to the flat slope at lower elevations. However, the wave-induced pressures are maximum for CPS-PSB followed by GS-PSB and VW-PSB. At lower elevations, the induced pressures due to short waves are the least for CPS-PSB due to the flatness of the profile. Though the run-up over CPS-PSB is high, there is no overtopping due to the presence of the recurve. Overtopping occurred over GS-PSB, thus necessitating an additional energy-dissipating mechanism on the seaward side. For a complete discussion, the reader can refer to Karthik Ramnarayan *et al.* (2020a, 2020b).

Jemi Jeya *et al.* (2021) investigated the hydrodynamic characteristics of a vertical (VPSB) and a quadrant face pile-supported breakwater (QPSB) under oblique waves and reported that the K_r for both models decreases with an increase in the angle of wave attack. The presence of a convex shape on the seaward side face of the QPSB model reduces K_r significantly. For any angle of the wave attack, K_r increases with an increase in the water depth for both models. In summary, it was reported that the QPSB satisfies the requirement of less reflection and transmission of incident wave energy, and the pressures exerted on the superstructure are also less compared to the VPSB. However, the run-up alone was found to be higher and hence the QPSB could be accommodated in locations of permissible occasional overtopping or included in the design by considering higher crest elevation.

7.5.1.2 *PSB as a berthing facility*

Pile breakwaters consist of closely spaced rigid, vertical piles. Vertical and horizontal slotted screens have also been used to form breakwaters. The following are examples of pile breakwaters in use:

Fig. 7.19. Plan view of a double-row piled breakwater with rocks at Mekong Delta.
Source: Schüttrumpf.

- Hanstholm, Denmark,
- Marsa el Brega, Libya,
- Port of Osaka, Japan,
- Pass Christian, USA,
- Pasir Panjang, Singapore,
- Mekong Delta, Vietnam (Fig. 7.19).

The details of the pile breakwater next to Pasir Panjang, Singapore, have been discussed by Liu *et al.* (2011).

A piled double-row concrete breakwater with rock filling has been used at the Mekong Delta. A gap of 1.5–2 m between the two rows was filled with natural riprap for the purpose of incident wave energy dissipation and permeability so as to allow the passage of sediments through the breakwater. The details of the piled double-row concrete breakwater have been discussed by Nguyen and Duong (2019).

As stated earlier, the effect of piled breakwaters on the shoreline is minimal or negligible as the longshore sediment transport passes through the structure without interception. The curved seaward side piled breakwaters have characteristics that allow their wider application in the marine

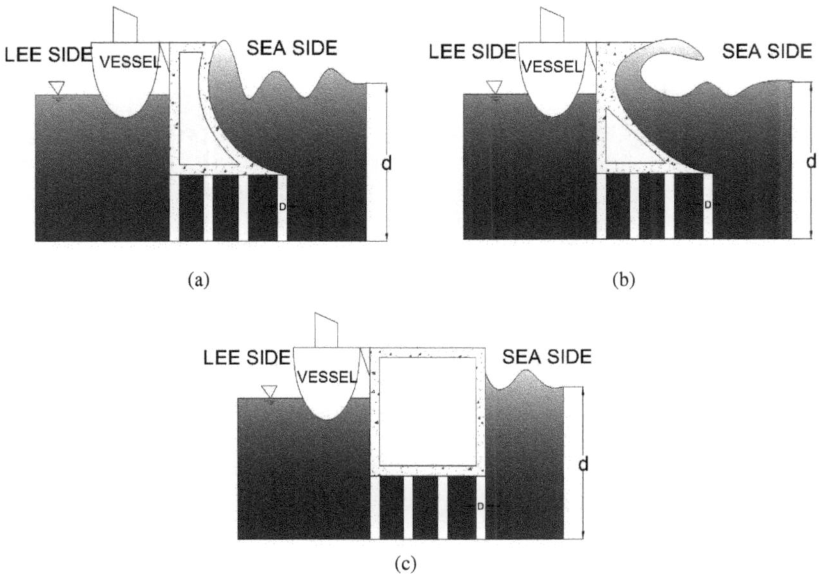

Fig. 7.20. Berthing facility with (a) CPS-PSB, (b) GS-PSB and (C) VP-PSB.

environment. The advantages of pile-supported curved wall breakwaters are as follows:

- Lesser reflection of incident wave energy,
- Minimal interference to the alongshore sediment transport,
- Significant dissipation of incident wave energy, thereby creating tranquillity on its leeward side,
- Facilitate berthing of vessels on the leeward side,
- More scenic compared to mound breakwaters,
- Suitable for islands and locations where natural rocks are scarce,
- Transportation of materials is easier and less cumbersome in the case of piled structures.

Illustrations of the three types of breakwaters, (a) CPS-PSB, (b) GS-PSB and (C) VP-PSB, with berthing facilities are shown in Fig. 7.20. A perspective view of a harbour basin with the proposed CP-PBW as the breakwater is shown in Fig. 7.21. Beyond coastal ports, there exist numerous river ports. The following are the ten largest river ports in the world:

- Port of Shanghai, China.
- Port of Rotterdam, Netherlands.

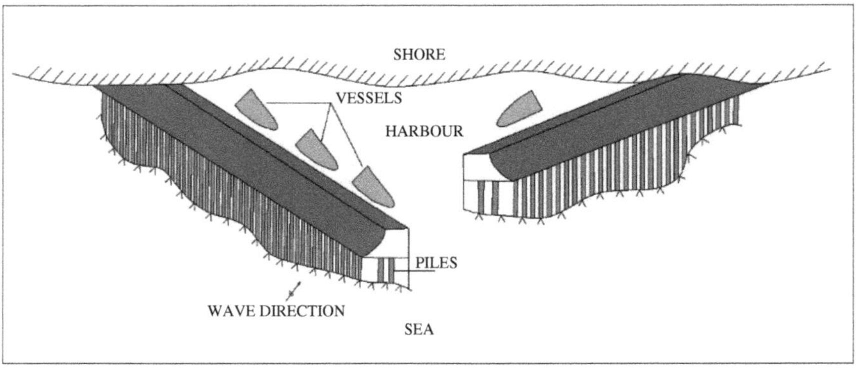

Fig. 7.21. A perspective view of a harbour basin with the proposed CP-PBW as the breakwater.

- Port of Hamburg, Germany.
- Port of Antwerp, Belgium.
- Port of New Orleans, U.S.A.
- Port of Osaka, Japan.
- Port of London, United Kingdom.
- Port of Kolkata, India.

In the case of riverine ports, the confluence of the river and the ocean can in certain locations create a wave climate most of the year that may permit only smaller crafts/vessels without posing challenges in navigation; hence, the operation of the port may be limited during some months of the year. In such scenarios, an approach trestle with a CPS-PSB or GS-PSB type of pile breakwater could serve as a berthing facility as shown in Fig. 7.22. Herein, the approach trestle will not intercept the passage of littoral drift as it is not shore connected, and it can thus be used for loading and unloading cargo. Supply crafts can operate between the riverine port and the deep-water berth, thereby increasing the number of operational days of the port. This concept would also be suitable for islands.

7.5.1.3 *Supporting structure for wave energy conversion*

Energy demand has been on the rise globally and energy from the ocean is looked upon as a major source. Samad *et al.*, (2021) have compiled a list of various types of wave energy converters. An oscillating water column (OWC) device, a type of wave energy convertor (WEC), is a relatively simple mechanism in which the incident wave energy is converted into water

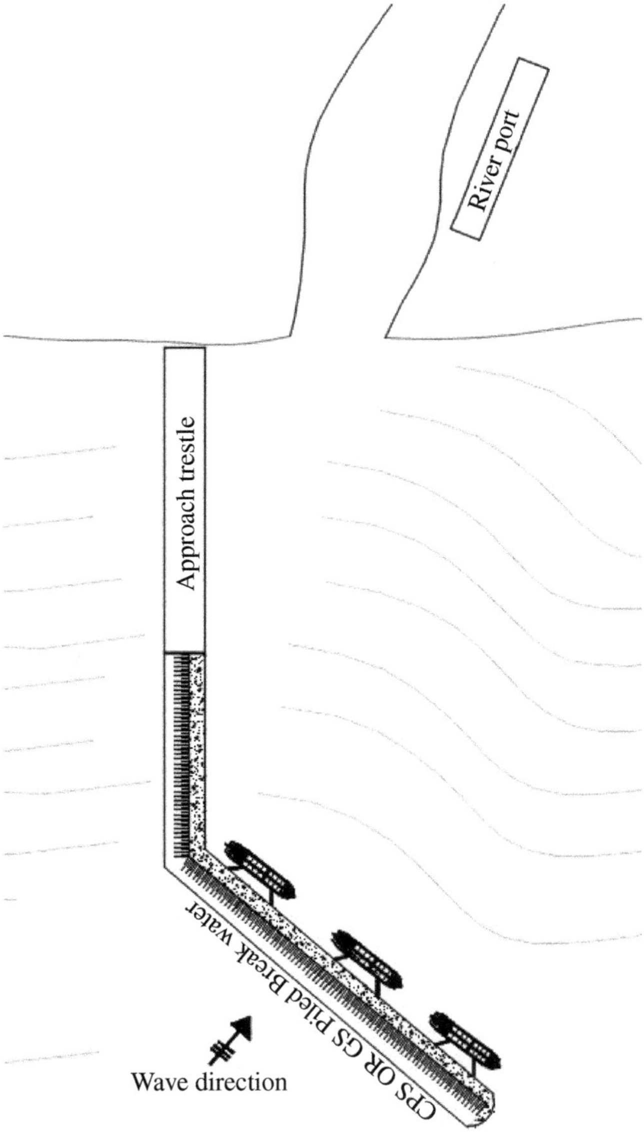

Fig. 7.22. A view of a deep-water berthing facility formed with CPS-PSB or GS-PSB.

column oscillations inside a chamber that creates the necessary airflow to run a turbine as shown in Fig. 7.23. As the waves rise and fall, they initiate a reciprocating motion of air within a chamber, driving turbines and

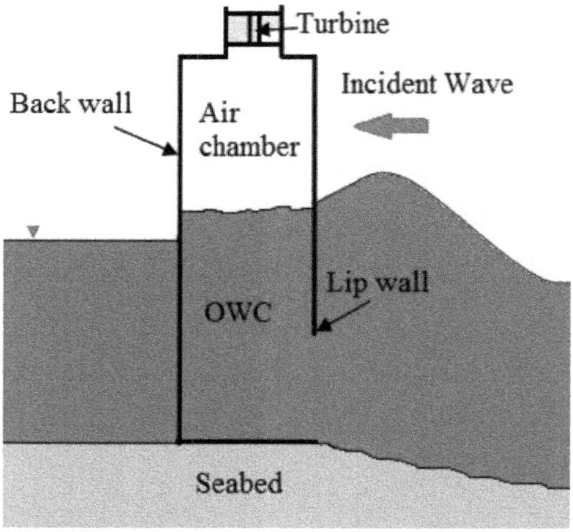

Fig. 7.23. Working principle of an OWC device.

generating electricity. Understanding the pressures and forces induced by ocean waves on these devices is necessary not only to maximize energy extraction but also to ensure structural resilience and longevity. In locations where the wave energy potential is moderate, standalone wave energy converters may not be feasible. Therefore, integrating additional functionality to harness a portion of the incident wave energy instead of dissipating it would be an ideal solution.

7.5.1.4 *Wave forces on an OWC WEC*

The study of wave loading on an OWC WEC presents difficult challenges when approached through analytical or numerical methodologies. These complexities arise predominantly from the structure's unique geometry, which features a frontal opening, a power take-off (PTO) device and a dual-phase chamber. Therefore, direct measurements are essential. Physical model tests are suitable and reliable in such cases. Ashlin *et al.* (2017) conducted a physical model study to gain a comprehensive understanding of the pressures and forces induced by non-breaking waves. These findings are crucial inputs for the stability check while designing the OWC WEC.

The study utilized a circular curve bottom OWC WEC model (Fig. 7.24), hydrodynamically optimized by Ashlin *et al.* (2016), to primarily focus on wave loading.

The expression for estimating the shoreward horizontal wave force acting on the vertical wall based on the theoretical method of Goda (1985),

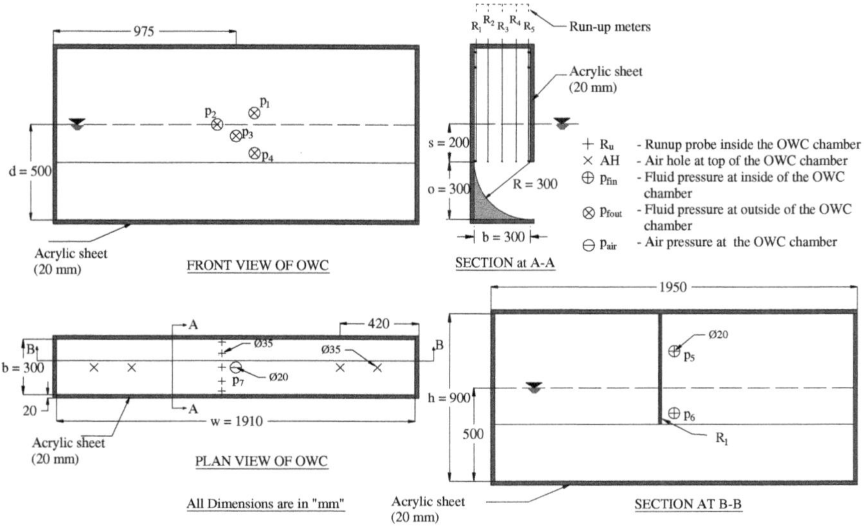

Fig. 7.24. Model setup inside the wave flume.
Source: Ashlin *et al.* (2017).

$(\mathrm{GM_{HZL}})^c$, is given as follows:

$$(\mathrm{GM_{HZL}})^C = \frac{(F_{\mathrm{HZL}})_{\max}}{0.5\rho g w d H} = \alpha_1 \left[\left(1 + \frac{1}{\cosh kd}\right) + \left(\frac{1.5\,H}{d}\right)\right], \quad (7.7)$$

$$\alpha_1 = 0.6 + 0.5\left(\frac{4\pi \frac{d}{L}}{\sinh 4\pi \frac{d}{L}}\right)^2, \quad (7.8)$$

where $k = 2\pi/L$. For comparison, the seaward horizontal wave force acting on the vertical wall based on the theoretical method of Goda (1985), $(\mathrm{GM_{HZL}})^t$, has been taken.

To understand the differences in force nature and magnitude between the OWC WEC and a vertical wall, horizontal wave forces from the study were compared with values from Goda's empirical formula (1985). The results are depicted in Fig. 7.25 as a function of d/L. Despite distinct hydrodynamic behaviours, efforts were made to quantify wave load differences at the two tested water depths.

The data revealed a consistent variation trend, indicating that, in general, Goda's method (1985) tended to overestimate shoreward peak forces, especially for lower d/L values. Conversely, the formula of Goda (1985) underestimated seaward peak forces. The high wave power absorption in the low-frequency region and the resulting partial standing wave in front of

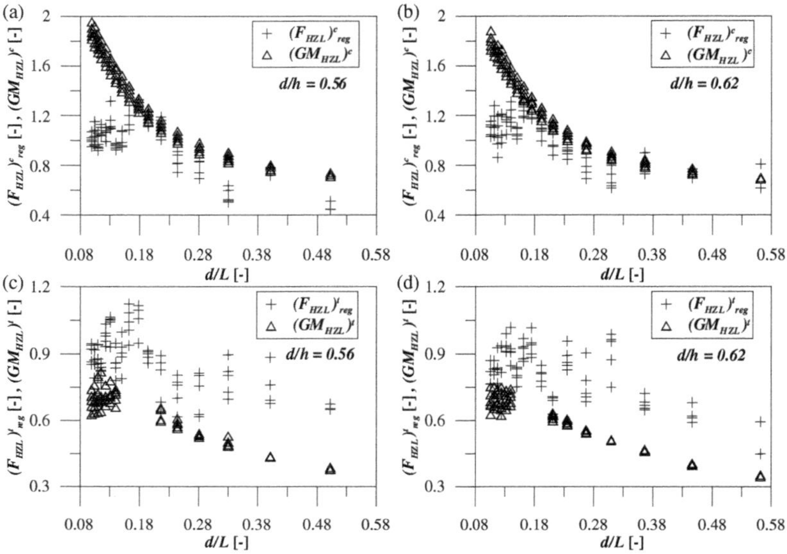

Fig. 7.25. Comparison of the measured forces with the formulation results from Goda (1985).
Source: Ashlin *et al.* (2017).

the OWC chamber reduced the shoreward horizontal wave force. The OWC system's efficiency peaked at 71% for $d/L < 0.15$, which might explain the overestimation by the Goda method. For $d/L > 0.15$, the agreement between the two sets of results was considerably close. The shoreward horizontal wave force for the two distinct water depths, when compared with $(\mathrm{GM_{HZL}})^c$, differed by roughly 46–90% for the lesser d/h of 0.56 and 27–80% for the larger d/h of 0.62. The seaward horizontal wave force consistently exceeded $(\mathrm{GM_{HZL}})^t$ by approximately 5–50%. With wave trough action, the flow transitioned from the chamber interior to its front, resulting in increased force on the rear side of the lip wall, a phenomenon absent in vertical walls due to the unobstructed wave trough receding. The reduction in horizontal load on the OWC device shows the advantages of using it in the construction of caisson-type breakwaters.

Sundar *et al.* (2010), through a comprehensive review of literature, have highlighted that the OWC is commonly linked with a caisson-type breakwater. A number of concepts on this aspect have been discussed. Karthik Ramnarayan *et al.* (2021) compared the hydrodynamic characteristics of an OWC mounted on a pile-supported breakwater (OWC-PSB) with those of VW-PSB, the details of which are shown in Fig. 7.26.

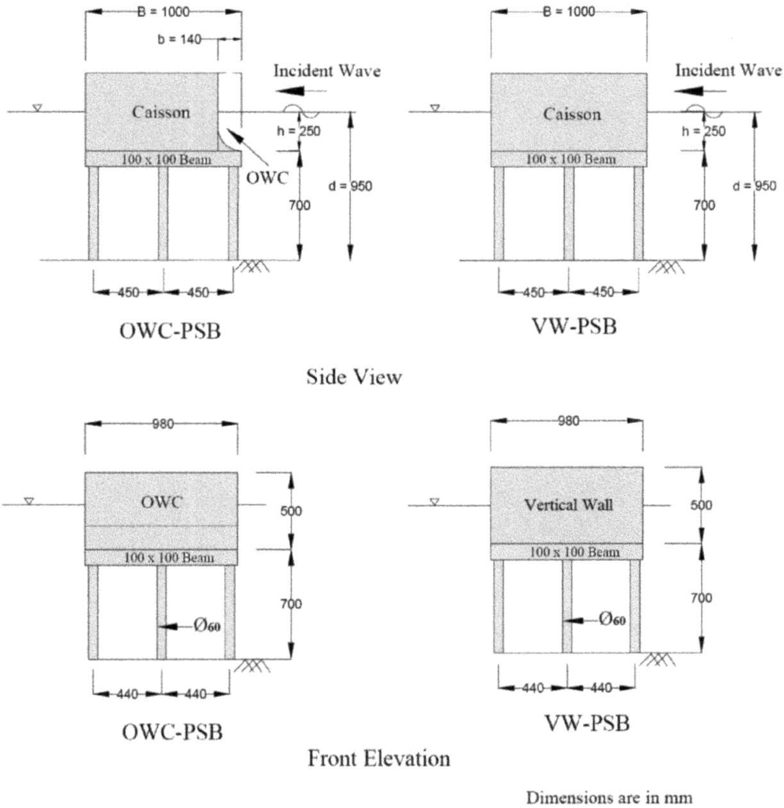

Fig. 7.26. Sectional view of the OWC-PSB and VW-PSB.
Source: Karthik Ramanarayanan *et al.* (2021).

The variations of K_t K_r and K_L for OWC-PSB and VW-PSB from random wave test results are depicted in Figs. 7.27(a), 7.27(b) and 7.27(c), respectively. The K_t for both the models in deep-water conditions is found to be identical but marginally lesser for OWC-PSB in intermediate-water depths as the wave energy absorption by the OWC device affects the transmission for lower d/L values. The K_t for OWC-PSB and VW-PSB varies from 0.05 to 0.6 and 0.05 to 0.6, respectively. The results indicate that the wave transmission past the structure is not significantly altered by housing an OWC in place of a vertical wall superstructure supported over piles.

The variation of K_r for both the models superposed shows that the range of K_r for OWC-PSB and VW-PSB is from 0.5 to 0.6 and 0.6 to 0.8, respectively. The variations of the energy loss coefficient K_L superposed

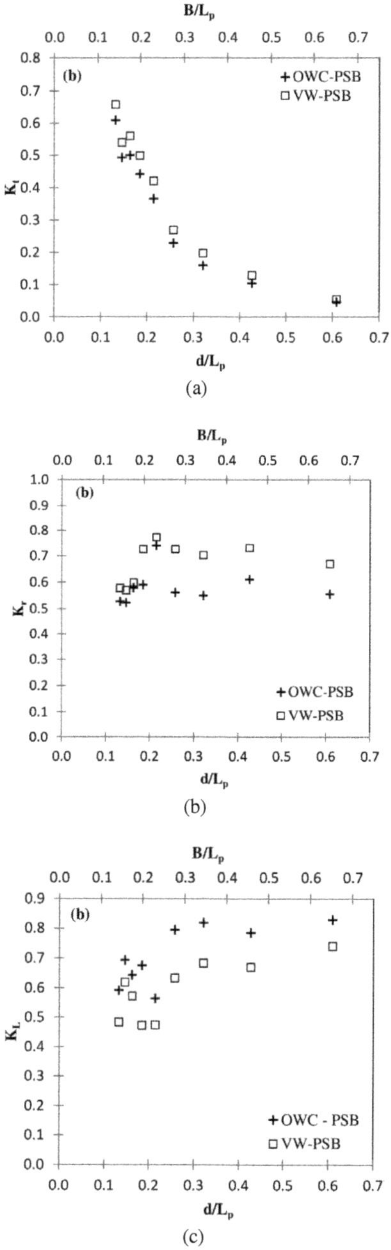

Fig. 7.27. (a) Variation of K_t with d/L_p. (b) Variation of K_r with d/L_p. (c) Variation of K_L with d/L_p.

for both models is found to be higher for OWC-PSB due to the conversion of the incident wave energy into water column oscillations in the OWC chamber. Beyond a d/L of 0.2, the variation in K_L is less as there is not much change in wave reflection. The range of K_L for OWC-PSB and VW-PSB under random waves is from 0.6 to 0.80 and 0.5 to 0.70, respectively.

Due to the OWC-PSB model absorbing the incident wave energy for energy conversion, it was reported to experience lesser reflection. Furthermore, the wave amplification factor for the OWC device was reported to be higher for long waves; however, the hydrodynamic efficiency was found to be higher for high-frequency waves with the efficiency being at a maximum close to 50% under deep-water conditions.

The wave amplification factor β, defined as the ratio of the water surface elevation inside the OWC chamber to the incident wave height, is expressed as $\beta = (R_u)_s/H_s$, where $(R_u)_s$ represents the significant value of the wave run-up shown in Fig. 7.28(a). The wave amplification factor is more for long waves as it is easy for the elliptical orbital motions to pass through the lip wall opening and provide sufficient time for the water surface oscillations inside the OWC chamber. In the case of short waves, the lip wall serves as an obstruction for the smaller circular orbital motions, thereby resulting in lesser water surface oscillations inside the OWC chamber.

The air pressure ratio is defined as the ratio of air pressure inside the OWC chamber to the incident wave-induced pressure and expressed as $(\mathrm{PR_{air}}) = (p_{\mathrm{air}})_s/\rho g H_s$, wherein $(p_{\mathrm{air}})_s$ represents the air pressure inside the OWC chamber measured from crest to trough and $(p_{\mathrm{air}})_s$ stands for the significant value of air pressure for random waves. The hydrodynamic efficiency is the ratio of pneumatic output power (P_{out}) to the incident wave power flux (P_{in}), defined as $\lambda = \frac{P_{\mathrm{out}}}{(P_{\mathrm{ave}})_{\mathrm{in}}}$, wherein $(P_{\mathrm{ave}})_{\mathrm{in}}$ is the average incident wave power flux.

The trend in the variations of air pressure ratio (Fig. 7.28(b)) is not similar to that of the wave amplification factor as it is also dependent on the pace of water column oscillations. As the wave amplification factor β increases with wavelength, the time period available for water column oscillation also increases, leading to less variation in the air pressure ratio over the entire d/L range. Smaller fluctuations in air pressure are generally desirable for the smooth operation of the turbine. $(\mathrm{PR_{air}})$ (Fig. 7.28(b)) is found to vary from 0.8 to 1. For details of this study, the reader can to refer to Karthik Ramnarayan (2021).

The hydrodynamic efficiency for random waves is given as

$$\lambda = {P_{\mathrm{out}}}\big/{(P_{\mathrm{ave}})_{\mathrm{in}}}, \tag{7.9}$$

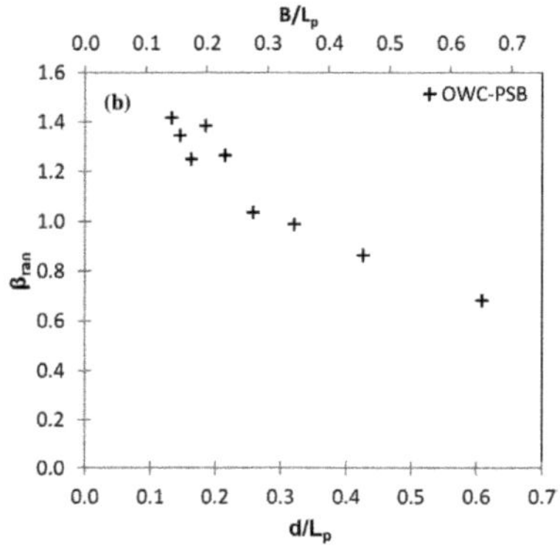

(a)

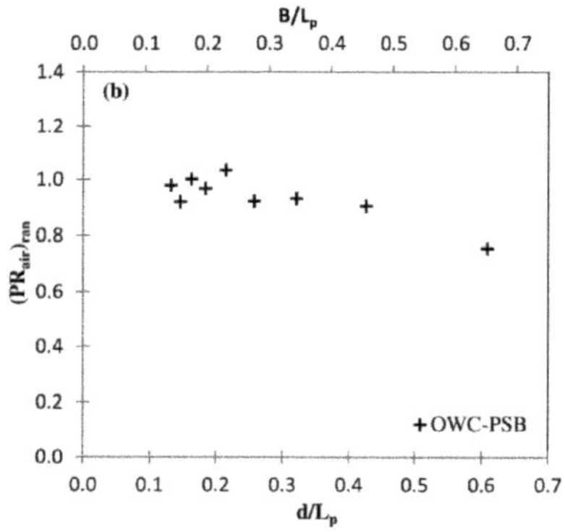

(b)

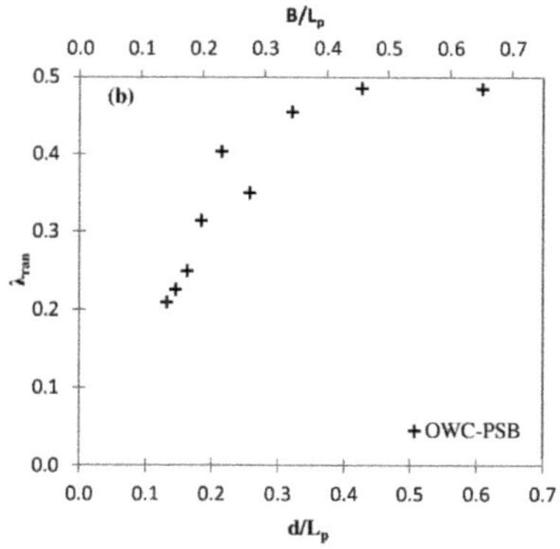

(c)

Fig. 7.28. (a) Variation of β with d/L_P. (b) Variation of $(\mathrm{PR_{air}})$ with d/L_P. (c). Variation of λ with d/L_P.

where $(P_{\text{ave}})_{\text{in}}$ is the average incident wave power flux evaluated as

$$(P_{\text{ave}})_{\text{in}} = w\rho g \sum_{i=1}^{N} S_i (C_g)_i \Delta f, \tag{7.10}$$

in which S_i refers to the spectral density at the i_{th} band, $(C_g)_i$ stands for the group celerity at the i_{th} band and Δf is the frequency interval considered. $(C_g)_i$ is given by the expression

$$(C_g)_i = \frac{C_i}{2}\left[1 + \frac{2k_i d}{\sinh(2k_i d)}\right], \tag{7.11}$$

wherein $k_i = L_i/T_i$ is the wave number corresponding to the ith band.

The variation of λ with d/L_p is given in Fig. 7.28(c). It is observed that, although the wave amplification factor reduces with an increase in d/L, the hydrodynamic efficiency increases. A maximum efficiency of about 50% is attained in deep-water conditions. The hydrodynamic efficiency depends on the phase difference between the water surface oscillations and the air pressures inside the OWC chamber. The hydrodynamic efficiency is mainly affected by the wave energy transmission past the structure. As the transmission is more for long waves and the wave energy tapped by the OWC is less, there is a steep reduction in its efficiency for $d/L < 0.2$. Since the power of long waves is not completely captured by OWC-PSB, the efficiency is less compared to an OWC resting on the seabed for which a maximum efficiency of about 65% is achieved in intermediate-water depths as reported by Ashlin *et al.* (2016). However, the combined advantage of a berthing facility and harvesting energy from ocean waves looks promising. Vasanthakumar *et al.* (2023) revealed that a vertical front pile-supported breakwater (VF-PSB) experiences maximum wave force at lower submergence levels, with the vertical force decreasing exponentially as relative depth increases through a comprehensive experimental investigation. For a circular cum parabola-shaped pile-supported breakwater (CPS-PSB), the variation of vertical force with relative depth is similar to the pattern of energy dissipation. Induced moments are higher for longer waves and greater submergence. CPS-PSB is subjected to lower forces and moments compared to VF-PSB.

7.5.2 *Bottom-seated type*

7.5.2.1 *OWC integrated with rubble mound type*

Detached offshore breakwaters have proven to be effective coastal protection measures. Building upon previous observations, integrating Oscillating

Water Column (OWC) technology with pile-supported breakwaters offers promising possibilities and advantages. Configuring OWC devices appropriately and arranging them in arrays with optimal spacing between each unit could constitute a highly effective hybrid integrated system. Such a system not only facilitates energy conversion but also plays a crucial role in protecting against coastal erosion. This concept would also be highly beneficial for islands. Realizing this, research on the hydrodynamic performance of an array of OWC with detached rubble mound breakwaters has been carried out (Ashlin *et al.*, 2018). Ashlin *et al.* (2016) conducted extensive experimental studies on the geometric and configuration aspects of OWC being exposed to the action of regular and irregular waves in isolated conditions. The convergence of waves towards the OWC device in integrated applications plays a critical role in the concentration of wave energy, thereby enhancing its performance. Through well-controlled laboratory investigations, it was concluded that an inner chamber with a curved bottom, an air vent ratio of 0.68% and a front lip wall submergence ratio of 60% could be effective in the hydrodynamic performance of the device. These results paved the way to study the characteristics of the OWC device after integration with detached rubble-mound offshore breakwaters, the setup of which is shown in Fig. 7.29(a) with a cross-sectional view of the details of the integrated system shown in Fig. 7.29(b). Harbour walls were adopted near the entrance of OWCs (for details, refer to Daniel Raj *et al.*, 2018), which could help in avoiding or minimizing the sliding of rubble materials.

Different parametric studies concluded that the hydrodynamic performance of OWC increases when integrated with detached rubble-mound breakwaters. The parametric studies included the effect of wavelength, wave height, wave steepness, wave interaction between devices, spacing between devices and the reflective nature of OWC with the breakwater structure. It is reported that the hydrodynamic performance increases with an increase in wavelength and then decreases once it reaches the resonance frequency. Waves with less amplitude produce higher efficiency than large-amplitude waves because of the reflection of wave energy. Low steep waves yield better hydrodynamic performance than high steep waves. Higher wave energy penetration inside the chamber of OWC was observed in low steep waves. From the study on multiple devices, the maximum efficiency was obtained with five OWC devices kept at a spacing of three times the width of a device. The effect of the wave obliqueness on the OWC system's performance has also been discussed and the normal wave incidence is more effective. The detailed results are presented in Ashlin *et al.* (2018).

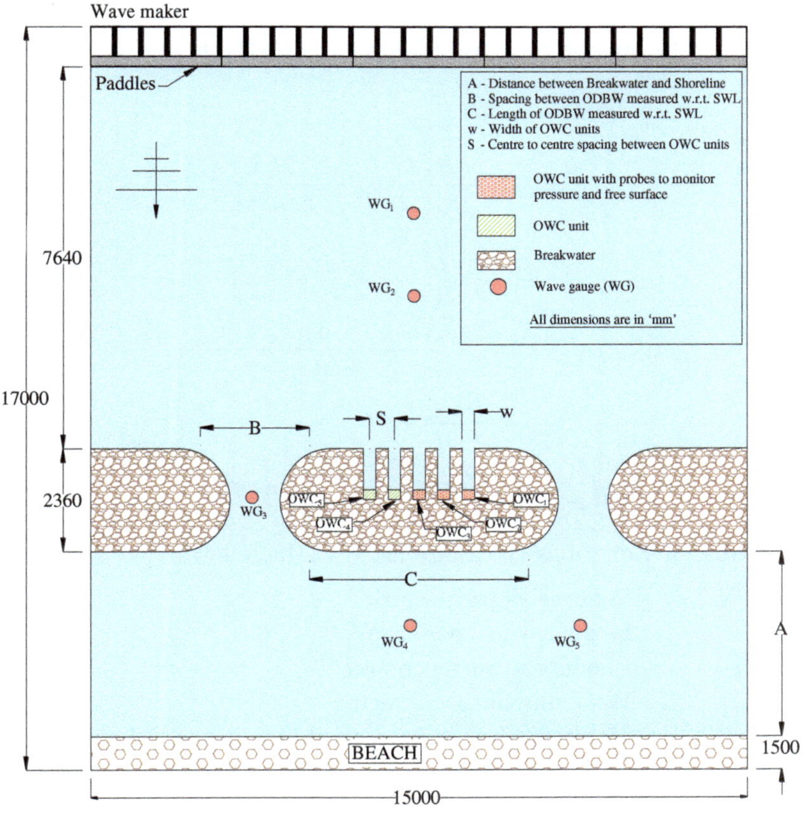

(a)

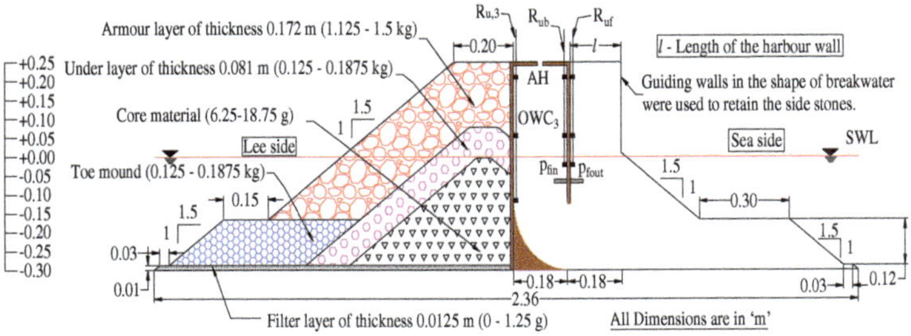

(b)

Fig. 7.29. (a) Schematic plan view of the experimental setup in the shallow-wave basin. (b) Cross-sectional view of OWCBW system with guiding walls on its adjacent side with measurement details.

Source: Ashlin *et al.* (2018).

For regular waves, the hydrodynamic performance of the array of OWC integrated with breakwaters can be estimated by the following equations:

$$\mathrm{RCW}_{i_{\mathrm{reg}}} = \frac{P_{\mathrm{out}_i}}{P_w * w}. \tag{7.12a}$$

$$P_w = \frac{\rho g H^2}{8}\frac{C}{2}\left(1 + \frac{2kd}{\sinh 2kd}\right). \tag{7.12b}$$

$$P_{\mathrm{out}} = \frac{1}{T}\int_t^{t+T} P_{\mathrm{air}} A V_w dt. \tag{7.12c}$$

$$V_w = \frac{2z_{j+1} + 3j - 6z_{j-1} + z_{j-2}}{6\Delta t}. \tag{7.12d}$$

$$\beta_{i_{\mathrm{reg}}} = \frac{R_{u,i_{\mathrm{max}}}}{H}. \tag{7.12e}$$

$$\mathrm{PR}_{\mathrm{air},i_{\mathrm{reg}}} = \frac{P_{\mathrm{air},i_{\mathrm{max}}}}{\rho g H}. \tag{7.12f}$$

Notations for Equations (7.12) are listed in the following:

$(\mathrm{RCW}_i)_{\mathrm{reg}}$	Relative capture width
P_w	Incident wave power [W]
$(P_{\mathrm{out}})_i$	Pneumatic output power [W]
$(\beta_i)_{\mathrm{reg}}$	Wave amplification factor
$(R_{u,i})_{\mathrm{max}}$	Maximum free surface oscillation inside the chamber [m]
$(\mathrm{PR}_{\mathrm{air},i})_{\mathrm{reg}}$	Non-dimensionalized air pressure inside the chamber
$(\mathrm{P}_{\mathrm{air},i})_{\mathrm{max}}$	Maximum air pressure inside the chamber [Pa]
i	denotes respective
ρ	OWC number density of the fluid [kg/m^3]
g	Acceleration due to gravity [m^2/s]
H	Incident wave height [m]
C	Wave celerity [m/s]
k	Wave number
d	Water depth [m]
P_{air}	Chamber air pressure [Pa]
A	The roof area of chamber [m^2]
V_w	The velocity of free surface elevation inside the chamber [m/s]
T	Wave period [s]
z_i	Free surface elevation inside the chamber at a different time step, j [m]
w	Width of the OWC model

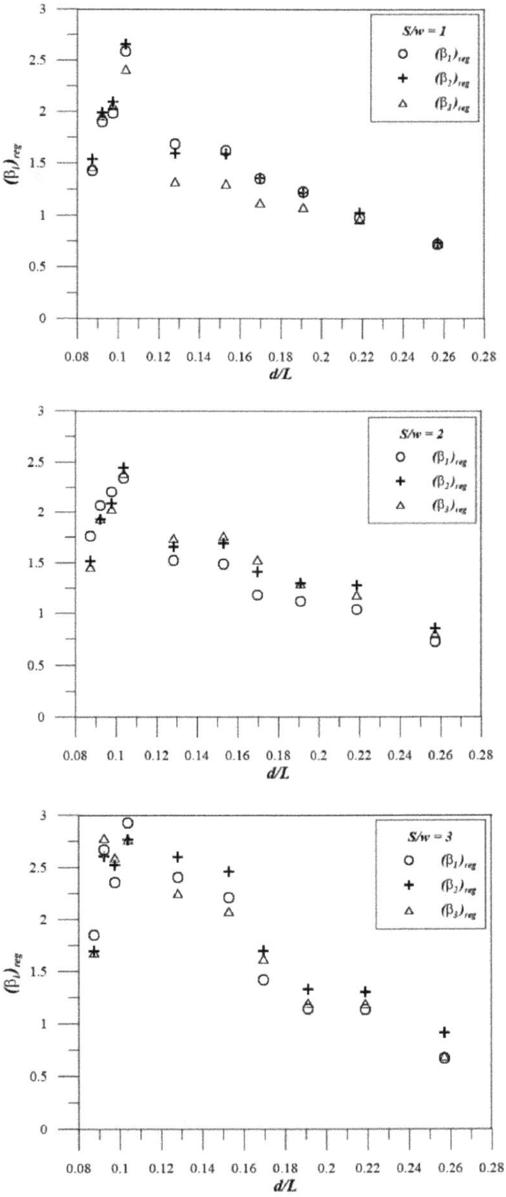

(a)

Fig. 7.30. The interaction among OWC models in an array influences the regularization parameters of the first three OWC devices: (a) $(\beta_i)_{reg}$, (b) $(PRair, i)_{reg}$, and (c) $(RCWi)_{reg}$ corresponds to d/L for various spacings ($H = 0.03$ m).

Source: Ashlin *et al.* (2019).

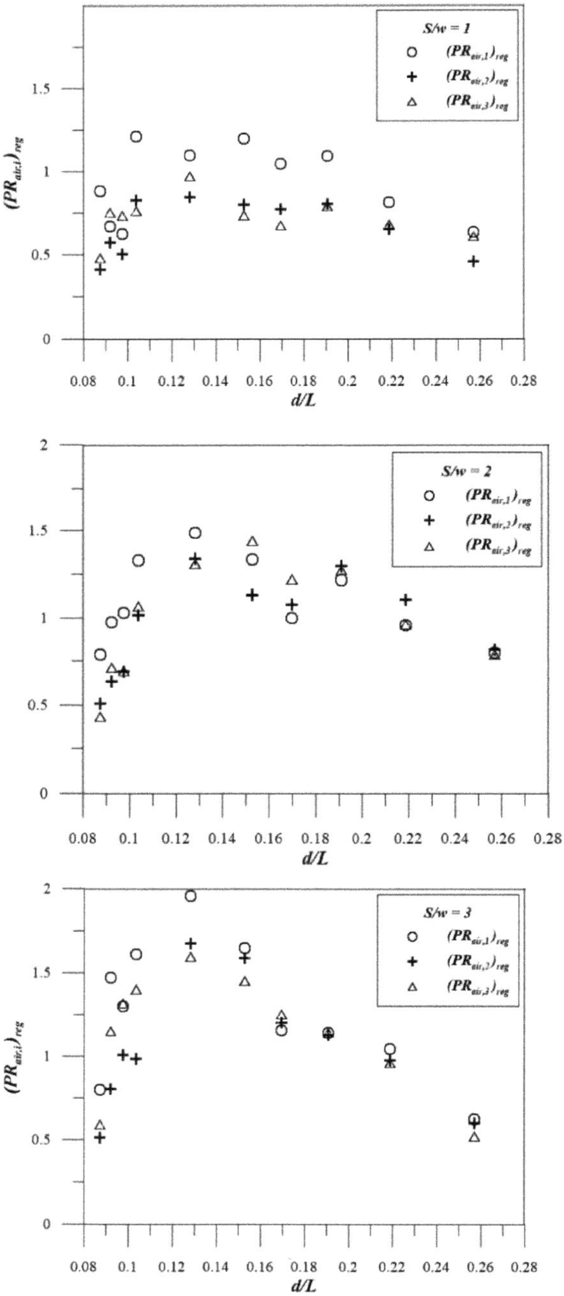

(b)

Fig. 7.30. *(Continued)*

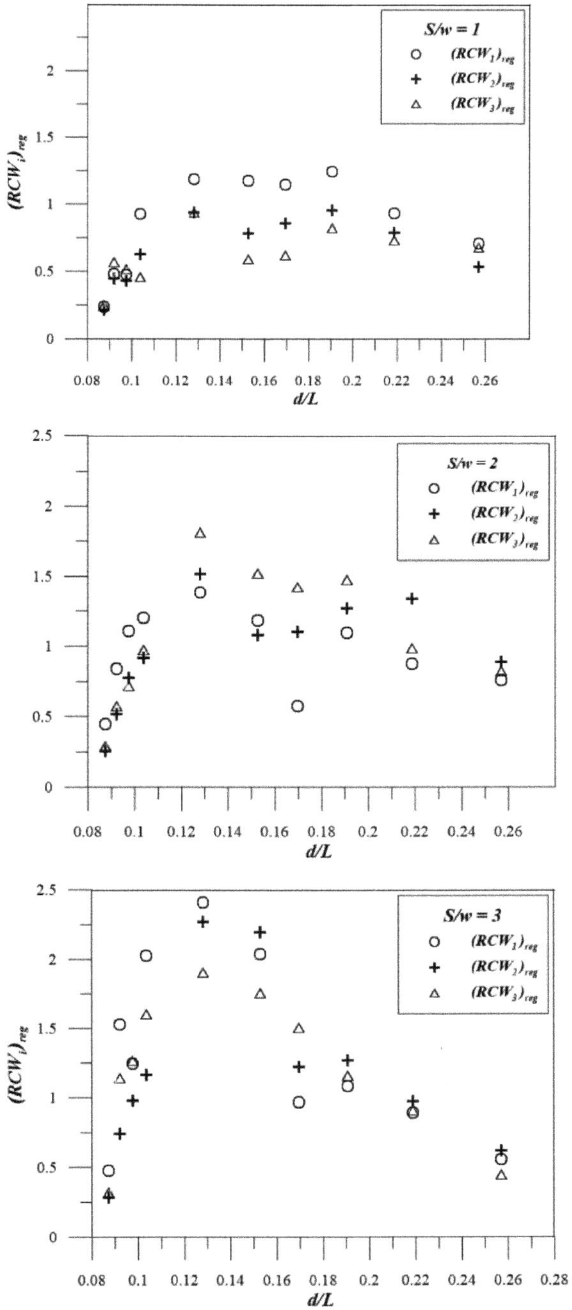

(c)

Fig. 7.30. *(Continued)*

The regular wave tests (Fig. 7.30) show that the maximum values of β and PR_{air} close to 3 occur for a spacing of three times the width, while a maximum RCW of 2.8 is seen. Extensive studies were also conducted on random waves, showing that a 17–22% reduction in average hydrodynamic performance was observed compared to regular waves. Thus, it is clear that due to three-dimensional effects, the convergence of water waves concentrates wave energy in front of the array of OWC chambers, increasing the absorbed power to a maximum of 2.2 times the input wave power. This demonstrates that the OWC system may absorb more power when placed in a network of devices than when installed in a single device due to energy concentration.

7.5.2.2 *Precast type*

Considering the installation cost of individual OWC devices with the benefit derived from the extraction of energy from waves, researchers in the recent past found that the OWC with integrated applications could be more beneficial. Ashlin *et al.* (2018) concluded that the hydrodynamic performance of OWC increases when integrated with detached offshore breakwaters. Although the integrated system is demonstrated to be more advantageous than an isolated OWC, there are inherent problems such as rolling down of armour stones during extreme events. This has led to the idea of considering precast elements of semi-circular breakwaters embedded with OWC. These can be a mound/solid structure or a hollow structure constructed with precast concrete materials. These can be installed onshore and as an offshore detached breakwater. Dhinakaran *et al.* (2010) explained that the hydrodynamic performance of semi-circular breakwater is more effective than rubble mound breakwaters in terms of wave reflection, stability of the structure and hydrodynamic characteristics.

Socrates *et al.* (2023) considered three different configurations of OWCs: forward duct at the bottom of OWC named Forward Duct OWC (FD-OWC), a curved wall at the front side of OWC with a radius equal to the height of OWC called Curved Front Lip Wall OWC (CFLW-OWC), and a hybrid of the curved and vertical wall at the front portion of OWC named Curved Vertical Front Lip Wall (CVFLW-OWC) as shown in Fig. 7.31.

The following equations (7.13) are required to estimate the hydrodynamic performance of the OWC:

$$\lambda_{\text{eff}} = \frac{P_{\text{out}}}{P_{\text{in}}}. \tag{7.13a}$$

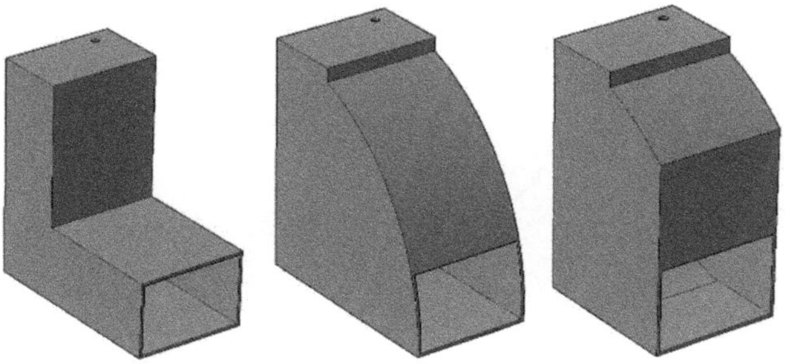

Fig. 7.31. (a)–(c) Different configurations of OWCs.
Source: Ashlin *et al.* (2018).

$$P_{\text{in}} = \frac{\rho g H^2}{8}\frac{C}{2}w\left(1 + \frac{2kd}{\sinh 2kd}\right). \tag{7.13b}$$

$$P_{\text{out}} = \frac{1}{T}\int_t^{t+T} P_{\text{air}} A V_w dt. \tag{7.13c}$$

$$\frac{2z_{i+1} + 3z_i - 6z_{i-1} + z_{i-2}}{6\Delta t}. \tag{7.13d}$$

$$\beta = \frac{\xi_{\max}}{H}. \tag{7.13e}$$

$$\text{PR}_{\text{air}} = \frac{P_{\text{air}_{\max}}}{\rho g H}. \tag{7.13f}$$

Notations:

P_{in}	Incident wave power [W]
P_{out}	Output power [W]
ρ	Density of the fluid [kg/m^3]
g	Acceleration due to gravity [m^2/s]
H	Incident wave height [m]
C	Wave celerity [m/s]
w	Width of OWC [m]
k	Wave number
d	Water depth [m]
P_{air}	Chamber air pressure [Pa]

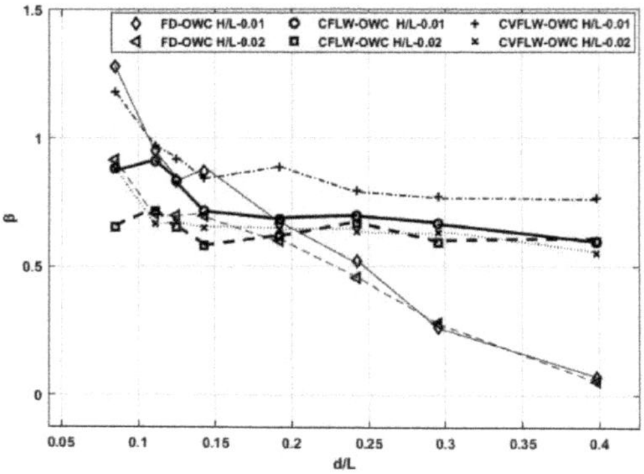

(a)

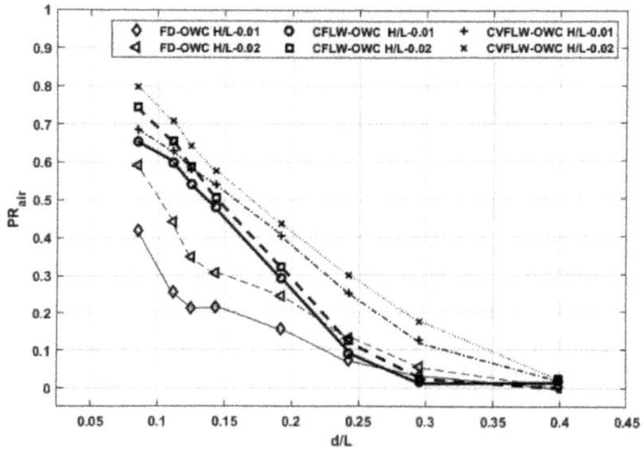

(b)

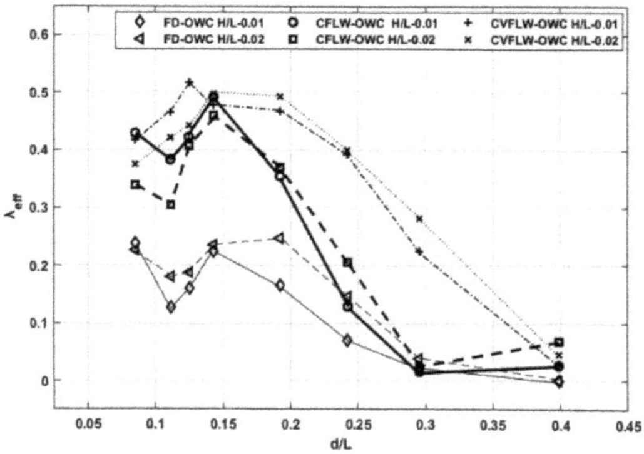

(c)

Fig. 7.32. (a) Variation of β with d/L. (b) Variation of PR_{air} with d/L. (c) Variation of λ_{eff} with d/L.
Source: Socrates *et al.* (2023).

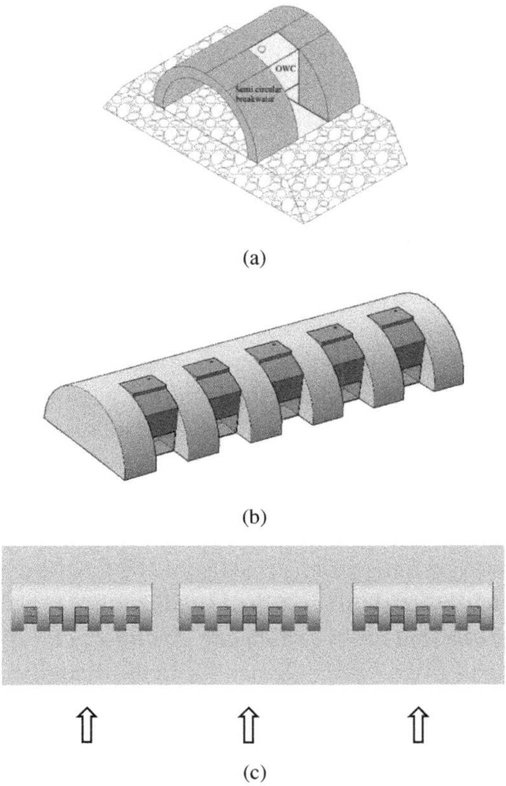

Fig. 7.33. (a) Typical semi-circular breakwater with OWC, (b) array of CVFLW-OWC integrated with semi-circular breakwater and (c) layout of the detached semi-circular breakwater.
Source: Ashlin *et al.* (2019).

A	Roof area of chamber [m^2]
V_w	The velocity of free surface elevation inside the chamber [m/s]
T	Wave period [s]
z_i	Free surface elevation inside the chamber [m]
ξ	Free surface oscillation inside the chamber [m]
β	Wave amplification factor
$\mathrm{PR}_{\mathrm{air}}$	Dimensionless air pressure inside the chamber
λ_{eff}	Hydrodynamic efficiency

The hydrodynamic performance of three isolated OWCs under the regular wave action was explored through numerical study by Socrates *et al.*

(2023). The variation of β, PR_{air} and λ_{eff} with d/L is shown in Figs. 7.32(a), 7.32(b) and 7.32(c), respectively. The β and PR_{air} increase with a decrease in d/L, which denotes that for short-period waves, the wave amplification and air pressure reduce and increase when exposed to long-period waves. This is because the fluid pressure inside and outside the chamber is high for long-period waves. CVFLW-OWC exhibits the maximum results of β of 1.2 and PR_{air} of 0.8. The λ_{eff} increases with a decrease in d/L, reaches the maximum at d/L of 0.125 and decreases further. This explains that the hydrodynamic efficiency increases from short-period waves to intermediate waves and then decreases once it reaches the resonance conditions. Maximum hydrodynamic efficiency of 52% was observed in CVFLW-OWC for low steep waves. As more wave energy dissipation was observed in FD-OWC due to ducts toward the seaward side, it yields the least hydrodynamic performance. Hence, this device is preferred for further study of its integration with semi-circular breakwaters.

There are few studies on the integration of OWC with semi-circular breakwaters. However, integrating an array of OWC within a single offshore breakwater among the offshore detached breakwaters to serve as a WEC as well as coastal protection looks promising. Ashlin *et al.* (2019) proved that optimum spacing of OWC will enhance its hydrodynamic performance. A schematic representation of an offshore detached semi-circular breakwater layout with an array of OWC is shown in Fig. 7.33. Hence, in the future, numerical and experimental studies should be conducted on OWC with semi-circular breakwaters with different parametric and wave conditions.

References

Samad, A., Sundar, V., Sannasiraj, S.A. and Halder, P. (2021). *Ocean Wave Energy Systems-Hydrodynamic Analysis, Power Take Off and Control Systems.* Springer, Cham, Switzerland.

Ackers, P. and White, W.R. (1973). Sediment transport: New approach and analysis. *Journal of the Hydraulics Division* (ASCE), HY11, 204–254.

Anand, K.V., Sundar, V. and Sannasiraj, S. A. (Nov, 2011). Hydrodynamic Characteristics of Curved Front Seawall Models Compared with Vertical Seawall under Regular Waves. *Journal of Coastal Research*, 27(6), pp 1103–1112, https://doi.org/10.2112/jcoastres-d-09-00169.1.

Andiroglu, E. and Suraneni, P. (2021). Multifunctional performance of coastal structures based on south florida coastal environs. *Journal of Coastal Research*, 1–40. DOI: 10.2112/JCOASTRES-D-20-00068.1.

Ashlin, S.J., Sundar, V. and Sannasiraj, S.A. (2016). Effects of bottom profile of an oscillating water column device on its hydrodynamic characteristics. *Renewable Energy*, 96, 341–353. https://doi.org/10.1016/j.renene.2016.04.091.

Ashlin, S.J., Sundar, V. and Sannasiraj, S.A. (2017). Pressures and forces on an oscillating water column type wave energy caisson breakwater. *Journal of Waterway, Port, Coastal, and Ocean Engineering*, 143, 1–18. DOI: 10.1061/(ASCE)WW.1943-5460.0000405.

Ashlin, S.J., Sannasiraj, S.A. and Sundar, V. (2018). Performance of an array of oscillating water column devices integrated with offshore detached breakwater. *Ocean Engineering Journal,* 163, 518–532. DOI: 10.1016/j.oceaneng.2018.05.043.

Ashlin S.J. (2019). Hydrodynamics of an oscillating water column device integrated with breakwaters, Ph.D thesis, Department of ocean engineering, IIT Madras, INDIA

Berkhoff, J.C.W. (1972). Computation of combined refraction — diffraction. *Coastal Engineering,* 471–490.

Bilskie, M.V., Hagen, S.C., Medeiros, S.C. and Passeri, D.L. (2014). Dynamics of sea level rise and coastal flooding on a changing landscape. *Geophysical Research Letters*, 41, 927–934.

Blacka, M.J., Shand, T.D., Carley, J.T. and Mariani, A. (2013). A review of artificial reefs for coastal protection in NSW. WRL Technical Report 2012/8, Watre Research Laboratory, University of New South Wales, June.

Brand, F.S. and Jax, K. (2007). Focusing the meaning(s) of resilience: Resilience as a descriptive concept and a boundary object. *Ecology and Society*, 12, 16 p.

Britannica. (2023). The Editors of Encyclopaedia (2023). "Atlantic City". Encyclopaedia Britannica, 28 June 2023. https://www.britannica.com/place/Atlantic-City-New-Jersey (Accessed 30 June 2023).

Brody, S.D., Highfield, W.E. and Kang, J.E. (2011). *Rising Waters: The Causes and Consequences of Flooding in the United States.* Cambridge; New York: Cambridge University Press.

Brossard, J., Druaux, A., Marin, F. and Aoul, E.H. (2003). Fixed absorbing semi-immersed breakwater. *Coastal Engineering*, 49, 25–41.

Cengiz, C., Cengiz, B. and Smardon, R.C.A. (2022). Bridge between coastal resilience and tourism-recreation: Multifunctional benefit of boardwalk design for sustainable development in the Western Black Sea Region. Turkey. *Water*, 14, 1434. https://doi.org/10.3390/w14091434.

CIRIA/CUR/CEFMEF. (2007). *The Rock Manual. The Use of Rock in Hydraulic Engineering*, 2nd edn., p. 1200.

Coghlan, I.R., Carley, J.T., Cox, R.J., Blacka, M.J., Mariani, A., Restall, S., Hornsey, W., and Sheldrick, S. (2009). Two-dimensional physical modelling of sand filled Geocontainers for coastal protection. In *Proceedings of Australasian Coasts and Ports Conference*, Wellington, pp. 295–301. https://search.informit.org/doi/10.3316/informit.862603013166460.

Daemen, I.E.R. (1991). Wave Transmission at Low Crested Breakwaters. Delf University of Technology, Delf, The Netherlands.

Daemrich, K. and Kahle, W. (1985). Schutzwirkung von Unterwasser Wellen brechern unter dem Einfluss unregelmässiger Seegangswellen. Eigenverslag des Franzius-Instituts fur Wasserbau und Küsteningenieurswesen.

Daniel Raj, D., Sundar, V. and Sannasiraj, S.A. (2018). Effect of harbour walls in an oscillating water column on its efficiency. *Journal of Waterway Port Coastal and Ocean Engineering* (ASCE), 144, 114(2), 1–16. Paper No: 04017043. https://doi.org/10.1061/(ASCE)WW.1943-5460.0000429.

Dassanayake, D.T. and Oumeraci, H. (2012). Engineering properties of geotextile sand containers and their effect on hydraulic stability and damage development of low-crested/submerged structures. *The International Journal of Ocean and Climate Systems*, 3(3), 135–150. DOI: 10.1260/1759-3131.3.3.135.

De Chowdhury, S., Anand, K.V., Sannasiraj, S.A. and Sundar, V. (2017). Non-linear wave interaction with curved front seawalls. *Ocean Engineering* 140, 84–96. https://doi.org/10.1016/j.oceaneng.2017.05.015.

Demirbilek, Z. and Panchang, V. (1998). CGWAVE: A Coastal Surface Water Wave Model of the Mild Slope Equation. Technical Report CHL-98-26.

Dhinakaran, G. (2002). Experimental studies on hydrodynamic characteristics of perforated semi-circular breakwaters. Ph.D Thesis, Indian Institute of Technology Madras, India.

Dhinakaran, G., Sundar, V., Sundaravadivelu, R. and Graw, K.U. (2010). Regular wave measurements on a submerged semi-circular breakwater. *Journal of Offshore Mechanics and Arctic Engineering*, 132(3), 034501-1–034501-6.

Dhinakaran, G., Sundar, V., Sundaravadivelu, R. and Graw, K.U. (2012). Performance of a perforated submerged semi-circular breakwater due to non-breaking waves. *Proceedings of the Institution of Mechanical Engineers Part M: Journal of Engineering for the Maritime Environment*, 226(1), 36–50.

Goda, Y. (1985). *Random Seas and Design of Maritime Structures.* University of Tokyo Press, Tokyo, Japan.

Harris, L.E. (2009). Artificial Reefs for Ecosystem Restoration and Coastal Erosion Protection with Aquaculture and Recreational Amenities.

Hornsey, W.P., Carley, J.T., Coghlan, I.R. and Cox, R.J. (2011). Geotextile sand container shoreline protection systems: Design and application, Geotext. *Geomembranes*, 29, 425–439. https://doi.org/10.1016/j.geotexmem.2011.01.009.

Hudson, R.Y. (1956). Laboratory investigation of rubble-mound breakwaters. *Journal of the Waterways and Harbors Division*, 85(3), 93–118. https://doi.org/10.1061/JWHEAU.0000142.

Jackson, L.A., Tomlinson, R., Corbett, B. and Strauss, D. (2012). Long-term performance of a submerged coastal control structure: A case study of the Narrow neck multi-functional artificial reef. *Coastal Engineering Proceedings,* 1, 54.

Jayanthi, M., Patterson Edward, J.K., Malleshappa, H., Gladwin Gnana Asir, N., Mathews, G., Diraviya Raj, K. and Sannasiraj, S.A. (2020). Perforated trapezoidal artificial reefs can augment the benefits of restoration of an island and its marine ecosystem. *Restoration Ecology*, 28(1), 233–243.

Jemi Jeya, T.J., Sriram, V. and Sundar, V. (2021). Hydrodynamic characteristics of vertical and quadrant face pile supported breakwater under oblique waves. *Journal of Engineering for the Maritime Environmentin.* https://doi.org/10.1177/14750902211031353.

Kamikubo, Y., Murakami, K., Irie, I. and Hamasaki, Y. (2000). Study on practical application of a non-wave overtopping type seawall. In *International Conference on Coastal Engineering, Proceedings of the Twenty-seventh International Conference*, Sydney, Australia, Vol. 126, pp. 2215–2228.

Karthik Ramnarayan. (2021). Hydrodynamic characteristics of curved front pile-supported breakwaters. Ph.D thesis, Indian Institute of Technology Madras, December.

Karthik Ramnarayan, S., Sundar, V. and Sannasiraj, S.A. (2020a). Hydrodynamic characteristics of concave front pile-supported breakwaters with a tubular wave screen. *Journal of Waterways, Oceans, Ports and Coasts*, 148, 1. https://ascelibrary.org/doi/10.1061/%28ASCE%29WW.1943-5460.000.

Karthik Ramnarayan, S., Sannasiraj, S.A. and Sundar, V. (2020b). Hydrodynamic characteristics of curved and vertical front face pile-supported breakwaters in regular waves. *Ocean Engineering,* 216, 108105, 1–11.

Karthik Ramnarayan, S., Sannasiraj, S.A. and Sundar, V. (2021). Hydrodynamic performance characteristics of an oscillating water column device integrated with a pile breakwater. *Journal of Ocean Engineering and Marine Energy* (Springer), 7, 229–241. https://doi.org/10.1007/s40722-021-00199-7.

Koraim, A.S., Heikal, E.M. and Rageh, O.S. (2011) Hydrodynamic characteristics of double permeable breakwater under regular waves. *Marine structures*, 24, 503–527.

Koraim, A.S., Iskander, M.M. and Elsayed, W.R. (2014). Hydrodynamic performance of double rows of piles suspending horizontal c shaped bars. *Coastal Engineering*, 81, 81–96.

Lenze, B., Heerten, G., Saathoff, F. and Stelljes, K. (2002). Geotextile sand containers — Successful solutions against beach erosion at sandy coasts and scour problems under hydrodynamic loads. In *Proceedings of the Sixth International Conference Littoral 2002, The Changing Coast.* EUROCOAST/EUCC, Porto, Portugal, pp. 375–381.

Liu, H., Ghidaoui, M.S., Huang, Z., Yuanb, Z. and Wang, J. (2011). Numerical investigation of the interactions between solitary waves and pile breakwaters using BGK-based methods. *Computers and Mathematics with Applications*, 61, 3668–3677.

Lokesha. (2019). Hydrodynamic characteristics of perforated and non-perforated submerged structures. Ph.D. Thesis, March, Department of Ocean Engineering, Indian Institute of Technology Madras, India.

Lokesha, Sundar, V. and Sannasiraj, S.A. (2013). Artificial reefs: A review. *International Journal of Ocean and Climate Systems (IJOS)*, 4(2), 117–124. https://doi.org/10.1260/1759-3131.4.2.117.

Mori, E., D'eliso, C. and Aminti, P.L. (2008). Physical modelling on geotextile sand container used for submerged breakwater. In *Proceedings of 2nd International Conference on the Application of Physical Modelling to Port and Coastal Protection*, Coastlab08, Bari, Italy.

Morris, R.L., Fest, B., Stokes, D., Jenkins, C. and Swearer, S.E. (2023). The coastal protection and blue carbon benefits of hybrid mangrove living shorelines. *Journal of Environmental Management*, 331, 117310.

Nakamura, T., Kohno, T. and Morita, Y. (2001). Performance of a double-walled barrier with a front wall of inclined plate array. In *Proceedings of 11th International Offshore and Polar Engineering Conference*, Stavanger, Norway, June, pp. 499–505.

Neelamani, S. and Rajendran, R. (2002). Wave interaction with T-type breakwaters. *Ocean Engineering*, 29, 151–175.

Neelamani, S. and Vedagiri, M. (2002). Wave interaction with partially immersed twin vertical barriers. *Ocean Engineering*, 29, 215–238.

Nguyen, H.H. and Duong, C.M. (2019). Stability analysis of a double-row pile breakwater subjected to combined loading in mekong delta. In *Proceedings of 10th International Conference on Asian and Pacific Coasts (APAC 2019)*, Hanoi, Vietnam, September 25–28, 2019APAC.

Oumeraci, H., Bleck, M., Hinz, M. and Kübler, S. (2002a). Large-scale investigations into the hydraulic stability of geotextile sand containers under wave loading. Leichweiss-Institute for Hydraulic Engineering and Water Resources, Report Nr. 878, Braunschweig, Germany (in German).

Oumeraci, H., Bleck, M., Hinz, M. and Möller, J. (2002b). Theoretische Untersuchungen zur Anwendung geotextiler Sandcontainer im Küstenschutz ("Theoretical studies on the use of geotextile sand containers in coastal protection"). Leichweiss-Institute for Hydraulic Engineering and Water Resources, Report Nr. 866, Braunschweig, Germany (in German).

Oumeraci, H., Hinz, M., Bleck, M. and Kortenhaus, A. (2003). Sand-filled geotextile containers for shore protection. In *Proceedings of 6th Conference of Coastal and Port Engineering in Developing Countries*. COPEDEC VI, Colombo, Sri Lanka, pp. 1–15.

Panin, N. and Nicolaev, S. (2002). Coastal erosion in the Black Sea. In *Sustainable Coastal Management: A Transatlantic and Euro-Mediterranean Perspective*. Springer, Dordrecht, The Netherlands, pp. 241–251.

Patterson, E.J.K., Sannasiraj, S.A., Malleshappa, H., Sundar, V. and Deepak, S.B. (2016). Integrated approach to protect and restore sinking Vaan Island in Gulf of Mannar, India. In *3rd International Conference on Coastal Zone Engineering and Management in Middle East*, pp. 81–83.

Pilarczyk, K.W. (2003). Alternative systems for coastal protection- An overview. In *International Conference on Estuaries and Coasts*, Hangzhou, China, November 9–11.

Powell, K.A. and Allsop, N.W.H. (1985). Low-Crest Breakwater, Hydraulic Performance and Stability.

Pradjoko, E., Bachtiarb, I., Matalattac, N. and Sugihartonoc, G. (2015). The submerged breakwater as prototype of coastal protection in Gili Trawangan, Lombok, Indonesia. In *Proceedings of 5th International Conference of Euro Asia Civil Engineering Forum (EACEF-5), Procedia Engineering*, 125, 284–290.

Rageh, O.S. and Koraim, A.S. (2010). Hydraulic performance of vertical walls with horizontal slots used as breakwater. *Coastal Engineering*, 57, 745–756.

Rageh, O.S., Koraim, A.S. and Salem, T.N. (2009). Hydrodynamic efficiency of partially immersed caissons supported on piles. *Ocean Engineering*, 36, 1112–1118.

Rahman, M.A. and Aysha, A. (2014). The effect of porosity of submerged and emerged breakwater on wave transmission. *International Journal of Environmental Science and Development*, 5(5), 473–478.

Ranasinghe, R., Hacking, N. and Evans, P. (2001). Multi-functional artificial surf breaks: A review. NSW Department of Land and Water Conservation. www.asrltd.co.nz.

Recio, J. (2007). Hydraulic stability of geotextile sand containers for coastal structures-effect of deformations and stability formulae. PhD thesis, Leichweiss-Institute for Hydraulic Engineering Water Resources, Braunschweig, Germany, pp. 167–174. DOI: 10.24355/dbbs.084-20080205 0100-0.

Recio, J., Oumeraci, H. and Mocke, G. (2010). Stability formula and numerical model for structures made with geotextile sand containers used for coastal stabilization. In *2nd International Conference on Coastal Zone Engineering and Management (Arabian Coast 2010)*, Muscat, Oman.

Schoonees, T., Gijón Mancheño, A., Scheres, B., Bouma, T.J., Silva, R., Schlurmann, T. and Schüttrumpf, H. (2019). Hard structures for coastal protection, towards greener designs. *Estuaries Coasts*, 42, 1709–1729.

Sasajima, H., Koizuka, T. and Sasayama, H. (1994). Field demonstration test of a semi-circular breakwater. In *International Conference on Hydro-Technical Engineering for Port and Harbor Construction*, Yokosuka, Japan, 19–21 October 1994, pp. 593–615.

Seelig, W.N. (1980). Two Dimensional Tests of Wave Transmission and Reflection Characteristics of Laboratory Breakwaters. US Army Crops of Engineers, Coastal Engineering Research Center.

Shin, E.C. and Oh, Y.I. (2007). Coastal erosion prevention by geotextile tube technology. *Geotextiles and Geomembranes*, 25(4–5), 264–277. https://doi.org/10.1016/j.geotexmem.2007.02.003.

Socrates, S.S., Sriram, V. and Sundar, V. (2023). Hydrodynamic performance of three different oscillating water column devices in regular waves. In *Proceedings of the 42nd International Conference on Ocean, Offshore, and Arctic Engineering (OMAE)*, June 11–16, Melbourne, Australia.

Suh, K.D., Shin, S. and Cox, D.T. (2006). Hydrodynamic characteristics of pile-supported vertical wall breakwaters. *Journal of Waterway, Port, Coastal, and Ocean Engineering*, 132(2), 83–96.

Sukop, M.C., Rogers, M., Guannel, G., Infanti, J.M. and Hagemann, K. (2018). High temporal resolution modelling of the impact of rain, tides, and sea level rise on water table flooding in the Arch Creek basin, Miami-Dade 666 County Florida USA. *Science of The Total Environment*, 616–617, 1668–1688.

Sundar, V. and SubbaRao, B.V.V. (2003). Hydrodynamic characteristics of quadrant face pile supported breakwater. *Journal of Waterway, Port, Coastal and Ocean Engineering* (ASCE), 129(1), 22–33.

Sundar, V. and Sundaravadivelu, R. (2007). Role of groyne field for coastal protection considering socio-economic aspects. In *Proceedings of 32nd International Association Hydraulic Engineering Research*, Venice, July 1–7.

Sundar, V., Moan, T. and Hals, J. (2010). Conceptual design of OWC wave energy converters combined with breakwater structures. *International Conference on Offshore Mechanics and Arctic Engineering*, Vol. 49118.

Sundar, V. and Sannasiraj, S.A. (2019). *Coastal Engineering-Theory and Practice.* Advanced Series on Ocean Engineering, Vol. 47, April 2019. https://www.worldscientific.com/worldscibooks/10.1142/11148. worldscientific.com.

Sundar, V., Sannasiraj, S.A., Sukanya, R. and Maiti, D.P. (2022). Submerged geosynthetic reef as shore protection measure for Islands. *Journal of Marine Science and Application*, 21, 128–139. https://doi.org/10.1007/s11804-022-00256-z.

Takahashi, S. (1994). Lecture Notes on Design of Breakwaters. Port and Harbour Research Institute, Japan International Cooperation Agency.

Teh, H.M. and Venugopal, V. (2013). Performance evaluation of a semi-circular breakwater with truncated wave screens. *Ocean Engineering*, 70, 160–176.

Van der Meer, J.W. (1988). Rock slopes and gravel beaches under wave attack. PhD Thesis, Delft University of Technology, Delft, (publication no. 396).

Van der Meer, J.W. (1990). Low Crested and Reef Breakwaters, Delft Hydraulics, Report H 986, The Netherlands.

Van der Meer, J.W. and d'Angremond, D.K. (1991). Wave transmission at low crested structures. *Coastal Structures and Breakwaters.* ICE, London, UK, pp. 25–42.

Vasanthakumar, S., Sannasiraj, S.A. and Sundar, V. (2023). Wave forces and moments on concave and vertical front caissons supported on piles. *Ships and Offshore Structures Journal.* https://doi.org/10.1080/17445302.2023.2233852.

Voorde, M.T., Do Carmo, J.S.A. and Da Graça Neves, M. (2009). Multi-functional artificial reefs for coastal protection. In Hudspeth, C.A. and Reeve, T.E. (eds.) *Agricultural Runoff, Coastal Engineering and Flooding.* Nova Science Publishers, Inc., Hauppauge, New York, USA.

Wang, Y.X., Wang, G.Y. and Li, G.W. (2006). Experimental study on the performance of the multiple-layer breakwater. *Ocean Engineering*, 33(13), 1829–1839.

Wang, G., Ren, B. and Wang, Y. (2016). Experimental study on hydrodynamic performance of arc plate breakwater. *Ocean Engineering*, 111, 593–601.

Womera, S.A. (2011). Experimental and Numerical Investigation on Wave Interaction with Submerged Breakwater. Bangladesh University of Engineering and Technology, Dhaka.

Wouters, J. (1998). Open Taludbekleidingen; Stabiliteit van Geosystems (Stability of Geosystems). Delft Hydraulics Report, No. H1930,Delft, The Netherlands.

Yoshioka, K., Kawakami, T., Tanaka, S., Koarai, M. and Uda, T. (1993). Design manual for artificial reefs, in coastlines of Japan I. In Nagao, Y. (ed.). *Proceedings of Coastlines of Japan*, pp. 93–107.

Chapter 8

Numerical Modelling Concepts

8.1 Introduction

This chapter deals with the diverse array of modelling tools and approximations employed in the realm of wave–structure interaction problems. The focus primarily centers on coastal engineering, emphasizing the critical importance of considering both large-scale and local-scale temporal and spatial factors. This involves addressing broad, long-term patterns and trends in coastal processes, as well as the specific, immediate impacts and dynamics at smaller, localized sites. Such a comprehensive approach ensures the effective design and management of coastal structures, erosion control measures, and habitat restoration projects. The different modelling approaches characterized by their varying levels of approximation or theoretical underpinnings, accompanied by elucidations of their underlying assumptions, are explored. Such exposition equips readers with a foundational grasp of theory while illuminating the nuanced constraints inherent in numerical tools, thereby fostering an informed interpretation of results. To facilitate comprehension, examples illustrating the application of each tool are furnished. Moreover, a burgeoning trend in recent years has been observed, wherein standalone tools are increasingly coupled to address diverse physical processes. This emergent practice is scrutinized comprehensively following our exposition on individual modelling tools, elucidating the evolution and synergies inherent in these modelling endeavours.

In contemporary times, numerical models are increasingly accessible in both open-source and commercial variants, employing diverse numerical methods that offer varying degrees of approximation in grid resolutions, stability, accuracy and computational efficiency. Recognizing this diversity, recent efforts within the numerical modelling community have been directed towards comparative analyses and benchmarking exercises, a topic explored in the present chapter. This enables readers to evaluate their own tools against established benchmarks, leveraging available theories and open-source experimental data, which are also catalogued herein.

In addition to providing an overview of existing tools, this chapter offers a systematic classification of models based on their applicability range, computational aspects and physical processes, alongside a comprehensive survey of the historical evolution of these developments aimed at beginners. It is important to acknowledge that while this chapter offers a broad understanding, the intricacies of individual models extend beyond its scope, and interested readers are encouraged to consult the corresponding literature for further details.

8.2 Different Levels of Approximations

A single numerical tool to address all classes of problems in coastal engineering is ideal; however, it is not possible due to the following reasons:

- a large sea area, having both a large spatial domain and time scale.
- highly nonlinear wave–structure interaction processes that are not only fluid but with the structure sometimes also behaving nonlinearly, such as vegetations, fenders or hydro-elasticity.
- waves coexist with nonlinear currents of various levels, sediment movements and other phenomena.
- viscosity, surface tension and turbulence.
- two-phase (air–sea) or multiphase (air–sea–oil or air–sea–sediment) process.
- violent wave impacts (during cyclonic storm surges, flooding) and aeration on rubble mound structures.

For the above-mentioned phenomena, one needs a large spatial/time scale to capture the wave propagation phenomenon as well as a small spatial/time scale to understand the wave–structure–soil interaction process. Thus, a single mathematical model may not always be a solution to this complex problem. Hence, researchers have developed various levels of approximations in mathematical modelling.

The levels of approximations in the mathematical modelling are based on two principles: (a) the physical process that is governing the problem at hand and (b) reducing the computational aspect in a numerical algorithm for industrial/practical needs. Now, let us look at the first principles, the process and the application in Table 8.1, which provide some applications in coastal engineering that require either large-domain or small-domain modelling. A typical example of large-scale modeling is wave propagation from offshore to nearshore and its interactions with harbor structures to understand its tranquillity, run-up, or inundations. On the other hand, local-scale modeling might involve studying the sediment transport and deposition around a specific coastal construction site, such as a seawall

Table 8.1. Physical processes required to be modelled for the various coastal engineering applications.

	Application				
	Large-domain modelling		Local/small-domain modelling		
Physical Process	Current/flow interactions with structure	Wave propagation & interactions	Steep wave and rigid structure	Extreme waves and rigid structure	Wave–soil–structure interactions
Surface tension	Not Required	Not Required	Not Required	Not Required	Not Required
Viscosity and turbulence	May or May not	May or May not	May or May not	May or May not	May or May not
Nonlinearity in fluid phase	Required	Required	Required	Required	Required
Nonlinearity in structure	Not Required	Not Required	Not Required	Not Required	Not Required
Two-phase modelling (air–water)	Not Required	Not Required	Not Required	Required	Not Required

or breakwater, to optimize its design and minimize localized erosion. For these applications, physical process such as surface tension, nonlinearity in structure is not important. Further, complete physical modelling of the air–sea process is not required; some empirical treatment would suffice. Thus, a full continuity and momentum equation can be simplified based on this approximation. Similarly, consider an application of small-domain modelling, wherein one is interested in quantifying the forces experienced by the coastal structures (such as caisson, seawall, rubble mound or scour) against operating or extreme sea state conditions. In this scenario, normally, researchers will carry out the physical model studies in experimental wave tanks. A similar process can be done using numerical modelling in a numerical wave tank. A numerical wave tank is a numerical tool that can reproduce the experimental facility as closely as possible. Thus, a detailed physical flow process is required by solving the continuity and momentum equations, only slightly reducing the physical approximations to be as close as possible reality. For example, in coastal engineering, surface tension, nonlinearity in structure, sediment-to-sediment interactions or rigid body (rubble mound) interactions can be relaxed.

So, for large-scale problems, one can use different approximations based on the wave characteristics and their applications compared to small-scale physical processes, thus leading to various levels of approximations and computational efficiency. This is further emphasized in the bubble plot showing the various environmental aspects in spatial and time scale for mathematical modelling efforts in coastal engineering, as shown in Fig. 8.1.

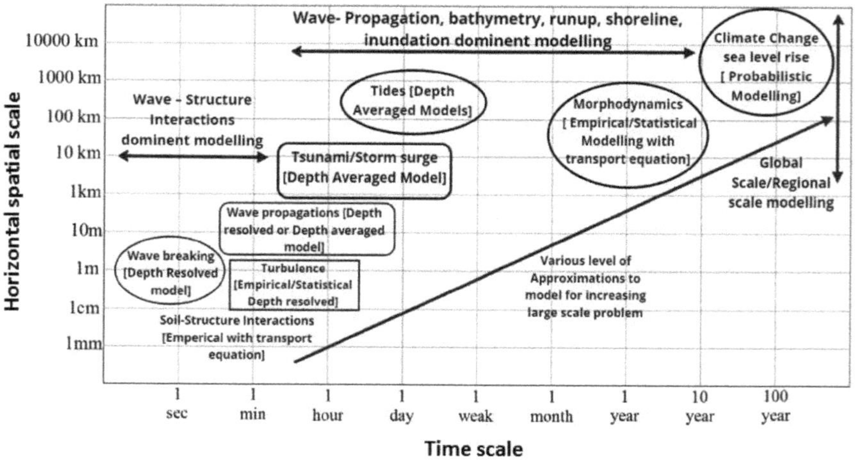

Fig. 8.1. Bubble plot variation to represent the scales of physical process and the modelling effort in coastal engineering.

The illustration clearly depicts a comprehensive range of processes, from climate change and its impact on sea level rise. It then covers morphodynamics, illustrating the changes in landforms and coastal structures. The illustration further includes the effects of tides, the occurrence of tsunamis, and the impact of storm surges. It also shows the propagation of waves over varying offshore bathymetry, leading up to the interactions between waves and structures. Each element is interconnected, demonstrating the complex interplay between natural phenomena and coastal environments. Each of these cases has a different horizontal spatial scale from 1 mm to more than 10,000 km and a time scale from less than 1s to 100 years. Further, which type of modelling is dominant is also represented in the figure, along with global-scale or regional-scale modelling. For large scale problems, the mathematical modelling approximations based on the depth average are predominant. This is assuming the shallow water conditions.

When the vertical flow motion is considered weak and not of significant importance, then depth-averaged horizontal velocities can be adopted. Such classifications of mathematical models are called depth-averaged models and, depending upon the approximations adopted in the horizontal velocities, different models are available. This will be discussed in the later part of this chapter. When the time scale and horizontal spatial scale are small, then wave–structure interaction is dominant; in those cases, depth-resolving models are normally adopted. This is solved based on Navier–Stokes (NS) equations with various simplified approximations based on applications. Depending upon the applications (such as porous structure, vegetation interactions, hydroelasticity or sediment transport), either microscopic or macroscopic modelling can be adopted within the NS framework to model the structure interaction process.

In recent decades, the primary motivation for employing various physical approximations to model different scales of a problem has been to mitigate computational time. Nevertheless, this approach often necessitates compromises on the fidelity of the physical representation.

The three main categories, i.e. depth averaging, depth resolving and hybrid models, that are widely recognized in coastal engineering are depicted in Fig. 8.2. These categories encompass a range of governing equations employing diverse approximations commonly used within the coastal engineering field. Despite the diversity, the fundamental governing equations typically revolve around solving continuity and momentum equations for fluid dynamics problems.

In coastal engineering, many structures (e.g. breakwaters, seawalls, pile structures) are fixed or stationary. Consequently, the primary focus is in representing the wave transformation process, encompassing phenomena

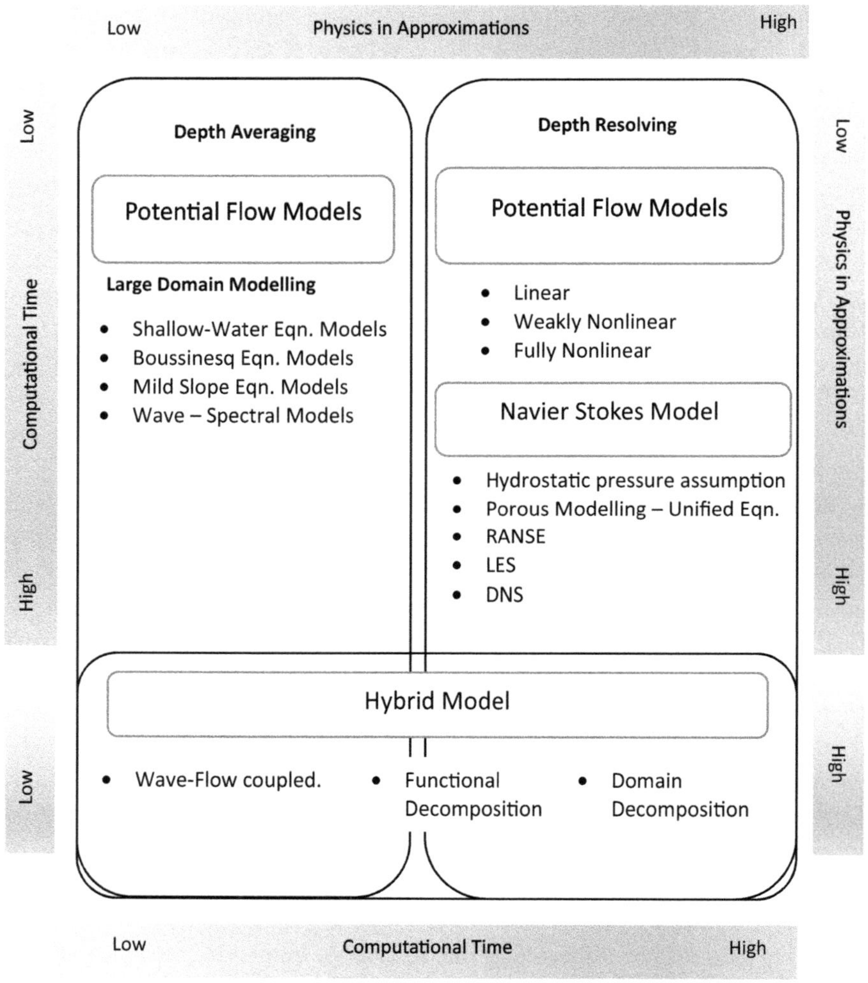

Fig. 8.2. Different types of the modelling approximations with respect to computational efficiency.

such as wave shoaling, diffraction, refraction, reflection, wave overtopping and wave breaking, along with their interactions with the structures.

Thus, the mathematical models are simplified based on the assumptions, which can be either in the treatment of the governing equations or boundary conditions. This is pictorially shown in Fig. 8.3 for an example problem of wave tranquillity, wherein the mathematical models are derived based on the depth-averaged assumptions, the bathymetric slopes are mild and the free surface is treated as linear or weakly nonlinear.

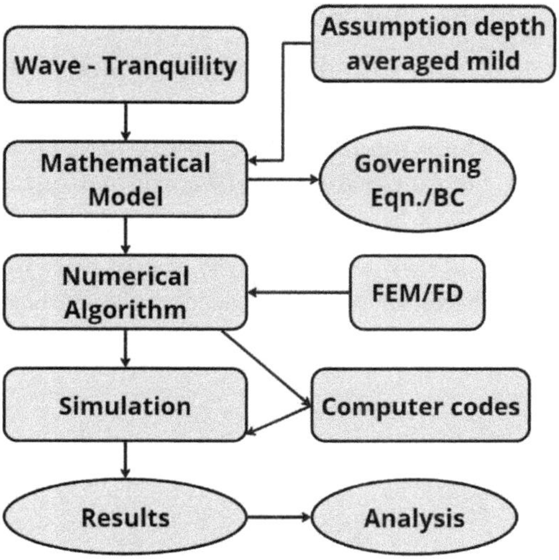

Fig. 8.3. Representation of the physical problem of wave tranquillity in coastal engineering and the development of numerical tools.

Following the formulation of mathematical models, appropriate numerical methods and algorithms are employed to solve them, leading to the execution of simulations. Subsequently, utilizing advanced post-processing tools like Paraview, Matlab or similar software, the results are meticulously analyzed and discussed.

Various numerical methods are currently in practise to solve the mathematical model. The numerical methods are broadly classified into strong and weak forms. Traditional methods, such as finite different, finite element method, boundary element method and finite volume method, and modern methods, such as particle/mesh free-based methods, are being employed in coastal engineering problems. Mostly, the choice of the numerical method is dependent upon the developers and one is not superior to the others as one would think. Each of these numerical methods has some advantages and disadvantages, and the overall goal is to reduce or minimize the disadvantages using numerical treatments/algorithms/schemes.

In the following sections, these mathematical models are discussed with their governing equations to handle physical problems, assumptions, implementation strategies and the adopted numerical methods along with their applications. The existing numerical efforts carried out worldwide are provided along with a detailed discussion on model development by IIT Madras.

8.3 Depth-Resolving Mathematical Models

Let's initiate the discussion with depth-resolving models, which find extensive application in addressing wave–structure interaction problems within coastal engineering. These models are primarily utilized to estimate wave loads and damping characteristics, particularly for problems characterized by small spatial and time scales. Depth-resolving approaches offer high-resolution representations of the physical processes involved, demanding significant computational power to accurately simulate such intricate phenomena.

8.3.1 *Full Navier–Stokes equations*

8.3.1.1 *Governing equations*

The governing equations for fluid motion consist of the basic principles of mass and force balance conservations. There are four different flow models: infinitesimally small element in fixed frame, infinitesimally small element in moving frame, fixed control volume and moving control volume. These forms lead to partial derivative forms or integral forms of equations. Further, these forms lead to conservative and non-conservative forms of equations. Each of these has its own advantages and disadvantages. More detailed information can be found in computational fluid dynamics (CFD) books, for example by Anderson (1995).

To summarize, the continuity and momentum equations for incompressible fluids are as follows in partial differential form:

$$\frac{\partial u_i}{\partial x_i} = 0, \tag{8.1}$$

$$\frac{\partial u_i}{\partial t} + u_i \frac{\partial u_i}{\partial x_i} = -\frac{1}{\rho}\frac{\partial p}{\partial x_i} + g_i + \frac{1}{\rho}\frac{\partial}{\partial x_j}\left(\mu \frac{\partial u_i}{\partial x_j}\right). \tag{8.2}$$

In integral form:

$$\int_{\mathcal{S}} \vec{V} \cdot \overrightarrow{\mathrm{d}\mathcal{S}} = 0, \tag{8.3}$$

$$\underbrace{\int_{\mathcal{V}} \frac{\partial \vec{V}}{\partial t} \mathrm{d}\mathcal{V}}_{\text{time}} + \underbrace{\int_{\mathcal{S}} (\vec{V}\vec{V}) \cdot \overrightarrow{\mathrm{d}\mathcal{S}}}_{\text{advection}} = \underbrace{-\frac{1}{\rho^*}\int_{\mathcal{V}} \vec{\nabla} p \, \mathrm{d}\mathcal{V}}_{\text{pressure}} + \underbrace{\frac{1}{\rho^*}\int_{\mathcal{S}} \{(\mu^* + \mu_{\mathcal{T}})\vec{\nabla}\vec{V}\} \cdot \overrightarrow{\mathrm{d}\mathcal{S}}}_{\text{diffusion}}$$

$$-\frac{2}{3}\int_{\mathcal{V}} \vec{\nabla} k \, \mathrm{d}\mathcal{V} + \int_{\mathcal{V}} \vec{g} \, \mathrm{d}\mathcal{V}, \tag{8.4}$$

where $\vec{V}$ is the Reynolds averaged or mean velocity field, p is the pressure, ρ^* and μ^* are the mixture density and viscosity, respectively, μ_T is the turbulent or eddy viscosity, k is the turbulence kinetic energy and $\vec{g}$ is the gravitational acceleration vector.

8.3.1.2 *Solving NSE*

The main difficulty in solving the above-mentioned equations is that pressure is a passive variable, which is not determined by the momentum equation. Instead, it depends on the instantaneous velocity field. Thus, the solution of the pressure can be obtained by incorporating the continuity equation in the momentum equation leading to the Pressure Poisson Equation (PPE). Since the PPE is an elliptic type, the numerical solution involves a large sparse matrix. So, iterative algorithms have been normally employed. There are three ways in which the NSE will be solved:

(1) Pressure–velocity iteration method and pressure correction technique;
(2) Projection method;
(3) Artificial compressibility method.

In the pressure–velocity iteration method, by enforcing zero divergence of the velocity field at both previous and present time steps, the pressure is solved in an iterative way. Based on updated pressure, the velocity information at the present time step can be obtained. This idea was first proposed by Harlow and Welch (1965) and it was followed in the later development of the NSE, such as SIMPLE (Semi-Implicit Method for Pressure Linked Equations) by Patankar and Spalding (1972) and SIMPLER (Revised version of SIMPLE) by Patankar (1981). Another improvement to the original solution procedure was the pressure–velocity coupling scheme adopted by Issa (1982) based on PISO (Pressure Implicit with Splitting of operators). The advantage of the PISO loop is solving the transient problem with less computational time but it has a limitation of the Courant number (Co $<$ 1) to maintain stability. However, the SIMPLE algorithm developed for steady-state simulation has no time step limitations. Combining the advantages of PISO and SIMPLE, the segregated iterative approach or PIMPLE was developed (Patankar and Spalding, 1983). The PIMPLE algorithm comprises two loops, with the inner loop solving the pressure velocity coupling like PISO and the outer loop following SIMPLE algorithm's idea to deal with velocity. Co can be much larger than 1 and no under-relaxation is required for a stable solution. This allows for larger time steps, but the computational effort for each time step is larger than with PISO. The second approach was introduced by Chorin (1968), in which the calculation is split into two steps. In the first step, the intermediate velocities are estimated by neglecting the pressure term and thus the velocity field carries the correct vorticity. In the second step, the pressure is estimated based on PPE and the final velocity field is satisfied by the continuity conditions. Later, an artificial compressibility method to solve the incompressible NSEs was proposed. This was suggested to overcome the difficulty of solving the elliptic PPE using iterative schemes.

8.3.1.3 *Boundary conditions – Wave/current generation and absorption*

The boundary conditions play an important role in modelling the free surface problem and must be specified at all domain and physical boundaries. Traditionally, inflow and outflow boundaries are called domain boundaries, whereas the free surface, bottom boundaries and boundaries on the solid or moving surfaces are called physical boundaries.

In this section, different boundaries such (a) as inflow boundary for flow or wave simulations; (b) outflow boundary considering as open boundary or sponge layers; (c) Free surface boundary treatments; (d) boundary conditions for solid surfaces with or without turbulence are discussed.

Inflow boundary: There are different methods of generating waves in NSE solvers: (a) Static boundary conditions, (b) Relaxation approach, (c) Moving wavemakers and (d) internal wavemakers.

(a) Static boundary conditions

In static boundary conditions, based on the wave and current theories, both free surface displacement and velocity distributions are prescribed as follows:

$$\phi = \phi_i, \tag{8.5}$$

where ϕ represents velocity or free surface displacement. The variables on the right side of the equation with the subscript i are the known values. The values can be prescribed from various theories such as linear, irregular, stokes, Cnoidal and solitary waves, and recently it became quite popular to prescribe fully nonlinear potential flow theories (this will be discussed in a later subsection of this chapter).

Apart from the generation of waves using the Dirichlet-type condition, there will be a reflection from the structure which will interfere with the inflow boundary. Hence, an absorption boundary condition is required. The simplest wave absorption is Sommerfeld's (1949) radiation condition:

$$\left(\frac{\partial}{\partial t} + c_{\text{out}}\frac{\partial}{\partial x}\right)\phi_{\text{out}} = 0, \tag{8.6}$$

where $\phi_{\text{out}} = \phi - \phi_i$ is the measured wave reflection that needs to be absorbed. In this case, if reflection is zero, then it leads to a static boundary condition (Equation (8.5)). Celerity can be obtained by different means such as using Wellens' (2012) digital filter or the approach of Schäffer and Klopman (2000). These are quite popular in the physical flume active wave absorption method and can be directly applied in a numerical approach.

The more popular and robust method is the relaxation zone approach, in which part of the computational domain is dedicated to performing wave generation with or without absorption.

(b) Relaxation approach

The relaxation zone method for generating and absorbing waves is based on the concept of dedicating a part of the CFD model domain to perform wave generation and/or absorption. It can be broadly summarized in the following equation, which, without loss of generality, is written in 1D form:

$$F(x_r) = F_b\sigma(x_r) + F_d(1 - \sigma(x_r)), \quad x_r \in [0, L_r], \tag{8.7}$$

where F is a model variable or parameter in broad terms, σ is a weighting function that ranges from 0 at the non-relaxed part to 1 at the boundary and can have various forms (e.g. linear, polynomial, exponential, cos or sin), x_r is the coordinate along the wave propagation axis, $x_r = 0$ at the interface of the numerical domain and the relaxation zone, and L_r is the length of the relaxation zone along the x_r-axis. Indices b (boundary) and d (domain) indicate that these are calculated or imposed at the boundary and the internal domain, respectively. It has been demonstrated that an appropriate form of Equation (8.7), in the context of wave absorption and generation problems, can be introduced to the flow equations or the numerical procedure to relax the solution towards the boundary condition within the relaxation zone domain. Given enough length, this transition becomes smooth and reflections of outgoing waves are minimized. These techniques may be referred to as "sponge layer," "numerical beach," or "generation or absorption layers." It is important to highlight that for classification, the relaxation zone must interface with the domain boundary; otherwise, it is considered part of the internal wavemaker techniques. Alternatively, for the absorption of waves, mesh stretching is also adopted (coarser mesh in the sponge layers), based on the numerical modeller's expertise.

(c) Moving wavemakers

Generating and absorbing waves in numerical models using moving wavemakers presents the advantage of replicating waves generated in an experimental facility. In the laboratory, two types of generations are quite common: flap-type or hinge-type and piston-type wavemaker. A piston-type wavemaker is ideal for the waves in shallow- and intermediate-water depths, whereas the flap type is preferred for the deep-water wave characteristics in the laboratory. The same approach can be implemented in a numerical model using the Neumann-type boundary conditions. This was quite successful in reproducing experimental measurements. However, one of the limitations of this approach is that the side boundary will be moving, hence proper treatment for mesh needs to be considered, particularly for domain-based mesh methods such as FEM.

(d) Integrated wavemakers (IWMs)

The IWM concept is usually implemented by creating a generation line or area inside the domain and adding source terms to the mass or momentum equation to generate waves. This line/area is expected to be transparent to the wave propagation. Therefore, full wave absorption can be achieved as long as passive absorption zones or absorbing boundary conditions are added to inlet/outlet boundaries. Initially, Larsen and Dancy (1983) proposed an internal generation scheme which is based on removing or adding to some volume of water from the domain by modifying the equation of mass in the Boussinesq approach. After the equations had been discretized, source terms were implemented. It was demonstrated that the IWM remained transparent to the propagation of nonlinear regular waves. It was also demonstrated that it was not suitable for more advanced formulations such as Boussinesq equations (Nwogu, 1993; Wei and Kirby, 1995). The overall concept was relatively successful as it was the starting point for further improvement and adaptation of the Boussinesq and CFD models.

In CFD models, Lin and Liu (1999) applied an IWM to their volume of fraction (VOF)-based model, initiating this concept in 2D NWT. The approach uses a relatively concise source region, which is placed internally in the flumes and below the free surface. The continuity equation is locally modified to include inflow and outflow fluxes that cause wave excitation. For example, the mass source function for a monochromatic wave can be considered as

$$s(t) = \frac{cH}{A}\cos(\omega t + \phi), \tag{8.8}$$

wherein ω, H, A and c are a wave angular frequency, wave height, area of the IWM and celerity, respectively. By using linear superimposition, this concept can be extended to cover irregular, random and directional waves.

Saincher and Banerjee (2017) further assessed the performance of Lin and Liu's (1999) IWM by placing an IWM in the middle of the tank and using absorption zones at the sides to dissipate the outgoing waves. It was concluded that steep waves generated in deep or shallow waters may experience excessive wave height damping, which is caused by vortices generated by the IWM. The arrival of these vortices to the free surface in deep-water waves was associated with wave height dampening. The problem is solved by adjusting the location of the IWM towards deeper water and increasing the size of the source area to reduce velocities. Normally, additional calibration of the IWM has to be performed due to the change of location and size of the IWM. In order to overcome the shallow-water wave limitations, the source size is increased to full depth. Although it can be argued that this method converges to the one proposed by Wei *et al.* (1999), as the source area is extended throughout the whole water column, this modification yielded good results.

Choi and Yoon (2009) introduced a novel momentum source approach within ANSYS Fluent, building upon the framework proposed by Wei *et al.* (1999), and tailored it for CFD applications. Their model enabled the creation of a 3D numerical wave tank capable of generating both planar and directional waves. This framework offers flexibility, allowing for the representation of higher-order, random and directional waves through linear superimposition of wave components.

Validation efforts showcased promising results, with the model demonstrating satisfactory agreement when compared against physical experiments, analytical solutions and benchmark cases simulated by Lin and Liu (1999) in both 2D and 3D scenarios of regular wave propagation. Notably, Choi and Yoon (2009) observed that within the context of intermediate- to shallow-wave conditions, the influence of evanescent modes originating from the vertical profile at the IWM is confined within a range of twice to thrice the water depths from the IWM location. However, this phenomenon primarily manifests under conditions where the assumption of constant velocity across depth holds true, such as in shallow-water scenarios, as opposed to deep-water wave settings.

Ha *et al.* (2013) coupled the Integrated Wavemaker (IWM) with passive absorption layers and conducted numerical tests involving random and multidirectional waves. Their findings indicated satisfactory performance overall, particularly for intermediate- and shallow-wave conditions. However, their approach encountered limitations in accurately reproducing deep-water wave behaviours. One potential explanation for this discrepancy lies in the utilization of a relatively coarse mesh in their simulations. While their results demonstrated accurate wave height generation at the IWM location, they observed rapid dissipation during wave propagation within the NEWTANK model. This suggests that the coarse mesh resolution may have hindered the model's ability to accurately capture the dynamics of deep-water waves. The IWM concept of Choi and Yoon (2009) was implemented by Liu *et al.* (2015) in the framework of the Incompressible Smooth Particle Dynamics (ISPH) model. Using mass source terms would require adding and removing SPH particles from the domain, thus causing difficulties and complications for numerical implementation and stability.

8.3.2 *Free surface tracking*

The accurate tracking of the free surface is of supreme importance for water wave models solving NSEs. There have been two types of approaches for tracking free surfaces, namely the Eulerian and Lagrangian methods. Although the Lagrangian method determines the exact free surface location within a computational cell, the Eulerian method provides the bulk fluid property in a computational cell, from which the free surface can be only approximately reconstructed. The Eulerian method is classified as an "interface capturing method," whereas the Lagrangian method is classified as an "interface tracking method."

The VOF method: In the Eulerian approach, the normalized density at the fixed location is tracked by solving the transport equation of the VOF function:

$$\frac{\partial F}{\partial t} + u_i \frac{\partial F}{\partial x_i} = 0, \tag{8.9}$$

where F is defined as the volume fraction of the fluid in the cell (control volume), which is equivalent to the normalized mean density of the liquid–air mixture in the cell. It is obvious that $F = 1$ in liquid, $F = 0$ in air and $0 < F < 1$ on the free surface. Hence, free surface geometry can be approximated and reconstructed with the updated information of F in all computational cells. This approach forms the basis of the so-called VOF method originally developed by Nichols *et al.* (1980) in their code SOLA-VOF for the simulation of free surfaces (breaking and broken). Saincher and Sriram (2022b) discussed and reviewed various VOF schemes, both geometric and algebraic.

Marker-and-cell method (MAC): This is a Lagrangian approach that follows the motion of particles on the free surface and in the interior domain based on ambient flow velocity information. This way of tracking philosophy is the basis of the MAC method which was initially developed by Harlow and Welch (1965):

$$\frac{dX_i}{dt} = u_i, \tag{8.10}$$

where X_i is the fluid particle position in the ith direction and u_i is the corresponding velocity at the particle location. The updated free surface information can be obtained at each time step by following these particles originally deployed on the free surface with an assumption that particles on the free surface will remain so for non-breaking waves. For breaking waves, the free surface may leave the surface during the breaking process. MAC can be regarded as an extension of the earlier particle-in-cell (PIC) method that was originally developed for plasma stimulation. Hirt *et al.* (1975) developed the SOLA method with the use of the MAC method to stimulate free surface flows.

Level set method: This method developed by Osher and Sethian (1988) became popular for capturing interface motion (e.g. Sussman *et al.*, 1994). In this method, the so-called level set function ϕ, which is designed to be zero on the interface but smoothly changing across the interface, is tracked by solving the following equation, instead of tracking the density, which has a sharp front on the free surface or two-fluid interface:

$$\frac{\partial \phi}{\partial t} + u_i \frac{\partial \phi}{\partial x_i} = 0. \tag{8.11}$$

The most intuitive way to design ϕ is to define it as the normal distance function to the free surface, which leads to $\phi < 0$ in air, $\phi > 0$ in water and $\phi = 0$ on the free surface.

Once the level set function is known, it will be used to determine the density at each cell through the Heaviside function as follows:

$$\gamma_\phi = \begin{cases} 0, & \text{if } \phi < 0, \quad f_p \notin \Gamma(p) \\ a_p, & \text{if } \qquad\quad\; f_p \in \Gamma(p) \\ 1, & \text{if } \phi > 0, \quad f_p \notin \Gamma(p) \end{cases}, \tag{8.12}$$

where $f_p \notin \Gamma(p)$ represents the non-free surface cell and $f_p \in \Gamma(p)$ the free surface cell. The coefficient a_p is defined as the volume ratio of liquid in the cell that can be calculated based on the tangential plane of the zero-level set contour ($\phi = 0$) across the cell (Gueyffier *et al.*, 1999). The mean fluid density for a particular cell is then calculated as follows:

$$\rho = \rho_a + (\rho_w - \rho_a)\gamma_\phi, \tag{8.13}$$

Constrained Interpolation Profile (CIP): The CIP method uses a colour function called density function similar to volume fraction in the VOF method, and this is used to find out the interface. In this method, the profile inside the grid cell can be well reconstructed using the spatial derivative of the density function without any additional computation cost. The main disadvantage of the CIP method is poor mass conservation.

8.3.3 *Turbulence modelling*

In wave–structure interaction, the turbulence model is important. However, whether this should be included in the list of models depends on the specific problem being addressed. Including turbulence in non-breaking waves and low viscous dominant problems leads to excess numerical dissipation and should be included with caution. Turbulence modelling can be categorized into various strategies: (a) based on the definition of velocity, (b) based on the decomposition of velocity and (c) based on the number of transport equations. Based on the definition of velocity, it can be unfiltered (e.g. Direct Numerical Simulation), spatially or temporally filtered (LES) and time-averaged (RANS). Based on the decomposition of velocity, it can be without any decomposition (DNS), resolved and subgrid scale (LES) and mean/fluctuation (RANS). Based on the number of transport equations, it can be without any equation (e.g. Mixing length, WALE), one equation (Spalart-Allmaras, Prandtl's Model), two equations (e.g. $k - \varepsilon$, $k - \omega$) or seven (e.g. RSM). Discussing these methods is beyond the scope of this chapter and the reader can refer to suitable textbooks on turbulence modelling.

8.3.4 *Numerical methods*

Based on the numerical method used to solve the governing equations discussed previously, numerical models in general can be grouped into mesh-based methods and meshfree (particle) methods. In these methods, the fluid flow is based on either Lagrangian, Eulerian or Arbitrary Lagrangian–Eulerian (ALE) formulation.

Mesh-based methods for free surface water wave problems mainly include the finite difference method, finite element method, finite volume method and boundary element method. In Eulerian formulation, popular numerical techniques such as the Marker and Cell (MAC) (Harlow and Welch, 1965), Volume of Fluid (VOF) (Hirt and Nichols, 1981), Constrained Interpolation Profile (CIP) (Takewaki and Yabe, 1985) and Level Set (LS) (Osher and Sethian, 1988) methods are used to track the free surface. Even though mesh-based methods produced impressive results, the success of these methods mainly depends on the quality of the mesh used. In order to obtain a promising result, the mesh should be updated frequently or be refined to capture the motion of the free surface and it needs to be maintained. It is a time-consuming task and costly, especially in cases of breaking waves. For the NS models with the Eulerian grid, considerable numerical diffusion caused by the treatment of the convective term in the momentum equations may be observed, if the mesh resolution is insufficient.

In scenarios characterized by intense free surface dynamics, such as violent free surface behaviour, accurately identifying the free surface particles poses a significant challenge within the Lagrangian formulation. Additionally, in the Eulerian formulation, mitigating numerical diffusion resulting from the convective term remains a key concern. The meshfree (or particle) method offers a solution to the challenges inherent in mesh-based approaches. Instead of discretizing the computational domain into grids, it treats the fluid domain as a collection of randomly distributed nodes. This eliminates issues related to mesh orientations, commonly encountered in finite element methods, as well as concerns about element connectivity.

In the meshfree method, as there is no connectivity between nodes, the aforementioned drawbacks of mesh-based methods are automatically circumvented. Particle methods typically employ the Lagrangian form of governing equations or boundary conditions for each particle, allowing them to track fluid motion without frequent updates. Consequently, particle methods exhibit considerable promise in modelling scenarios involving violent wave–structure interactions. Various meshfree methods reported in the literature are diffusion element method, element free Galerkin method, smooth particle hydrodynamics, reproducing kernel particle method, moving particle semi-implicit method, meshless local Petrov–Galerkin method and particle finite element method (Sriram and Ma, 2021). Among meshfree methods, smoothed particle hydrodynamics (sph), moving particle semi-implicit (mps), meshless local Petrov–Galerkin (MLPG) and particle finite element method (PFEM) are the usual methods for solving water wave problems. In general, particle methods can be categorized into weakly compressible and incompressible variants. The weakly compressible particle method involves solving the equation of state to estimate pressure, whereas the incompressible particle method solves the pressure Poisson equation for pressure calculations.

However, it is essential to recognize that both mesh and meshfree methods possess their own sets of advantages and disadvantages. Therefore, numerical modellers must exercise discretion in selecting the most appropriate method for their specific application.

To overcome the problems associated with Lagrangian and Eulerian formulations, a new formulation was suggested by Hirt *et al.* (1974) called Arbitrary Lagrangian–Eulerian formulation (ALE), which combines the advantages of both formulations and alleviates their drawbacks. In this method, the computational grid is not required to adhere to the fluid particles or be fixed in space, but it can be moved arbitrarily. While using ALE, the grids inside the domain will move arbitrarily to optimize the shape of the elements/cells, whereas the mesh on the boundaries and interfaces of the domains will move along with the material point to track the boundaries and interfaces of a multiphase system. The main objective of the ALE formulation is to make the mesh independent of the material so that mesh distortion can be minimized. However, the governing equations need to be modified to include the moving mesh velocity. The ALE method does not work effectively if huge free surface deformation occurs. So, recently researchers adopted this scheme in meshfree methods such as SPH.

8.4 Modelling Sediment Dynamics

With the rapid development of CFD and computational power, numerical studies of local scour have become increasingly popular in recent years.

Numerical models for the simulation of the scour usually comprise two distinct elements:

i. A hydrodynamic module to solve the fluid flow;
ii. A morphological module to deal with sediment transport and bed erosion.

The hydrodynamic module solves the (Reynolds Averaged) Navier–Stokes equations for the flow field. The sediment-laden flow can be treated as a single-phase flow, in which the two phases are modelled as a mixture or as a two-phase flow, which includes the water phase and the sediment phase.

The two-phase flow model can be classified into an Euler–Euler type (Chen *et al.*, 2011), Lagrange–Lagrange type (Zubeldia *et al.*, 2018) and an Euler–Lagrange type (Li *et al.*, 2014). In the Euler–Euler approach, the dispersed phase is described in a similar manner to that used for the continuous phase. Efforts have been made to develop such sediment transport models, representative examples of which are the models of Jha and Bombardelli (2010) and Chen *et al.* (2011). A new type of Euler–Euler approach has been developed more recently that makes use of two different intergranular stress models, the kinetic theory of granular flows and the dense granular flow rheology (Chauchat *et al.*, 2017; Mathieu *et al.*, 2019). However, the Euler–Euler type of two-phase models is typically complicated (Chen *et al.*, 2011).

In the Euler–Lagrange type of model, the sediment phase is treated as representing the motion of a certain number of individual particles. This approach succeeds in capturing the individual and collective dynamics of natural sand grains. Many good numerical models of the Euler–Lagrange type have been proposed recently such as those by Li *et al.* (2014), Finn *et al.* (2016), and Sun and Xiao (2016). However, this approach translates into a huge demand for computational resources since it requires a large number of particles to simulate practical problems. In hydraulic engineering applications, including scour estimations, single-phase models are typically used since in two-phase models, the continuity and momentum equation needs to be solved for both the phases. The most difficult part is to properly describe the turbulence characteristics of the two phases.

The most popular single-phase approach solves the volumetric concentration of the dispersed phase using an advection–diffusion model and neglects the effects of the dispersed phase on the continuous phases. The advection–diffusion model has been successfully applied to many numerical studies of scour problems, for example, the work of Liang *et al.* (2005) and Roulund *et al.* (2005). However, it should be noted that the advection–diffusion model is only valid for dilute problems where the suspended load concentration is not large enough to influence the principal properties of the flow field since it neglects the effects of dispersed phases.

The morphological module for scour simulations calculates the bed level change and hence includes a sediment conservation equation (the Exner equation) and a sediment transport model. There are many possible choices for the sediment transport model, which usually includes both suspended load and bed load. As mentioned earlier, the suspended load in scour problems is often solved by the advection–diffusion model with a reference concentration given at a certain reference height.

The scour FOAM is a state-of-the-art three-phase model for free surface flow over a mobile sediment bed; the air and water phases are combined together as a mixed continuum phase and solved by the existing Open FOAM® inter FOAM solver using a single momentum equation over the whole computational domain (Jasak, 1996). In order to model the scour, the sediment phase is introduced as Lagrangian particle parcels, whose motions are coupled to the local flow field. It is assumed that the size of the particle parcel is small in comparison with the mesh size and hence the presence of the particle is considered at a sub-grid scale. The incompressible Navier–Stokes equations resolve the dynamics of the fluid phase, i.e. the mixture of water and air, which are modified with the two-fluid method. The sediment phase is treated as individual discrete particles, and their motion is solved based on Newton's law of motion. In order to describe the turbulent characteristics at the various scales involved in the scour process, the Large Eddy Simulation (LES) with the k-equation sub-grid-scale model of Yoshizawa (1993) is normally adopted. Further, a suitable model is normally chosen to represent the drag force from the fluid on each particle (Andrews and O'Rourke, 1996).

8.5 Unified Navier–Stokes Equations to Model Porous and Vegetal Flows

In earlier years, either potential flow theory or shallow water equations (SWEs) (Kobayashi and Wurjanto, 1990) were used as the base model in earlier numerical studies of wave-induced porous flows that incorporate damping terms in porous media. These models will be discussed in a subsequent section. However, incorporating porous and vegetal flows in the NSEs is computationally expensive. There are two types of modelling approaches: microscopic and macroscopic. The more rigorous and accurate porous flow models in connection with NSE solvers were developed later (van Gent, 1995). The NSEs were averaged over space to have a new set of equations that have similar characteristics to NSEs but the additional friction in porous media is included. This is a macroscopic approach, which will be discussed further.

The numerical studies on the wave porous structure can be carried out in two different ways:

(i) Coupling of pure fluid and porous flow equations, in which the fluid flow was solved for Navier–Stokes equation and porous flow with different porous flow models, after which the interface was coupled by matching the flow properties. It can be explicit, implicit or iterative in nature.
(ii) Based on unified or single governing equations to model both porous structure and fluid flow.

The first approach is computationally expensive as one needs to solve this iteratively at each time step. Lin and Liu (1998) used volume-averaged Navier-Stokes equations, whereas the porous flow velocity was averaged with respcect to volume (VARANS) and the fluid flow was Reynolds averaged (RANS). The second approach can be classified into three main categories based on (a) the governing equation for porous flow, (b) the linear and nonlinear coefficients used, and (c) numerical methods and algorithms. The governing equation for porous flow was initially suggested by the classical work of Darcy (1856) in which the pressure gradient of the flow was found to be directly proportional to the flow velocity by introducing the permeability coefficient, which for higher value means a higher fluid flow in the porous region. Darcy's law was valid for laminar flows having less permeability. Later, Forchheimer (1901) introduced a nonlinear term to account for turbulent flows in which a nonlinear drag force is directly proportional to the square of fluid velocity. The porous flow model was extended by incorporating the virtual mass coefficient in the local acceleration term (Polubarinova-Kochina, 1962). Karunarathna and Lin (2006) considered the transitional flow by incorporating the drag force proportional to the square root of velocity. Thus, for modelling the wave porous structure, in the governing equations, additional resistance terms such as the linear drag coefficient representing the laminar flow, nonlinear drag coefficient representing the turbulent flow, the coefficient for the transitional flow and the virtual mass coefficient for inertia terms were incorporated.

In this chapter, the derivation of the saturated porous flow equations is discussed. To simplify the derivation, we assume that the porous medium is of spherical shape with the mean diameter d_{50}. The original NSEs are still valid in the fluid domain. Now, consider a control volume (filter space); the volume taken by the porous material and the pore space filled with fluid are V_s & V_f, respectively. The porosity is defined as $n_w = V_f/(V_f + V_s)$.

Let us now introduce the spatial average operation as follows:

$$\overline{\overline{\phi}} = (1/V_f) \int_{V_f} \phi \, dV. \tag{8.14}$$

We obtain the new set of equations by applying the divergence theorem by taking the spatial average of the NSEs within the control volume for the fluid only. Now, assume that the operator is commutable for integration and differentiation of velocities:

$$\frac{\partial \overline{\overline{u_i}}}{\partial x_i} = 0, \tag{8.15}$$

$$\frac{\partial \overline{\overline{u_i}}}{\partial x_t} + \overline{\overline{u_j}} \frac{\partial \overline{\overline{u_i}}}{\partial x_j} + \frac{\partial \overline{\overline{\ddot{u}_i \ddot{u}_j}}}{\partial x_j} = -\frac{1}{\rho V_f} \int_{S_c+S_p} p n_i \, dS + g_i + \frac{1}{\rho V_f} \int_{S_c+S_p} \tau_{ij} n_j \, dS, \tag{8.16}$$

where S_c is the control surface of the fluid part for the chosen control volume, S_p is the total surface of the porous material inside the control volume and n_i is the outward unit normal vector from the fluid to the solid surface. In the Equation (8.16), we have two terms requiring proper closures: (1) the 3rd term on the left side of the equation, i.e. the additional stress term in space $\overline{\overline{\ddot{u}_i \ddot{u}_j}}$ (this is similar to Reynolds stress for turbulent flows) and (2) the additional forcing terms on the right side of the equation, i.e. from the surface integration of the reaction force from the porous materials within the control volume. This is not difficult to understand because the surface of the porous material coincides with the surface of integrated fluid volume, through which the porous material acts as the source of external force.

We can now relate the control volume mean velocity S_c with the spatial-averaged velocity in the fluid domain $\overline{\overline{u_i}}$ by $\bar{u}_i = n\overline{\overline{u_i}}$ (based on the mass conservation argument). By further assuming that the porous material shares the same pressure and stresses in fluid, we can apply the divergence theorem again in the entire control volume and simplify the 1st and 3rd terms on the right side of Equation (8.16):

$$\frac{1}{\rho V_f}\int_{S_c} pn_i\,dS = -\frac{1}{\rho}\frac{\partial \bar{p}}{\partial x_i}, \tag{8.17}$$

$$\frac{1}{\rho V_f}\int_{S_c} \tau_{ij}n_j\,dS = -\frac{v}{n}\frac{\partial^2 \overline{u_i}}{\partial x_j \partial x_j}. \tag{8.18}$$

If we define the friction term

$$\bar{f}_i = \frac{1}{\rho V_f}\int_{S_p} pn_i\,dS - \frac{1}{\rho V_f}\int_{S_p} \tau_{ij}n_j\,dS \tag{8.19}$$

and substitute Equations (8.17)–(8.19) into Equation (8.16), we have the new set of governing equations for the mean flow averaged over the entire control volume:

$$\frac{\partial \overline{\overline{u_i}}}{\partial x_i} = 0. \tag{8.20}$$

$$\frac{1}{n_w}\frac{\partial \overline{u_i}}{\partial t} + \frac{\overline{\overline{u_i}}}{n_w^2}\frac{\partial \overline{u_i}}{\partial x_j} = -\frac{1}{\rho}\frac{\partial \bar{p}}{\partial x_i} + g_i + \frac{v}{n_w}\frac{\partial^2 \overline{u_i}}{\partial x_j \partial x_j} - \bar{f}_i - \frac{1}{n_w^2}\frac{\partial \overline{\ddot{u}_i\ddot{u}_j}}{\partial x_j}. \tag{8.21}$$

The flow inside and outside the porous media can be directly linked by the same velocity by using the control volume mean velocity, which is continuous across the interface. As a result, there is no need to impose flow boundary conditions on the porous interface when the flow inside and outside of the porous domain is solved as one fluid flow system.

The main issue now is how to model the friction term $\bar{f}_i$ and the turbulence stress term $-\rho\overline{\ddot{u}_i\ddot{u}_j}$. While the latter term can be modelled by the same type of modified turbulence eddy viscosity model, i.e. $-\overline{\ddot{u}_i\ddot{u}_j} = v_{tp}(\frac{\partial \overline{u_i}}{\partial x_j} + \frac{\partial \overline{u_j}}{\partial x_i})$, the former can be modelled by the well-known Morison equations for general and steady flow:

$$\bar{f}_i = \bar{f}_I + \overline{f_D}, \tag{8.22}$$

where $\bar{f}_I$ and $\overline{f_D}$ are the inertial and drag force contributions, respectively. The inertial force is caused by the resistance of the rigid and fixed porous particles to the ambient flow acceleration. The following expressions were derived by Lin and Karunarathna (2007) assuming that the control volume is filled with uniform spheres:

$$\bar{f}_I = \frac{1}{\rho V_f}\left(N\rho C_M V_p \frac{1}{n_w}\frac{\partial \overline{u_i}}{\partial t}\right)$$
$$= \frac{NV_p}{V_f}\left(C_M \frac{1}{n_w}\frac{\partial \overline{u_i}}{\partial t}\right) = \frac{C_M(1-n_w)}{n_w^2}\frac{\partial \overline{u_i}}{\partial t}, \tag{8.23}$$

where N is the total number of spheres in the control volume, V_P is the volume of 1 sphere and C_M is the virtual mass coefficient. C_M is also called the inertial coefficient, which is the sum of the Froude–Krylov force coefficient 1.0 and the added mass force coefficient C_a, i.e.

$C_M = 1 + C_a$. The Froude–Krylov (F–K) force results from the pressure difference in an accelerating fluid flow and behaves in a similar way to buoyancy force. The additional force resulting from the presence of the body is the added mass force and it is the function of body shape. For a single sphere, $C_M = 1.5$ and $C_a = 0.5$. The suggested values of C_M by van Gent (1995) and Gu and Wang (1991) are $C_M = 0.34$ and $C_M = 0.46$, respectively. The reduction of the F–K force coefficient is due to the porous materials that reduce the effective cross-sectional area which is exposed to the flow direction. The reduced F–K force coefficient in porous media can be explained by the reduced buoyancy for a composite cylinder sitting on the bottom since the F–K force has the same expression and similar justification as the buoyancy force. The added mass coefficient is also reduced besides the F–K force coefficient when we consider the porous materials.

The drag force can be expressed as the following based on the Morison equation:

$$\overline{f_{Di}} = \frac{1}{\rho V_f}\left(N\frac{1}{2}\rho C_D A \frac{\overline{u_c}}{n_w}\frac{\overline{u_i}}{n_w}\right) = \frac{N}{V_f}\left(\frac{1}{2}C_D\frac{\pi}{4}d_{50}^2\frac{\overline{u_c}}{n_w}\frac{\overline{u_i}}{n_w}\right)$$
$$= \frac{NV_p}{V_f}\left(\frac{1}{V_p}\frac{1}{2}C_D\frac{\pi}{4}d_{50}^2\frac{\overline{u_c}}{n_w}\frac{\overline{u_i}}{n_w}\right) = \frac{V_s}{V_f}\left(\frac{1}{\frac{\pi}{6}d_{50}^3}\frac{1}{2}C_D\frac{\pi}{4}d_{50}^2\frac{\overline{u_c}}{n_w}\frac{\overline{u_i}}{n_w}\right)$$
$$= \frac{3}{4}C_D\frac{(1-n_w)}{n_w^3}\frac{1}{d_{50}}\overline{u_c}\bar{u}_i, \tag{8.24}$$

where $\overline{u_c}$ is the characteristic velocity that can be estimated by $\overline{u_c} = \sqrt{\bar{u}_i\bar{u}_i}$, $A = \pi d_{50}/4$ is the maximum cross-sectional area of the sphere and C_D is the drag force coefficient.

The drag coefficient formula was proposed by Fair *et al.* (1968):

$$C_D = \frac{24.0}{\text{Re}} + \frac{3.0}{\sqrt{\text{Re}}} + 0.34, \tag{8.25}$$

where the Reynolds number $Re = \overline{u_c} d_{50}/v$. Porous media have various shapes, surface roughness and packaging patterns. The wake flows behind particles also have an influence on the drag force of the following particles. All these would have an impact on the value of C_D. Lin and Karunarathna (2007) suggested the following modification for C_D in simulating wave-induced flows in porous materials by employing a large amount of experimental data:

$$C_D = c_1 \frac{24.0}{\text{Re}} + c_2 \left(\frac{3.0}{\sqrt{\text{Re}}} + 0.34 \right) \left(1 + \frac{7.5}{\text{KC}} \right), \tag{8.26}$$

where $c_1 = 7.0$ and $c_2 = 1.8$–4.2 are found in their study, with the smaller c_2 better describing the flow in a small porous medium. In the Keulegan–Carpenter number, $\text{KC} = U_{\max} T/(n d_{50})$, $U_{\max}$ is the maximum water particle velocity and T is the wave period. This term $(1 + \frac{7.5}{\text{KC}})$ is included to account for the additional influence of the flow and steadiness on the turbulent boundary layer around the porous particles.

The well-known Darcy's law for laminal flows in porous media where the mean velocity is linearly proportional to the mean pressure gradient can be adopted and is derived as follows:

Consider a 1D steady flow in the x-direction. We have

$$\begin{aligned} 0 &= -\frac{1}{\rho} \frac{\partial \bar{p}}{\partial x} - \frac{3}{4} C_D \frac{(1 - n_w)}{n_w^3} \frac{1}{d_{50}} \overline{u_c} \bar{u}_i \\ \Rightarrow -\frac{1}{\rho g} \frac{\partial \bar{p}}{\partial x} &= I = \frac{3}{4} C_D \frac{(1 - n_w)}{n_w^3} \frac{1}{g d_{50}} \bar{u} \bar{u}. \end{aligned} \tag{8.27}$$

For small Re (e.g. Re < 10), the porous flow is laminar and $C_D = c_1 \frac{24.0}{Re}$. Substituting it into Equation (8.27),

$$-\frac{1}{\rho g} \frac{\partial \bar{p}}{\partial x} = I = \frac{3}{4} \frac{24 c_1 v}{\bar{u} d_{50}} \frac{(1 - n_w)}{n_w^2} \frac{1}{g d_{50}} \bar{u} \bar{u} = \frac{(1 - n_w)}{n_w^3} \frac{18 c_1 v}{g (d_{50})^2} \bar{u} = \frac{\bar{u}}{K}, \tag{8.28}$$

where K (m/s) is the permeability.

When Re is large, $C_D = 0.34$ becomes a constant and Equation (8.28) is reduced to

$$-\frac{1}{\rho g}\frac{\partial \bar{p}}{\partial x} = I = \frac{3}{4}0.34c_1\frac{(1-n_w)}{n_w^3}\frac{1}{gd_{50}}\bar{u}\bar{u} = 0.255c_1\frac{(1-n_w)}{n_w^3}\frac{1}{gd_{50}}\bar{u}\bar{u}. \tag{8.29}$$

The drag is proportional to the mean velocity square and is called nonlinear friction, which prevails when the porous flow is fully turbulent (e.g. Re > 1000).

Only the linear and nonlinear terms are employed to describe flow resistance to porous media, neglecting the transitional flow range, which may be important for intermediate Re (e.g. $10 <$ Re < 1000). For example, Forchheimer (1901) proposed the following formula:

$$-\frac{1}{\rho g}\frac{\partial \bar{p}}{\partial x} = I = a\bar{u} + b\bar{u}|\bar{u}|. \tag{8.30}$$

Based on different theoretical assumptions and experimental data, there have been many proposals for the expression of a and b. Burcharth and Andersen (1995) compared different equations, which is extended here by using the following more general equation:

$$-\frac{1}{\rho g}\frac{\partial \bar{p}}{\partial x} = I = a\bar{u} + b\bar{u}|\bar{u}| + c\bar{u}\sqrt{|\bar{u}|}, \tag{8.31}$$

where I is the pressure gradient, a is the linear drag coefficient dominating in laminar flows, b is the nonlinear drag coefficient dominating in turbulent flows and c is the transitional coefficient dominating in transitional flows.

The linear and nonlinear drag coefficients depend on Reynold's number and the type of flow. Linear drag is dominant for laminar flows, whereas a nonlinear coefficient is dominant for turbulent flows. The general form of this coefficient is given as follows:

$$a = \alpha\frac{\vartheta_w(1-n_w)^2}{n_w^3 D_{50}^2} \quad b = \beta\frac{1-n_w}{n_w^3 D_{50}} \quad c = \alpha\frac{\vartheta_w^{1/2}(1-n_w)}{n_w^3 g D_{50}^{3/2}}. \tag{8.32}$$

Various coefficients for linear and nonlinear drag have been reported in the literature from experiments. Ergun (1952) suggested a linear coefficient of 150 and a nonlinear coefficient of 1.75 based on experiments. On the other hand, Engelund (1953) suggested a linear coefficient of 780–1500 and a nonlinear coefficient of 1.8–3.6 based on his experiments. The most widely adopted linear and nonlinear coefficients were suggested by Van Gent (1995), who used a linear coefficient of 1000 and a nonlinear coefficient of

1.1. Liu *et al.* (1999) suggested linear and nonlinear coefficients of 200 and 1.1.24, respectively. Karunarathna and Lin (2006) have included both linear and nonlinear coefficients with the transitional coefficient. This transitional coefficient considers the square root of flow velocity, which takes into consideration the transitional flow. Karunarathna and Lin (2006) have used 1.26 as the linear coefficient, 0.51 as the nonlinear coefficient and 4.5 as the transitional coefficient. The power of linear coefficient "a" varies as per the literature. The nonlinear coefficient "b" follows the same functional form in almost all the literature. Both "a" and "b" are lesser for Karunarathna and Lin (2006) than all other models since they account for an additional transitional term.

Similar to incorporating the porosity, vegetation interactions can be studied in the NSEs by treating them as force terms, considering them as cylinders. Su and Lin (2005) considered the vegetation to be a collection of uniform, rigid and smooth cylinders in a control volume (V), with the inertia and drag force over the cylinder obtained by Morison's equation as follows:

$$F_D = \frac{1}{\rho V_f}\left(\frac{1}{2}\rho C_D A \frac{\bar{u}_c}{n_v}\frac{\bar{u}_i}{n_v}\right), \tag{8.33}$$

$$F_I = \frac{1}{\rho V_f}\left(\rho C_m V_p \frac{1}{n_v}\frac{\partial \overline{u_i}}{\partial t}\right), \tag{8.34}$$

where F_D is the drag force, F_I is the inertia force obtained for a single cylinder with a cross-sectional area A perpendicular to the flow, ρ is the water density, V_f is the volume of fluid in the control volume, V_p is the volume of a single cylinder, n_v is the porous rate of vegetation, $\bar{u}_c$ is the characteristic velocity $= \sqrt{\overline{u_i}\overline{u_j}}$ in which i and j are the direction of flow, C_D is drag force coefficient and C_m is inertia force coefficient. Equation (8.33) and (8.34) are then applied for N number of cylinders present in the control volume, and we get

$$F_D = \frac{1}{\rho V_f}\left(N\frac{1}{2}\rho C_D A \frac{\bar{u}_c}{n_v}\frac{\bar{u}_i}{n_v}\right), \tag{8.35}$$

$$F_I = \frac{1}{\rho V_f}\left(N\rho C_m V_p \frac{1}{n_v}\frac{\partial \overline{u_i}}{\partial t}\right). \tag{8.36}$$

Now, Equation (8.35) is taken to arrive at the drag force for a group of cylinders representing the vegetation. Multiply and divide by V_p in Equation (8.35) and apply the relation $NV_p = V_s$. Then, incorporate the volume and area of a cylinder of diameter "D_v" and height "h" in the arrived equation to get the final form as follows:

$$F_{D\_c} = \frac{2}{\pi}\left(\frac{(1-n_v)}{n_v^3}C_D\frac{1}{D_v}\bar{u}_c\bar{u}_i\right) = f_{d\_c}\bar{u}_c\bar{u}_i \tag{8.37}$$

In case, if the vegetation is assumed to be a collection of strips of width "b_v" and thickness "t_v" then the value V_s and A is changed accordingly in the Equation (8.37), and the final form is as follows,

$$F_{D\_s} = \frac{(1-n_v)}{n_v^3}\left(\frac{C_D}{2.t_v}\bar{u}_c\bar{u}_i\right) = f_{d\_s}\bar{u}_c\bar{u}_i. \tag{8.38}$$

To get a generalized form for the porous rate "n," a control volume of dimension $\Delta x.\Delta y.\Delta z$ is taken, where N number of cylinders is present. Applying the relation of the porous rate $n_v = V_f/(V_f + V_s)$ in the generalized form for the cylinder (Equation (8.37)) and strip/square (Equation (8.38)) vegetation, the porous rate can be given as follows:

$$n_v = 1 - \left(\frac{\pi D_v^2}{4}N_D\right), \tag{8.39}$$

$$n_v = 1 - (b_v t_v N_D), \tag{8.40}$$

where N_D is the vegetation density, which is the number of vegetation stems per unit area.

On arrives at inertia force for a group of cylinders/strips from Equation (8.36) by applying the relation $V_p = V_s$; the final form is as follows:

$$F_I = C_m\frac{(1-n_v)}{n_v^2}\frac{\partial\overline{u_i}}{\partial t}. \tag{8.41}$$

The arrived inertia and drag force term is added to the original momentum equation to get the interaction of vegetation.

The drag force f_d for the vegetation based on the Equations (8.37) and (8.38) for cylinder and strips, respectively, is given as follows:

$$f_{d\_c} = \frac{2}{\pi}.C_D.\frac{(1-n_v)}{n_v^3}.\frac{1}{D_v}, \tag{8.42}$$

$$f_{d\_s} = C_D\cdot\frac{(1-n_v)}{n_v^3}.\frac{1}{2t_v}. \tag{8.43}$$

Various numerical methods were used for solving the porous flow equation. Both mesh and meshless methods are used. In mesh-based methods, the finite element method was used by McCorquodale (1972), the boundary integral method was used by Ijima *et al.* (1976) and the finite difference method was used by Lin and Liu (1998). Karim *et al.* (2003) developed a two-phase model (fluid and air) based on the volume of fluid (VOF) method that was extended to include interaction with porous media. Further, developments were reported by Karim *et al.* (2009) and Hieu and Tanimoto (2006). The model presented by Liu *et al.* (1999) was developed by Losada *et al.* (2008) based on VOF, where overtopping of rubble mound breakwaters was investigated. A detailed description of model formulation and validation was reported by del Jesus *et al.* (2012). Examples of the use of OpenFOAM for wave–structure interactions such as breakwaters and modelling of scour around structures can be found in studies conducted by Lara *et al.* (2012) and Stahlmann and Schlurmann (2012). Further, the application of OpenFOAM in coastal engineering was presented in Higuera *et al.* (2013).

All the above-mentioned methods are based on the Eulerian framework. On the other hand, meshless methods are quite popular nowadays as they avoid the use of mesh. Notable works were carried out using incompressible smooth particle hydrodynamics (ISPH) by Shao (2010), where porous flow is solved separately, using incompressible smooth particle hydrodynamics in porous medium (ISPHP) by Akbari and Namin (2013), and using weakly compressible smooth particle hydrodynamics (WCSPH) by Ren *et al.* (2014).

In handling the governing equations using any numerical methods, interface treatment plays an important role, since one needs to satisfy the boundary conditions between the pure fluid and porous region. When using two sets of governing equations for the fluid and porous region, the flow properties, such as velocity and pressure, need to be matched. Notable works using SPH include Shao (2010). In particle-based methods, while using the unified governing equations, Akbari and Namin (2013) used background nodes with porosity values and then interpolated them based on SPH for each and every particle. The background nodes are made finer so that more porosity information can be obtained. At each and every time step, the position and porosity information were updated. Ren *et al.* (2014) suggested a transition zone in which the porosity value of the background nodes is linearly varying. Although the method is simple to incorporate, it requires special treatment for porosity or velocity using higher-order interpolation. Sometimes, these approximations reduce the accuracy of the results and compromise the flow properties near the interface. Further, background node usage throughout the domain may increase the computation time. The lower-order SPH interpolation carried out for each and every particle to update the porosity value at every time step leads to an accumulation of error. The meshless methods attempted till date have been based on the strong formulation for the unified equation; however, these may not effectively handle discontinuous flow. To overcome this, Divya and Sriram (2017) adopted the Petrov–Galerkin approach and proposed novel treatments to satisfy interface continuity.

8.6 Navier–Stokes Equation with Hydrostatic Pressure Assumption

For the flow motion whose horizontal characteristic length is much larger than local water depth, e.g. long waves, storm surges, tides, bores, ocean circulation and ocean currents, the hydrostatic pressure assumptions are valid:

$$\rho = \rho g(\eta - z). \tag{8.44}$$

The above-mentioned expression is the solution to the vertical momentum equation in the original NSEs after neglecting all inertial and viscous terms. The assumption of hydrostatic pressure also suggests that the free surface maintains a gentle slope, which is a single function of the horizontal plane, i.e. $\eta = \eta(x, y)$.

Let us consider a 3D case where $g_1 = g_2 = 0$ and $g_3 = -g$. By substituting (8.44) into the horizontal momentum equations in NSEs, we have the following:

$$\frac{\partial u}{\partial t} + \frac{\partial y}{\partial x} + v\frac{\partial u}{\partial y} + w\frac{\partial u}{\partial z} = -g\frac{\partial \eta}{\partial x} + \frac{1}{\rho}\frac{\partial \tau_{xx}}{\partial x} + \frac{1}{\rho}\frac{\partial \tau_{xy}}{\partial y} + \frac{1}{\rho}\frac{\partial \tau_{xz}}{\partial z}, \tag{8.45}$$

$$\frac{\partial v}{\partial t} + \frac{\partial v}{\partial x} + v\frac{\partial v}{\partial y} + w\frac{\partial v}{\partial z} = -g\frac{\partial \eta}{\partial y} + \frac{1}{\rho}\frac{\partial \tau_{yx}}{\partial x} + \frac{1}{\rho}\frac{\partial \tau_{yy}}{\partial y} + \frac{1}{\rho}\frac{\partial \tau_{yz}}{\partial z}. \tag{8.46}$$

The continuity equation takes the same form as in the original NSEs:

$$\frac{\partial u}{\partial x} + \frac{\partial v}{\partial y} + \frac{\partial w}{\partial z} = 0. \tag{8.47}$$

This equation can be used to calculate w, when u and v are solved from (8.45) and (8.46).

On further integrating the continuity equation in the vertical direction from the bottom to the free surface and applying the kinematic free surface boundary condition and the bottom boundary condition, we obtain the so-called height function transport equation:

$$\frac{\partial \eta}{\partial t} + \frac{\partial}{\partial x}\int_{-h}^{\eta} u dz + \frac{\partial}{\partial y}\int_{-h}^{\eta} v dz = 0. \tag{8.48}$$

The aforementioned equation offers a means to monitor the motion of water wave free surface displacement, provided the surface remains stable. By assuming hydrostatic pressure, the governing equations are simplified in contrast to the original NSEs, eliminating the need for iterative pressure solutions. Subsequently, the temporal term facilitates deriving vertical velocity via the continuity equation. A model tackling this PDE system is commonly termed a quasi-3D model, typically employed for simulating extensive phenomena like tides, tsunamis or large-scale currents.

Johns and Jefferson (1980) were the early explorers of this approach, though their model was only 2D in a vertical plane. Blumberg and Mellor (1987) developed a 3D ocean circulation model to study various ocean circulation and mixing problems. Casulli and Cheng (1992) also presented a 3D model of this type and used it to simulate coastal flooding by tides.

Most of these quasi-3D models are constructed on the σ-coordinate, which maps the physical space between a wavy free surface and an uneven bottom to a regular space:

$$\sigma = \frac{z+h}{H}, \tag{8.49}$$

where $H = h + \eta$ and σ is the new coordinate in the vertical direction that varies from 0 (when $z = -h$) to 1 (when $z = \eta$). The boundary conditions can be applied exactly on the free surface and bottom with the introduction of σ coordinate transformation. This introduction modifies the original NSEs to the following forms:

$$\frac{\partial u}{\partial x} + \frac{\partial u}{\partial \sigma}\frac{\partial \sigma}{\partial x^*} + \frac{\partial v}{\partial y} + \frac{\partial v}{\partial \sigma}\frac{\partial \sigma}{\partial y^*} + \frac{\partial w}{\partial \sigma}\frac{\partial \sigma}{\partial z^*} = 0, \tag{8.50}$$

$$\begin{aligned}\frac{\partial u}{\partial t} &+ u\frac{\partial u}{\partial x} + v\frac{\partial u}{\partial y} + w\frac{\partial u}{\partial \sigma}\\ &= -g\frac{\partial \eta}{\partial x} + \frac{\partial \tau_{xx}}{\partial x} + \frac{\partial \tau_{xx}}{\partial \sigma}\frac{\partial \sigma}{\partial x^*} + \frac{\partial \tau_{xy}}{\partial y} + \frac{\partial \tau_{xy}}{\partial \sigma}\frac{\partial \sigma}{\partial y^*} + \frac{\partial \tau_{xz}}{\partial \sigma}\frac{\partial \sigma}{\partial z^*},\end{aligned} \tag{8.51}$$

$$\begin{aligned}\frac{\partial v}{\partial t} &+ u\frac{\partial u}{\partial x} + v\frac{\partial u}{\partial y} + w\frac{\partial u}{\partial \sigma}\\ &= -g\frac{\partial \eta}{\partial y} + \frac{\partial \tau_{yx}}{\partial x} + \frac{\partial \tau_{yx}}{\partial \sigma}\frac{\partial \sigma}{\partial x^*} + \frac{\partial \tau_{yy}}{\partial y} + \frac{\partial \tau_{yy}}{\partial \sigma}\frac{\partial \sigma}{\partial y^*} + \frac{\partial \tau_{yz}}{\partial \sigma}\frac{\partial \sigma}{\partial z^*},\end{aligned} \tag{8.52}$$

where

$$\omega = \frac{D\sigma}{Dt^*} = \frac{\partial\sigma}{\partial t^*} + u\frac{\partial\sigma}{\partial x^*} + v\frac{\partial\sigma}{\partial y^*} + w\frac{\partial\sigma}{\partial z^*}$$

and

$$\frac{\partial\sigma}{\partial t^*} = -\frac{\sigma}{D}\frac{\partial D}{\partial t}, \tag{8.53}$$

$$\frac{\partial\sigma}{\partial x^*} = \frac{1}{D}\frac{\partial h}{\partial x} - \frac{\sigma}{D}\frac{\partial D}{\partial x}, \tag{8.54}$$

$$\frac{\partial\sigma}{\partial y^*} = \frac{1}{D}\frac{\partial h}{\partial x} - \frac{\sigma}{D}\frac{\partial D}{\partial y}, \tag{8.55}$$

$$\frac{\partial\sigma}{\partial z^*} = \frac{1}{D}. \tag{8.56}$$

The above-mentioned equations are valid assuming $x = x^*$ and $y = y^*$.

The free surface tracking equation then becomes

$$\frac{\partial\eta}{\partial t} + \frac{\partial}{\partial x}\left[(\eta + h)\int_0^1 u\, d\sigma\right] + \frac{\partial}{\partial y}\left[(\eta + h)\int_0^1 v\, d\sigma\right] = 0. \tag{8.57}$$

The reason for conversion to the mapped coordinates system is that it is easy to numerically solve the problem with the fixed grid using the finite difference or finite element method.

8.7 Potential Flow Theory (Linear, Weakly Nonlinear and Fully Nonlinear)

8.7.1 *General*

The exploration of fully or weakly nonlinear potential flow began in the late 1970s, gaining significant traction with substantial advancements between the 1990s and 2000s. Today, these models are firmly established and widely utilized in both research and practical applications with a high degree of confidence.

Pioneering work in fully nonlinear modelling can be credited to Longuet-Higgins and Cooklet (1976), whose approach, subsequently modified by others, combined elements of both Eulerian and Lagrangian frameworks. In this mixed approach, Laplace equations are addressed within the Eulerian framework, while the evolution of the free surface is tracked in the Lagrangian framework. This comprehensive model is commonly referred to as fully nonlinear potential flow theory (FNPT).

Kim *et al.* (1999) conducted a thorough review of research and development in nonlinear wave simulation, focusing primarily on boundary element method (BEM) implementations, methods of wave generation and absorption. FNPT models offer distinct advantages over Navier–Stokes (NS) equations, notably in terms of computational cost and the mitigation of numerical damping, crucial for long-duration simulations.

Ensuring the quality of generated waves is essential for accurate wave–structure interaction simulations. These solvers must be more efficient and accurate than viscous CFD solvers in computing incident waves to enhance overall computational efficiency.

8.7.2 *Governing equations and boundary conditions*

Three-dimensional fluid motion is defined with respect to the fixed Cartesian coordinate system, $Oxyz$, with the z axis positive upwards. The water depth h is assumed to be constant. The fluid is assumed to be incompressible and the flow is irrotational. Viscous forces are neglected. This simplifies the flow problem to be defined with Laplace's equation involving velocity potential $\Phi(x, y, z, t)$ given by

$$\nabla^2\Phi = 0. \tag{8.58}$$

A potential flow in a rectangular flume with a wavemaker at one end and the nonlinear free surface boundary condition at the free surface is considered.

The boundary conditions on the solid smooth surface are

$$\frac{\partial\Phi}{\partial n} = \vec{U}.n, \tag{8.59}$$

rear side of the wall. This can be for fixed or moving structure. Dirichlet boundary conditions are specified at the free surface with respect to the instantaneous velocity potential $\Phi(x, y, z)$ which is obtained from the nonlinear kinematic and dynamic free surface boundary conditions. The nonlinear dynamic free surface condition to be satisfied at the air–water interface can be written as

$$\frac{\partial\Phi}{\partial t} + \frac{1}{2}\nabla\Phi\nabla\Phi + g\eta = 0 \quad \text{at } z = \eta(x, y, t). \tag{8.60a}$$

The kinematic boundary condition considering no flow through the free surface can be written as

$$\frac{\partial\Phi}{\partial z} = \frac{\partial\eta}{\partial t} + \frac{\partial\Phi\partial\eta}{\partial x\partial x} + \frac{\partial\Phi\partial\eta}{\partial y\partial y} \quad \text{at } z = \eta(x, y, t). \tag{8.60b}$$

The difficulty with the above-mentioned free surface boundary conditions is that the location of the free surface is not known beforehand which makes it an implicit form of solution. Hence, rewriting the solution in terms of Lagrangian or semi-Lagrangian form and one can march in time to get a new position. In the simulation of nonlinear free surface waves, the second term in Equation (8.60a), usually called convective term (i.e. the variation of the field variable in space), plays a major role.

8.7.2.1 *Lagrangian form*

The Lagrangian form of nonlinear free surface boundary condition was given by Longuet-Higgins and Cokelet (1976):

$$\frac{D\vec{x}}{Dt} = \nabla\Phi,$$
$$\frac{D\Phi}{Dt} = \frac{1}{2}\nabla\Phi\nabla\Phi - g\eta. \tag{8.61}$$

The above-mentioned equations are obtained by substituting the substantial or material derivative $\frac{D}{Dt} = \frac{\partial}{\partial t} + \nabla\Phi.\nabla$ in Equations (8.60a) and (8.60b).

8.7.2.2 *Semi-Lagrangian form*

This form of the free surface boundary condition restricts the movement of the nodes in the horizontal direction but allows movement only in the vertical direction and is given by

$$\frac{\delta\Phi}{\delta t} + \frac{1}{2}\nabla\Phi\nabla\Phi + g\eta + \frac{\partial\Phi}{\partial z}\frac{\partial\eta}{\partial t} = 0$$
$$\frac{\partial\eta}{\partial t} = \frac{\partial\Phi}{\partial z} - \nabla\Phi\nabla\eta. \tag{8.62}$$

Upon expansion, these can be written as

$$\frac{\partial\eta}{\partial t} = \frac{\partial\Phi}{\partial z} - \frac{\partial\eta}{\partial x}\frac{\partial\Phi}{\partial x} - \frac{\partial\eta}{\partial y}\frac{\partial\Phi}{\partial y},$$
$$\frac{\delta\Phi}{\delta t} = -\frac{1}{2}\left[\left(\frac{\partial\Phi}{\partial x}\right)^2 - \left(\frac{\partial\Phi}{\partial y}\right)^2 - \left(\frac{\partial\Phi}{\partial z}\right)^2\right]$$
$$- g\eta - \frac{\partial\Phi}{\partial z}\frac{\partial\eta}{\partial x}\frac{\partial\Phi}{\partial x} - \frac{\partial\Phi}{\partial z}\frac{\partial\eta}{\partial y}\frac{\partial\Phi}{\partial y}. \tag{8.63}$$

In the above-presented equations, the kinematic free surface boundary condition retains its Eulerian form, whereas the dynamic boundary condition deviates strictly from the Eulerian framework. The derivative $\delta/\delta t$

differs from $\partial/\partial t$ (local derivative) and D/Dt (substantial derivative) in that it allows node movement solely in the vertical direction over the time rate of change. Consequently, this formulation is recognized as the semi-Lagrangian or semi-Eulerian approach, owing to the constraint on nodal movement in the horizontal direction. Detailed derivations of both forms are documented in Sriram (2008).

Compared to the Lagrangian form, this method offers the advantage of eliminating the need for regridding. Further insights into regridding processes will be addressed subsequently.

Depending on the problem considered, one can choose a suitable method. Due to the free movement of nodes at the free surface, the Lagrangian approach is more suitable for floating bodies or breaking waves. For fixed structure, such as submerged obstacles or multiple cylinders under non-breaking waves, a semi-Lagrangian approach is preferred compared to the Lagrangian approach. There are two probable reasons for many researchers preferring the Lagrangian approach rather than a true Eulerian approach for handling wave–structure interactions: the Lagrangian points have the tendency to concentrate on the rapid region of variations and for a non-single value free surface, there is no projection available in the Eulerian manner. Thus, in either case, the Lagrangian method of updating seems to be simple and leads to higher numerical accuracy even though solving the Dirichlet problem as proposed by Longuet-Higgins and Cokelet (1976) at every step is non-physics-based updating. A truly Eulerian-based updating method has been attempted by combining the BEM and level set method by Garzon and Sethian (2006), wherein the velocity is computed using the higher-order BEM and the free surface nodes are moved using the level set method and the model is validated for the overturning waves. The paper states that merging of the overturning waves with the free surface is also possible using this method, whereas the Lagrangian method will break down. The formulation based on nonlinear boundary conditions solving Laplace's equation is called fully nonlinear potential theory (FNPT). To summarize the FNPT-based MEL, the boundary value problem (BVP) associated with Laplace's equation is split into two steps:

1. **Eulerian Framework:** For the given domain, the instantaneous velocity potential inside the domain is obtained based on the prescribed forcing or rigid boundaries.
2. **Lagrangian Framework:** Based on the solution sought in the previous step, the velocity potential and the free surface wave elevation are

updated by numerically integrating the free surface boundary condition (Equations (8.61) or (8.63)).

Further, Equation (8.61) or (8.63) can also be approximated as various orders, leading to weakly nonlinear models.

8.7.3 *Treating breaking waves and viscous effects*

The phenomenon of wave breaking remains an enigma in mechanics, with many of its intricacies yet to be fully elucidated. A significant challenge lies in preemptively identifying which wave events will result in a broken wave. While the mechanisms leading to this event appear to be generally universal, it has not been definitively established whether a simple parameter exists to predict its onset.

Wave breaking involves one time scale and two length scales that are intrinsic to the phenomenon, constituting an energy-dissipating mechanism. In the absence of a well-defined theory, estimates of energy dissipation rates have traditionally relied on dimensional analysis and empirical evidence. The temporal nature of wave breaking commences with the identification of local instabilities, leading to the onset or incipient wave-breaking geometry, culminating in the residual stage of wave breaking. The collective duration of this process is termed the time duration of wave breaking ($T_{\rm br}$). Spatially, this disturbance can be roughly observed over a horizontal length characterized as $L_{\rm br}$. Consequently, the presence of these two scales enables the establishment of a horizontal propagation velocity scale for this instability/disturbance.

In fully or weakly nonlinear models, the primary objective is to detect when a wave is on the verge of overturning using numerical approaches. Subsequently, viscous effects, or artificial damping, are applied to simulate the breaking characteristics.

One notable approach is the eddy viscosity-based wave-breaking model proposed by Tian *et al.* (2010). In this model, energy dissipation correlates with the turbulence intensity generated by the plunging of the wave crest into the surrounding water or by the breakdown of the toe of the bulge in the case of a spilling breaker. In both scenarios, a small area of whitecap coverage forms downstream of the breaking crest. This region subsequently integrates into the propagating waveform, entraining air bubbles that collapse and further dissipate energy (Tian *et al.*, 2010). Estimating an integral length scale of turbulence for this area presents a significant challenge in reality. Therefore, the wave-breaking height ($H_{\rm br}$) can serve as a surrogate for this length scale, as the depth of penetration into the water is closely related to this height. Building on these insights,

Tian *et al.* (2010) estimate the rate of shear in the turbulence region as $O(c_{\mathrm{br}}/H_{\mathrm{br}})$, where c_{br} is representative of characteristic breaking speeds. In this way, if this turbulence is characterized by an eddy viscosity (ν), then the energy dissipation rate is $O(\nu\frac{c_{\mathrm{br}}^2}{H_{\mathrm{br}}^2}H_{\mathrm{br}}L_{\mathrm{br}})$. Equating it to the kinetic energy of the propagating disturbance $O(c_{\mathrm{br}}^2 L_{\mathrm{br}}\frac{L_{\mathrm{br}}}{T_{\mathrm{br}}})$ that is dissipated, the following expression is obtained:

$$\nu = \alpha\frac{H_{\mathrm{br}}L_{\mathrm{br}}}{T_{\mathrm{br}}}, \tag{8.64}$$

where α is a proportionality constant dependent on the ambient conditions of wave breaking like prior air entrainment. For standalone breaking events, $\alpha \sim 0.02$ according to Tian *et al.* (2010). However, multiple violent events in close proximity need to be adjusted for a value of 0.035. This selection of proportionality constant is, thus, a subjective tuning parameter in the breaking model. A better model for α based on parameterization of air entrainment in breaking events is, thus, needed. Thus, the computation of eddy viscosity is dependent upon the geometry of the wave-breaking event that determines the relevant parameters H_{br}, L_{br}, T_{br}.

This approach was implemented in the higher-order spectral method by Seiffert and Ducrozet (2018). In Hasan *et al.*'s study (2019), the eddy viscosity model is extended to incorporate second-order wave-making signals into the existing semi-empirical model. This is followed by testing Barthelemy's parameter as a trigger for this model over a large number of values for intermediate water depths. The hypothesis is that if the parameter is universal in nature, then perhaps there would be a unique value where the experiments best fit the results. However, this turns out to be rather incorrect for varied reasons. Therefore, the best outcome of such testing is the calibration of this so-called Tian–Barthelemy eddy viscosity model against empirical evidence.

Eddy viscosity wave-breaking models have been proven to be very effective in relation to shallow-water dynamics (Kennedy *et al.*, 2000; Chen *et al.*, 2000). As such, Tian *et al.* (2010) extended this approach into a semi-empirical model in their paper that relied on local wave slope as a trigger mechanism for eddy viscosity. Seiffert *et al.* (2017) instead combined this model with Barthelemy's parameter as a threshold to trigger the eddy viscosity and gave it the name Tian–Barthelemy wave-breaking model. Furthermore, Seiffert *et al.* (2017) showed that, compared with Chalikov and

Sheinin (2005), the Tian–Barthelemy model remains simpler to implement and does not depend a priori on the knowledge of the physical wave-breaking event. More details can be found in Seiffert *et al.* (2017) or Hasan (2020).

8.8 Shallow-Water Equations

In shallow regions, where the depth is comparatively smaller compared to the horizontal scale, depth-averaged equations can be used to get rid of vertical dimension. Based on this assumption, we start from Euler equations to derive shallow-water equations:

Boundary conditions:

Free surface condition:

$$p = 0, \quad \frac{d\eta}{dt} = \frac{\partial \eta}{\partial t} + v.\nabla\eta = \mathrm{w}. \tag{8.65}$$

Bottom boundary condition:

$$u.\nabla(\mathrm{z} + \mathrm{h}(\mathrm{x}, \mathrm{y})) = 0. \tag{8.66}$$

We integrate continuity equation

$$\begin{aligned}
0 &= \int_{-h}^{\eta} [\nabla.u] dz \\
&= \int_{-h}^{\eta} \left[\frac{\partial u}{\partial x} + \frac{\partial v}{\partial y} + \frac{\partial w}{\partial z}\right] dz \\
&= \frac{\partial}{\partial x}\int_{-h}^{\eta} \mathrm{u}dz - u|_{z=\eta}\frac{\partial \eta}{\partial x} + u|_{z=-\mathrm{h}}\frac{\partial(-\mathrm{h})}{\partial x}, \\
&\quad + \frac{\partial}{\partial y}\int_{-h}^{\eta} \mathrm{v}dz - v|_{z=\eta}\frac{\partial \eta}{\partial y} + v|_{z=-\mathrm{h}}\frac{\partial(-\mathrm{h})}{\partial y}, \\
&\quad + w|_{z=\eta} + w|_{z=-\mathrm{h}}.
\end{aligned} \tag{8.67}$$

Applying the free surface and bottom boundary condition to Equation 8.67, the equation becomes

$$\frac{\partial \eta}{\partial t} + \frac{\partial}{\partial x}\int_{-h}^{\eta} \mathrm{u}dz - \frac{\partial}{\partial y}\int_{-h}^{\eta} \mathrm{v}dz = 0. \tag{8.68}$$

The momentum equation is as follows:

$$\frac{d\mathrm{u}}{dt} + \frac{1}{\rho}\nabla\mathrm{p} + \mathrm{gz} = 0. \tag{8.69}$$

By assuming the relative depth (ratio between water depth and wavelength) is small, vertical acceleration is neglected from the above-mentioned momentum equation. The pressure can be found by integrating the momentum equation:

$$\int_{-h}^{\eta} \frac{\partial p}{\partial z} dz = \int_{-h}^{\eta} \rho g dz, \tag{8.70}$$

$$p(x, y, \eta, t) - p(x, y, z, t) = -\rho g(\eta(x, y, t) - z). \tag{8.71}$$

Assuming no vertical variations in u and v and using free surface condition, $p(x, y, \eta, t)$ hydrostatic pressure is given by

$$p(x, y, z, t) = \rho g(\eta(x, y, t) - z). \tag{8.72}$$

Using the above-mentioned equation in horizontal momentum equations,

$$\frac{\partial u}{\partial t} + u\frac{\partial u}{\partial x} + v\frac{\partial u}{\partial y} + g\frac{\partial \eta}{\partial x} = 0, \tag{8.73}$$

$$\frac{\partial v}{\partial t} + u\frac{\partial v}{\partial x} + v\frac{\partial v}{\partial y} + g\frac{\partial \eta}{\partial y} = 0. \tag{8.74}$$

Equation (8.65) becomes

$$\frac{\partial \eta}{\partial t} + \frac{\partial}{\partial x}(\eta + h)u + \frac{\partial}{\partial y}(\eta + h)v = 0. \tag{8.75}$$

Equations (8.73) to (8.75) are shallow-water equations.

The SWEs are a system of hyperbolic/parabolic PDEs governing fluid flow in the oceans, coastal regions, estuaries, rivers and channels. The SWEs can be used to predict tides, storm surge levels and coastline changes from hurricanes, and ocean currents, and to study dredging feasibility. Various schemes and numerical techniques are present to discretize a physical domain, like the finite element, finite difference, finite volume or a mesh-free technique. Various numerical storm surge models have been developed for many regions in the world. Scientists utilize 2D depth-averaged shallow-water equations as their governing equations. One of the key ideologies in these models is that the coast is also responsible for the generation of a storm surge (using a moving boundary) and it experiences the subsequent effects of the storm surge (Madsen and Jakobsen, 2004).

These models are well developed and have been used for real-time forecasting of cyclone and storm surges by many countries. The available models that are widely used for cyclone inundations are Delft3D-FLOW, TELEMAC, SLOSH and ADCIRC. A typical comparison of the observed

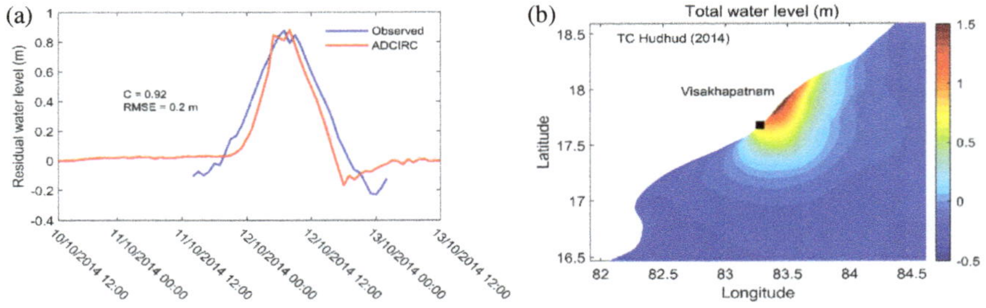

Fig. 8.4. (a) Validation of computed residual water levels with observed surge residuals obtained from the in-situ tide gauge located at Visakhapatnam. (b) ADCIRC model computed total water level (m) at the landfall location. The black square dot represents the tide gauge station located at Visakhapatnam (83.28°E, 17.68°N).

Source: Shashank *et al.* (2024).

surge and the ADCIRC model for Tropical Cyclone Hudhud is shown in Fig. 8.4.

8.8.1 *The Boussinesq models*

8.8.1.1 *The standard Boussinesq models*

The Boussinesq equations are intended to describe the irrotational motion of an incompressible homogeneous inviscid fluid in the long-wave limit. The goal of this type of modelling is to reduce 3D problems to two-dimensional (2D) ones. This is done by assuming a polynomial (usually linear) vertical distribution of the flow field while considering non-hydrostatic effects. This is the principal physical difference with nonlinear shallow-water (NSW) equations.

The Boussinesq equations describe a solitary wave in water of constant depth: Compared with the Airy equations, they include an extra term due to the effect of the vertical acceleration of the water on the pressure but they are limited to small-amplitude waves. These equations corresponding to Boussinesq's as per Peregrine (1967) for water of variable depth using an expansion similar to the one proposed by Keller (1948). These equations possess a hyperbolic structure (the same as in the nonlinear shallow-water equations) combined with high-order derivatives to model wave dispersion. There are two important parameters associated with long waves: One is the ratio of amplitude to depth and the other is the ratio of depth to wavelength, represented by ε and μ, respectively.

Non-dimensional variables are used as follows in this section:

$$\begin{aligned}(x,y,z) &= (1/h)^*(x^*,y^*,z^*), \quad t = t^*g/h, \quad p = p^*/gh,\\ (u,v,w) &= (gh\wedge(-1/2))(u^*,v,^*,w^*), \end{aligned} \tag{8.76}$$

where * indicates a dimensional variable and h is a length representing the depth of water. The z-axis is taken vertically upwards, the free surface is $z = (x,y,t)$ and the bottom is $z = -H(x,y)$. Wave amplitude η is given by $\varepsilon\zeta_1 + \varepsilon^2\zeta_2$.

The Boussinesq equations are given by

$$\frac{\partial\eta}{\partial t} + \nabla.[H+\eta]u + \frac{1}{2}\nabla.\{H^2\nabla[\nabla.(Hu)]\} + \frac{1}{3}\{H^3\nabla[\nabla.(u)]\} = 0, \tag{8.77}$$

$$\frac{\partial u}{\partial t} + (u.\nabla)u + \nabla\eta = 0, \tag{8.78}$$

where u is the mean velocity.

8.8.1.2 *Nwogu's model*

Nwogu (1993) used the velocity u_α at arbitrary elevation z_α, instead of using the bottom velocity or the depth-averaged velocity as the velocity variable. The velocities and the pressures are expressed in terms of u_α. The new set of equations to model horizontal propagation of irregular, multidirectional waves of varying depth is given as

$$\begin{aligned}&\frac{\partial\eta}{\partial t} + \nabla.[H+\eta]u_\alpha + \mu^2\nabla\\ &\quad .\left\{\left(\frac{z_\alpha^2}{2} - \frac{h^2}{6}\right)h\nabla[\nabla.(u_\alpha)] + \left(z_\alpha - \frac{h}{2}\right)h\nabla[\nabla.(hu_\alpha)]\right\} = 0,\end{aligned} \tag{8.79}$$

$$\frac{\partial u_\alpha}{\partial t} + \nabla\eta + \varepsilon(u_\alpha.\nabla)u_\alpha + \mu^2\left\{\left(\frac{z_\alpha^2}{2}\right)\nabla[\nabla.(u_\alpha)] + (z_\alpha\nabla)[\nabla.(hu_\alpha)]\right\} = 0. \tag{8.80}$$

The above-mentioned new set of equations may be regarded as a class of equations containing most known forms of Boussinesq-type equations with the elevation of z_α. Since the equations approximate fully dispersive and nonlinear problems, one can select a velocity variable to minimize the errors introduced by the approximation.

The numerical model showed good agreement for the shoaling of regular waves from intermediate to shallow water and irregular waves from deep to shallow water. The numerical model was able to reproduce several nonlinear effects that occur during shoaling such as generation of longer- and shorter-period waves, an increase in horizontal and vertical asymmetries, and the evolution of wave groups.

8.8.1.3 *Madsen and his co-workers*

The numerical model is based on two-dimensional equations of the Boussinesq type using a flux formulation with improved linear dispersion characteristics. The model is extended to include the possibility of wave breaking of regular and irregular waves in two horizontal dimensions and to consider a moving shoreline, allowing for a study of the surf zone and swash zone.

The incorporation of wave breaking in the two-dimensional Boussinesq model is based on the surface roller concept for spilling breakers using a geometrical determination of the instantaneous roller thickness at each point and modelling the effect of wave breaking using an additional convective momentum term. This is a function of the local wave celerity, which is determined interactively.

The modified version of the Boussinesq equation is as follows:

$$\frac{\partial \eta}{\partial t} + \frac{\partial u}{\partial x} + \frac{\partial v}{\partial y} = 0, \tag{8.81}$$

$$\frac{\partial \mathrm{u}}{\partial t} + \frac{\partial}{\partial x}\left(\frac{u^2}{d}\right) + \frac{\partial}{\partial y}\left(\frac{uv}{d}\right) + \frac{\partial R_{xx}}{\partial y} + \frac{\partial R_{xy}}{\partial y} + gd\frac{\partial \eta}{\partial x} + \varphi_x + \frac{\tau_x}{\rho} = 0, \tag{8.82}$$

$$\frac{\partial \mathrm{v}}{\partial t} + \frac{\partial}{\partial x}\left(\frac{uv}{d}\right) + \frac{\partial}{\partial y}\left(\frac{v^2}{d}\right) + \frac{\partial R_{yy}}{\partial y} + \frac{\partial R_{xy}}{\partial y} + gd\frac{\partial \eta}{\partial y} + \varphi_y + \frac{\tau_y}{\rho} = 0. \tag{8.83}$$

The terms denoted φ_x and φ_y are dispersive Boussinesq-type terms, which in shallow water may be taken from Peregrine (1967). Here, (u, v) is the depth-integrated velocity (the volume flux) in the Cartesian coordinate system (x, y), $d = \eta + h$ is the instantaneous depth and η is the surface elevation.

8.8.1.4 *Gobbi and his co-workers*

The O (μ^2) model describes wave propagation in shallow- to intermediate-water depth over a particular range, but still has its limitations in terms of the accuracy of water particle kinematics. The type of model of the order O (μ^2) becomes less useful for studies pertaining to seabed kinematics at water depths far smaller than that implied by the accuracy of the dispersion relation, unless an artificial means is employed to estimate near-bed velocities from surface values. As a result, higher-order approximations in the Boussinesq model are required to increase the range of water depths and to study seabed kinematics. Gobbi *et al.* (2000) approched this problem by constructing a velocity potential as a weight average at two Nwogu-type reference levels. The extension to O (μ^4) of the vertical structure of the velocity potential provides a dramatic enhancement in the prediction of velocity components.

8.8.1.5 *Liu and his co-workers*

Lynett and Liu (2004) offered an alternative approach in which they incorporated the multi-layer formulation with the fully nonlinear and weakly dispersive wave approximations. In their model, the water column is first divided into several layers with both the horizontal velocity and the pressure being matched at the interfaces between layers. A quadratic horizontal velocity profile and linear vertical velocity profile are employed by assuming it as weak and horizontal vorticity of O (μ^4) is also assumed in each layer. More accurate velocity profiles in the water column and better representation of linear frequency dispersion relation can be obtained by increasing the number of layers. For example, the two-layer model exhibits accurate linear and nonlinear characteristics up to $kd = 6.5$. A multi-layer irrotational BT model over a slow-varying bottom is proposed by generalizing the approach proposed by Liu and Fang (2016) in their two-layer BT model. The approach divides the simulation domain into multiple layers (Lynett and Liu, 2004), replaces the velocities at an arbitrary water depth with pseudo-velocities (Madsen *et al.*, 2002), and decouples problems into linear and nonlinear parts (Agnon *et al.*, 1999). An important attribute of the proposed model is that it simulates linear and nonlinear properties with high accuracy, thereby increasing its applicability for highly nonlinear and extremely dispersive water waves. Recently, Yang and Liu (2022) developed the multi-layer model based on rotational flow, instead of the irrotational flow assumption.

8.8.2 *Applications of the BSQs*

Boussinesq equations are widely used in coastal and ocean engineering. One example is tsunami wave modelling. These equations can also be applied to model tidal oscillations, and these types of wave motion are perfectly described by the Navier–Stokes equations, but currently it is impossible to solve fully three-dimensional (3D) models in any significant domain.

Thus, approximate models such as the Boussinesq equations must be used. There has been tremendous growth in the development and application of Boussinesq-type models (or BTMs) (Brocchini, 2013), with simulations of surface wind waves over domains with dimensions of tens of kilometres or of tsunami waves at global scale now being readily feasible.

Depth-averaged models are also used for addressing porous flow by incorporating the porosity and resistance terms as briefly described in the following:

Porosity is uniform and the continuity equation in porous media is given by

$$\nabla . U = 0, \tag{8.84}$$

where $U = (u, v, w)$ is the seepage velocity vector and ∇ is the gradient operator. The momentum equation in porous media is given by

$$\frac{\gamma \partial \mathrm{U}}{\partial t} + \frac{1}{\rho}\nabla(p + \rho g z) + D + I = 0, \tag{8.85}$$

where λ is the porosity, p is the pore pressure, ρ is water density, g is the gravitational acceleration, D is the drag resistance term and I is the inertial resistance term. It should be noted that the continuity equation and the momentum equation are expressed in terms of the seepage velocity instead of the discharge velocity. The drag resistance term is specified in different studies. Since the 1990s, in order to simulate waves in non-porous water, several researchers have developed extended Boussinesq equations which can be applied to deeper waters (Madsen and Sørensen, 1992; Nwogu, 1993). Also, for waves over a porous bed, Cruz *et al.* (1997) and Hsiao *et al.* (2002) followed the approaches of Madsen and Sørensen (1992) and Nwogu (1993), respectively, to extend their model for deeper waters. A small tuning parameter was first suggested by Madsen *et al.* (1991) and Madsen and Sørensen (1992) to improve the dispersion and shoaling properties of the model equations. In the momentum equations, the dispersive terms were added and multiplied by an optimum value of the tuning parameter. Later, Cruz *et al.* (1997) used the tuning parameter for waves over porous beds. The laminar and turbulent drag resistances depend on several parameters such as the porosity, the shape and the grade of the porous material, the Reynolds number and the flow characteristics. Various numerical models were calibrated to discover optimal values of the drag coefficients for the models (del Jesus *et al.*, 2012; Ma *et al.*, 2014; Shagun, 2022). An example of solitary wave propagation in the porous region is shown in Fig. 8.5 for the split-up nature as well as the transmission

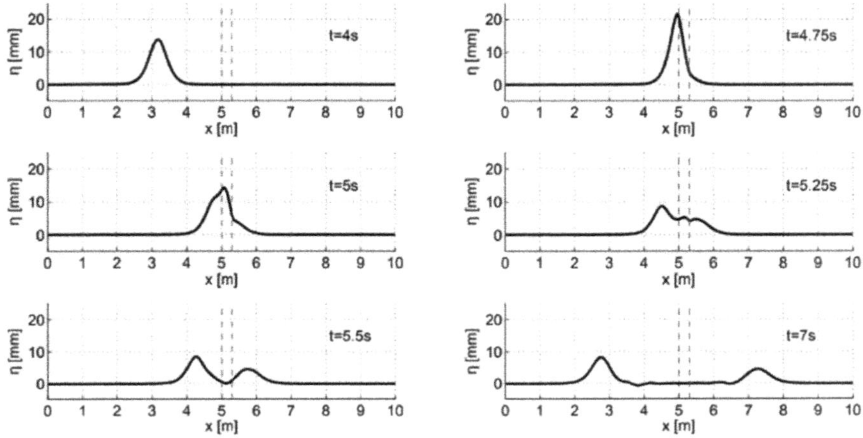

Fig. 8.5. Snapshots of surface elevation at different time instants depicting splitting of a solitary wave through the porous breakwater.
Source: Shagun (2022).

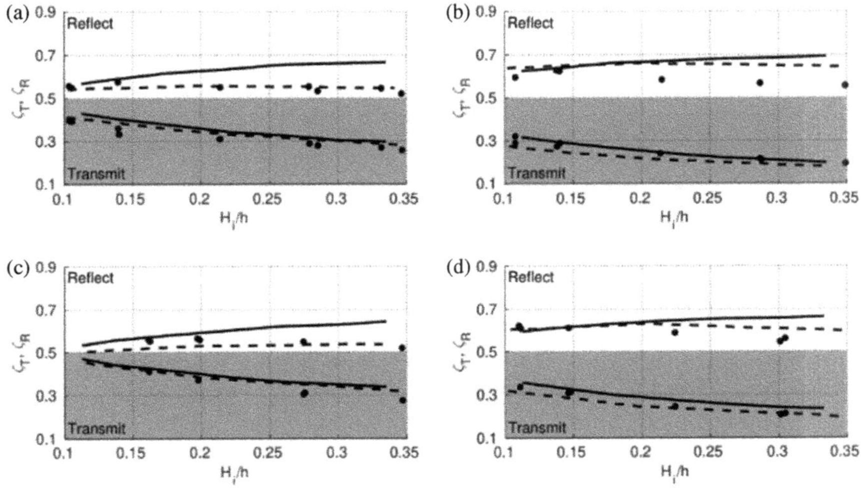

Fig. 8.6. Plots showing estimated transmission and reflection coefficient for 1D solitary waves through porous breakwater; shaded regions show transmission.
Source: Shagun (2022).

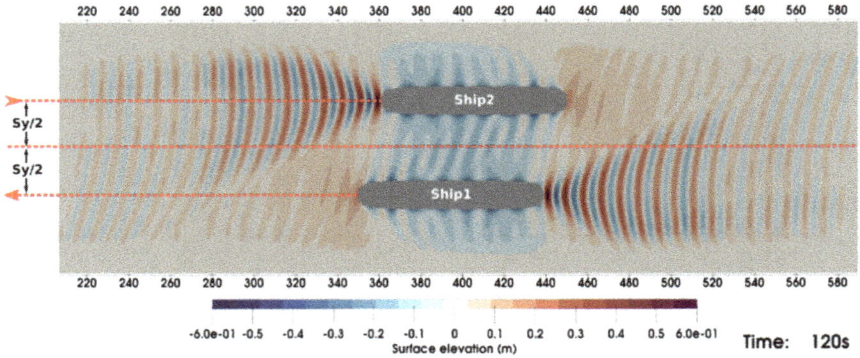

Fig. 8.7. Snapshot of the simulation results for two vessels passing each other while moving in opposite directions with a separation distance of 45 m.
Source: Shagun (2022).

and reflection coefficient as shown in Fig. 8.6. Recently, the Boussinesq model has been increasingly used to study ship wakes in inland waterways or near shallow-water region to understand bank effects (see Fig. 8.7); interested readers can see the works of Shagun (2022) and Chakraborty *et al.* (2022).

8.9 Mild Slope Equation Models

Based on the assumptions of linear waves and slowly varying bottoms, the mild slope equation (MSE) was originally derived. It can be used to describe combined wave refraction and diffraction in both deep and shallow waters. Even though it can be applied to irregular waves by summing up different wave harmonics, the MSE model is mostly employed to study monochromatic waves. The application of this model is limited to the region from an offshore location to a nearshore location some distance away from the shoreline before wave nonlinearity becomes strong. One exception is its application in harbour resonance modelling because water depth in a harbour is usually deep even along the boundaries. The MSE has three different formulations: the parabolic MSE (Radder, 1979) for a simplified steady-state wave field, the elliptic MSE (Berkhoff, 1972) for a steady-state wave field and the hyperbolic MSE (Lee *et al.*, 1998) for a time-dependent wave field that has a primary wave propagation direction.

8.10 Wave Spectral Models

The wave action spectral model or wave energy spectral model is the most commonly employed wave model in modelling large-scale wave motion. It is constructed on the assumption that a random sea state is composed of an infinite number of linear waves whose wave height is a function of wave propagation direction and wave frequency. The rate of change of wave energy flux for an individual wave train is balanced by the wave energy transfer among different wave components in different directions and different frequencies as well as energy input and dissipation. The representative wave energy spectral models are WAM (WAve prediction Model) (Hasselmann *et al.*, 1988) and WaveWatch III (Tolman, 1999), which can be used to model large-scale variations of wave heights in deep oceans. A wave energy spectral model connected with an atmospheric model is able to predict global ocean wave climate. When the current effect is considered, the wave action spectral model can be formulated and used instead to simulate combined wave–current interactions over a large-scale nearshore region. One such model is SWAN (Simulating WAves Nearshore; Ris *et al.*, 1999). The wave spectral models can use computational meshes much larger than a wavelength with the wave phase information being filtered out, thus allowing the models to be applied over a very large area. However, without wave phase information, the model is unable to represent wave diffraction that is phase related. Hence, this kind of model is usually used to provide far-field wave information only.

8.11 Hybrid Models

In the preceding section, we explored various models designed to address the challenges encountered in coastal engineering. Given the expansive spatial and temporal scales inherent in coastal engineering endeavours, the notion of relying on a singular model to tackle all problem classes is deemed ideal. One strategy to achieve this is through the integration of diverse modelling tools developed over time. This involves leveraging information from one approximation model to inform others. For instance, combining potential flow modelling for distant regions with Navier–Stokes solvers for nearby areas facilitates a comprehensive understanding of wave–structure interactions.

Traditionally, in coastal engineering, large-scale problems are dealt with as it is difficult to solve waves and currents together; hence, normally, flow models and wave models, such as SWAN and ADCIRC or TELEMAC tools, are integrated.

In CFD, there are two approaches: One is domain decomposition and the other is functional decomposition. The domain decomposition (DD) strategy divides the computational domain into parts and applies different mathematical models in each part. In most DD-based wave–structure interaction (WSI) problems, the computational domain is decomposed into a viscous inner sub-domain and a potential/rotational outer sub-domain. The information (velocity, pressure and surface elevation) will be transferred through either relaxation zones or a sharp interface. In addition, based on how the information between the solvers is being transferred, it can be either one-way coupling (weak coupling) or two-way coupling (strong coupling). In one-way coupling, information is transferred only from the potential solver into the viscous solver, but in two-way coupling, the information is transferred both ways, from depth-averaged or depth-resolving irrotational models to full NS and vice-versa. Two-way coupling is advantageous since it allows for a significantly smaller computational region for the viscous solver. However, it necessitates an iterative process or an implicit approach between the two models on a shared interface, which might increase the computational costs. The advantage of one-way coupling is that no such iterations are needed, but it needs a longer viscous domain to avoid the reflection from outer boundaries. This method is suitable when one needs to analyze the kinematics of the breaking waves in deep water or depth-induced breaking in the shallow water region. Grilli and Horrillo (1999) and Grilli *et al.* (2004) coupled the 2D HOBEM-FNPT with the NS model based on SL-VOF.

Extension of the FNPT (FEM) code coupled with the NS model was carried out by Clauss *et al.* (2004, 2005); for the NS model, commercial software such as FLUENT, CFX or COMET was adopted. The coupling approach was tested by investigating deep-water wave breaking (breaking of freak waves) and comparing it with experimental measurements. Yan and Ma (2011) coupled the QALE-FEM with the commercial software STAR-CD to study the wind effects on breaking waves. Hildebrandt *et al.* (2013) coupled the FEM with the commercial software ANSYS to model the wave impacts with a tripod structure. Narayanaswamy *et al.* (2010) used one-way coupling of the Boussinesq model with the SPH method for solitary wave simulations without feedback from the SPH to the Boussinesq model, with a fixed overlapping zone being considered to transfer the information. Recently, this was improved by Agarwal *et al.* (2022)

by coupling Boussinesq with the MLPG (Meshless Local Petrov Galerkin) method.

However, if one needs to analyze the wave structure interactions in the presence of floating bodies or fixed structures, then strong coupling of the two models is required, wherein the radiated waves will propagate from the NS model to depth-averaged or depth-resolved irrotational models. In strong coupling, the computational domain will be divided into two parts: In one part, the generation and propagation of waves are considered while, in the other part, the structure/breaking region will be present. The modelling of the first part of the domain will be carried out using depth-averaged or depth-resolved irrotational models, and then the boundary conditions (velocity and pressures) are fed into the NS model at the same time steps to study the remaining part of the domain. Then, the velocity from the NS model is again fed back to the depth-averaged or depth-resolved irrotational model domain for the next time steps. Thus, in general, strong coupling is needed to couple the models both in space and time domains. For coupling in the space domain, the following four methods have been employed as pointed out by Sriram *et al.* (2014): (a) fixed boundary interface, (b) moving boundary interface, (c) fixed overlapping zone and (d) moving overlapping zone.

Studies by Colicchio *et al.* (2006), Greco *et al.* (2007), Sitanggang (2008) and Sriram *et al.* (2014) explored the full capability of strong coupling and highlighted the feasibility through detailed analysis. However, to date, strong coupling has mostly been realized only in 2D problems, and, to the best of the authors' knowledge, strong coupling in 3D is yet to be attempted. The following are typical examples of simulations performed based on one-way coupling using the codes developed by the authors and their co-workers: IITM-RANS3D Saincher and Sriram (2022a), HOS-NWT-foamstar (using depth-resolved potential and viscous models, Aliyar *et al.*, 2022) and FEBOUSS-MLPG (using depth-averaged potential and viscous models, Agarwal *et al.*, 2022). Sriram and Ma (2021) reviewed the worked carried out in MLPG and the different coupling approaches that exist in practice. However, one of the theoretical issues in the existing domain decomposition-based coupling is that, traditionally, irrotational flow model assumption results are given as an input to the rotational flow models. Thus, there is a mathematical discontinuity in the velocity field, particularly for strong coupling methods.

The fundamental concept for functional decomposition (FD) is to use Helmholtz decomposition to separate the velocity field into rotational and

Table 8.2. Comparative studies carried out through different numerical models.

Experiments	Publication referred	Benchmark test case description
Benchmark test cases for numerical wave absorption	Clément (1999)	Test case for fully nonlinear boundary conditions in order to simulate the open sea problem using wave absorption. Different treatments were reported; however, comparison between them was not reported.
Benchmark test cases of radiation problem	Tanizawa and Clément (2000)	Sinusoidal heaving wedge was taken as benchmark. FEM and BEM codes are mostly used. Comparison between different codes is provided but no discussion on the results obtained from different codes.
Sloshing benchmark test	Loysel *et al.* (2012)	A simple tank geometry, excitation conditions and measurement of pressure and free surface elevation.
Focused wave interactions with a fixed FPSO-like structure	Ransley *et al.* (2019)	This is a blind test exercise, meaning the experiment details are only provided initially; the results of the experiments have not been released to the participants. Comparison between free surface elevation, run-up and pressures with different numerical models is reported.
Focused wave interactions with floating structures	Ransley *et al.* (2020, 2021)	Blind test on focused wave with floating structures. Low-fidelity linear models to high-fidelity Navier–Stokes (NS) solvers were reported. The numerical results for heave, surge, pitch and mooring load are compared against corresponding physical data.
Two tests: a hemispherical-bottomed buoy and a truncated cylinder with a moon pool		
Nonlinear interaction between a focusing wave and cylinder	Sriram *et al.* (2021),	Comparisons between 20 different numerical solvers on wave elevation and pressure at different locations on a fixed cylinder. A detailed discussion with the maximum, minimum and median relative errors in different numerical solvers is reported.
Nonlinear interaction between a focusing wave and moving cylinder	Agarwal *et al.* (2021)	This is a challenging problem, as the cylinder moves over 40 m and interacts with the focusing waves. Performances of the different solvers are discussed with errors for two different speeds of the cylinder.
Breaking waves interaction with vertical wall retrofitted with recurved parapet in small and large scale	Saincher *et al.* (2023).	The interaction between breaking waves and a vertical wall with a recurved parapet for two different scales (1:8 and 1:1) is reported for the first time. Pressure on the vertical and recurve portion, peak impacts are compared with 10 different solvers. Each of the solver has different two phase flow modelling approaches.

irrotational parts to investigate the free surface flow (Dommermuth, 1993). In FD, there are two categories: (1) The wave–structure problem is split into a potential component and a viscous part, i.e. a complete problem is initially solved by a potential solver and then rectified by adding the viscous correction (Kim *et al.*, 2005; Edmund *et al.*, 2013; Rosemurgy *et al.*, 2016; Robaux and Benoit, 2021). (2) The total unknown is decomposed into the incident part and the complementary part. Only the incident flow is modelled in the incident part (wave only), leaving all the interaction with structure, calculated by the viscous solver, as the complementary part (Ferrant *et al.*, 2002; Gentaz *et al.*, 2004; Li *et al.*, 2021; Kim, 2021; Aliyar *et al.*, 2022). Although computationally faster than the full Navier–Stokes equations (Aliyar *et al.*, 2022), this method inherits all the challenges associated with potential flow, including stability issues in steep waves and breaking induced by complex interactions with structures. More details about the methods can be found in the cited text.

8.12 Comparative Studies

In recent years, comparative studies have become quite common. There are many initiatives taken to make experiments available in the public domain and encourage researchers to validate their developed or existing numerical models. Table 8.2 provides a list of such available experiments. There may be other reliable experiments that are available in the literature that have been widely used by researchers to validate numerical models; however, Table 8.2 lists only the publications that report a comparative study from several numerical models.

References

Agarwal, S., Saincher, S., Sriram, V., Yan, S., Xie, Z., Schlurmann, T., Ma, Q., ... and Ferrant, P. (2021). A comparative study on the nonlinear interaction between a focusing wave and cylinder using state-of-the-art solvers: Part B. *International Journal of Offshore and Polar Engineering*, 31(1), 11–18. https://doi.org/10.17736/ijope.2021.jc832.

Agarwal, S., Sriram, V. and Murali, K. (2022). Three-dimensional coupling between Boussinesq (FEM) and Navier–Stokes (particle based) models for wave structure interaction. *Ocean Engineering*, 263, 112426. https://doi.org/10.1016/j.oceaneng.2022.112426.

Agnon, Y., Madsen, P.A. and Schüffer, H.A. (1999). A new approach to high order Boussinesq models. *Journal of Fluid Mechanics*, 399, 319–333.

Akbari, H. and Namin, M.M. (2013). Moving particle method for modelling wave interaction with porous structures. *Coastal Engineering*, 74, 59–73.

Aliyar, S., Ducrozet, G., Bouscasse, B., Bonnefoy, F., Sriram, V. and Ferrant, P. (2022). Numerical coupling strategy using HOS-Open FOAM-Moor Dyn for OC3 Hywind SPAR type platform. *Ocean Engineering*, 263, 112206. https://doi.org/10.1016/j.oceaneng.2022.112206.

Anderson, J.D. (1995). *Computational Fluid Dynamics, The Basics with Applications.* McGraw-Hill Education, United states of America.

Andrews, M. and O'Rourke, P. (1996). Tdupithe multiphase particle-in-cell (MP-PIC) method for dense particulate flows. *International Journal of Multiphase Flow*, 22, 379–402.

Berkhoff, J.C.W. (1972). Computation of combined refraction diffraction problem. In *Proceedings of 13th International Conference in Coastal Engineering*, pp. 796–814.

Blumberg, A.F. and Mellor, G.L. (1987). A description of a three-dimensional coastal ocean circulation model. In Heaps, N. (ed.), *Three Dimensional Coastal Ocean Models.* American Geophysical Union, Washington, DC, p. 208.

Burcharth, H.F. and Andersen, O.H. (1995). On the one dimensional steady and unsteady porous flow equations. *Coastal Engineering*, 24, 233–257.

Brocchini, M. (2013). A reasoned overview on Boussinesq-type models: The interplay between physics, mathematics and numerics. *Proceedings of the Royal Society A: Mathematical, Physical and Engineering Sciences*, 469(2160). https://doi.org/10.1098/rspa.2013.0496.

Casulli, V. and Cheng, R.T. (1992). Semi implicit finite difference methods for three-dimensional shallow water flow. *International Journal of Numerical Methods in Fluids*, 15, 629–648.

Chakraborty, M., Sriram, V. and Murali, K. (2022). Impact of ship-generated waves on the sediment re-suspension in restricted waterways. *ASCE Journal of Waterway, Port, Coastal and Ocean Engineering*, 149(4). https://doi.org/10.1061/JWPED5.WWENG-1951.

Chalikov, D. and Sheinin, D. (2005). Modeling extreme waves based on equations of potential flow with a free surface. *Journal of Computational Physics*, 210(1), 247–273.

Chauchat, J., Cheng, Z., Nagel, T., Bonamy, C. and Hsu, T. (2017). SedFoam-2.0: A 3-D two-phase flow numerical model for sediment transport. *Geoscientific Model Development Discussions,* 10(12), 4367–4392.

Chen, Q., Kirby, J.T., Dalrymple, R.A., Kennedy, A.B. and Chawla, A. (2000). Boussinesq modeling of wave transformation, breaking, and runup. II: 2D. *Journal of Waterway, Port, Coastal, and Ocean Engineering*, 126(1), 48–56.

Chen, X., Li, Y., Niu, X., Li, M., Chen, D. and Yu, X. (2011). A general two-phase turbulent flow model applied to the study of sediment transport in open channels. *International Journal of Multiphase Flow*, 37(9), 1099–1108.

Choi, J. and Yoon, S.B. (2009). Numerical simulations using momentum source wave-maker applied to RANS equation model. *Coastal Engineering*, 56, 1043–1060.

Chorin, A.J. (1968). Numerical solution of the Navier-Stokes equations. *Mathematics of Computation*, 22, 745–762.

Cruz, E.C., Isobe, M. and Watanabe, A. (1997). Boussinesq equations for wave transformation on porous beds. *Coastal Engineering*, 30, 125–156.

Clément, A.H. (1999). *Benchmark Test Cases for Numerical Wave Absorption: 1st Workshop of ISOPE Numerical Wave Tank Group*, Montréal, May 1998. In *Proceedings of 9th International Offshore and Polar Engineering Conference*, Brest, ISOPE.

Clauss, G.F., Hennig, J. and Schmittner, C. (2004). Modelling extreme wave sequences for the hydrodynamic analysis of ships and offshore structures. In *Proceedings of 9th Symposium on Practical Design of Ship and other Floating Structures*, Lübeck Travemünde, Germany, September 12–17, pp. 1–9.

Clauss, G.F., Schmittner, C.E. and Stück, R. (2005). Numerical wave tank: Simulation of extreme waves for the investigation of structural responses. In *ASME 2005 24th International Conference on Offshore Mechanics and Arctic Engineering*. https://doi.org/10.1115/omae2005-67048.

Colicchio, G., Greco, M. and Faltinsen, O.M. (2006). A BEM-level set domain-decomposition strategy for non-linear and fragmented interfacial flows. *International Journal for Numerical Methods in Engineering*, 67(10), 1385–1419.

Darcy, H. (1856). Les Fontaines Publiques De La Ville De Dijon. Dalmont, Paris.

Del Jesus, M., Lara, J.L. and Losada, I.J. (2012). Three-dimensional interaction of waves and porous coastal structures. Part I: Numerical model formulation. *Coastal Engineering*, 64, 57–72.

Divya, R. and Sriram, V. (2017). Wave-porous structure interaction modelling using improved meshless local Petrov-Galerkin method. *Applied Ocean Research*, 67, 291–305.

Dommermuth, D.G. (1993). The laminar interactions of a pair of vortex tubes with a free surface. *Journal of Fluid Mechanics*, 246, 91–115. https://doi.org/10.1017/s0022112093000059.

Edmund, D.O., Maki, K.J. and Beck, R.F. (2013). A velocity-decomposition formulation for the incompressible Navier–Stokes equations. *Computational Mechanics*, 52(3), 669–680.

Engelund, F.A. (1953). On the Laminar and Turbulent Flows of Ground Water through Homogeneous Sand. Danish Academy of Technical Sciences, Denmark.

Ergun, S. (1952). Fluid flow through packed columns. *Chemical Engineering Progress*, 48, 89–94.

Fair, G.M., Geyer, J.C. and Okum, D.A. (1968). *Waste Water Engineering. Vol. 2, Water Purification and Wastewater Treatment and Disposal.* Wiley, New York.

Ferrant, P., Gentaz, L. and Touzé, D.L. (2002). A new RANSE/potential approach for water wave diffraction. In *Numerical Towing Tank Symposium, NuTTS'2002*, Pornichet, France. https://hal.science/hal-01155923.

Finn, J.R., Li, M. and Apte, S.V. (2016). Particle based modelling and simulation of natural sand dynamics in the wave bottom boundary layer. *Journal of Fluid Mechanics*, 796, 340–385.

Forchheimer, P. (1901). Wasserbewegung durch boden. *Zeitschrift des Vereines Deutscher Ingenieure*, 45, 1782–1788.

Garzon, M. and Sethian, J.A. (2006). Wave breaking over sloping beaches using a coupled boundary integral-level set method. In Figueiredo, I.N., Rodrigues, J.F. and Santos, L. (eds.) *Free Boundary Problems*. International Series of Numerical Mathematics, Vol. 154. Birkhäuser, Basel. https://doi.org/10.1007/978-3-7643-7719-9_19.

Gentaz, L., Luquet, R., Alessandrini, B. and Ferrant, P. (2004). Numerical simulation of the 3D viscous flow around a vertical cylinder in non-linear waves using an explicit incident wave model. In *23rd International Conference on Offshore Mechanics and Arctic Engineering*, Vol. 1, Parts A and B. https://doi.org/10.1115/omae2004-51098.

Gobbi, M.F., Kirby, J.T. and Wei, G. (2000). A fully nonlinear Boussinesq model for surface waves Part 2. Extension to O(kh). *Journal of Fluid Mechanics*, 405, 181–210.

Gueyffier, D., Li, J., Nadim, A., Scardovelli, R. and Zaleski, S. (1999). Volume-of-fluid interface tracking with smoothed surface stress methods for three-dimensional flows. *Journal of Computational Physics*, 152 (2), 423–456.

Gu, Z. and Wang, H. (1991). Gravity waves over porous bottoms. *Coastal Engineering*, 15, 497–524.

Grilli, S.T. and Horrillo, J. (1999). Shoaling of periodic waves over barred-beaches in a fully nonlinear numerical wave tank. *International Journal of Offshore and Polar Engineering*, 9(04), 257–263.

Grilli, S.T., Gilbert, R.W., Lubin, P., Vincent, S., Astruc, D., Legendre, D., Duval, M., Kimmoun, O., Branger, H., Devrard, D., Fraunié, P. and Abadie, S. (2004). Numerical modeling and experiments for solitary wave shoaling and breaking over a sloping beach. In *4th International Offshore and Polar Engineering Conference IGARSS, 2004*, Toulon, France, Vol. 3, pp. 306–312.

Greco, M., Colicchio, G. and Faltinsen, O.M. (2007). Shipping of water on a two-dimensional structure. Part 2. *Journal of Fluid Mechanics*, 581, 371–399.

Ha, T., Lin, P. and Cho, Y.S. (2013). Generation of 3D regular and irregular waves using Navier–Stokes equations model with an internal wave maker. *Coastal Engineering*, 76, 55–67.

Hsiao, S.C., Liu, P.L.-F. and Chen, Y. (2002). Nonlinear water waves propagating over a permeable bed. *Proceedings of the Royal Society of London. Series A, Mathematical and Physical Sciences*, 458, 1291–1322.

Harlow, F.H. and Welch, J.E. (1965). Numerical calculation of time dependent viscous incompressible flow. *Physics of Fluids*, 8, 2182–2189.

Hasan, S.A. (2020). Nonlinear interactions among water wave systems. PhD thesis, Indian Institute of Technology Madras.

Hasan, S.A., Sriram, V. and Panneer Selvam, R. (2019). Evaluation of an eddy viscosity type wave breaking model for intermediate water depths. *European Journal of Mechanics-B/Fluids*, 78, 115–138.

Hasselmann, S., Hasselmann, K., Bauer, E., Janssen, P.A.E.M. and Komen, G.J. (1988). The WAM model- a third generation ocean wave prediction model. *Journal of Physical Oceanography*, 18, 1775–1810.

Hieu, P. and Tanimoto, K. (2006). Verification of VOF based two phase flow model for wave breaking and wave structure interaction. *Ocean Engineering*, 33, 1565–1588.

Higuera, P., Lara, J.L. and Losada, I.J. (2013). Simulating coastal engineering processes with OpenFOAM. *Coastal Engineering*, 71, 119–134.

Hirt, C.W. and Nichols, B.D. (1981). Volume of fluid (VOF) method for the dynamics of free boundaries. *Journal of Computational Physics*, 39, 201–225.

Hirt, C.W., Amsden, A.A. and Cook, J.L. (1974). An arbitrary Lagrangian Eulerian computing method for all speeds. *Journal of Computational Physics*, 14, 227–253.

Hirt, C.W., Nichols, B.D. and Romero, N.C. (1975). SOLA: A Numerical Solution Algorithm for Transient Fluid Flows. Los Alamos National Laboratory Report LA5852.

Hildebrandt, A., Sriram, V. and Schlurmann, T. (2013). Simulation of focusing waves and local line forces due to wave impacts on a tripod structure. In *The Twenty-Third International Offshore and Polar Engineering Conference*, Anchorage, Alaska.

Ijima, T., Chou, T.R. and Yoshida, A. (1976). Method of analysis for two-dimensional water wave problems. In *Proceedings of 15th Coastal Engineering Conference*, ASCE, pp. 2717–2736.

Issa, R.I. (1982). Solution of the Implicit Discretized Fluid Flow Equations by Operator Splitting. Mechanical Engineering Report FS/82/15, Imperial College, London.

Jasak, H. (1996). Error analysis and estimation for the finite volume method with applications to fluid flows. Ph.D. thesis, Imperial College, University of London.

Jha, S.K. and Bombardelli, F.A. (2010). Toward two-phase flow modelling of non-dilute sediment transport in open channels. *Journal of Geophysical Research: Earth Surface*, 115, F03015, doi:10.1029/2009JF001347.

Johns, B. and Jefferson, R.J. (1980). The numerical modeling of surface wave propagation in the surf zone. *Journal of Physical Oceanography*, 10, 1061–1069.

Karim, M.F., Tanimoto, K. and Hieu, P.D. (2003). Simulation of wave transformation in vertical permeable structure. In *Proceedings of the 13th International Offshore and Polar Engineering Conference (2003)*, pp. 729–735.

Karim, M.F., Tanimoto, K. and Hieu, P.D. (2009). Modeling and simulation of wave transformation in porous structures using VOF based two-phase flow model. *Applied Mathematical Modelling*, 33, 343–360.

Karunarathna, S.A.S. and Lin, P. (2006). Numerical simulation of wave damping over porous sea beds. *Journal of Coastal Engineering*, 53, 845–855.

Keller, J.B. (1948). The solitary wave and periodic waves in shallow water. *Communications in Pure and Applied Mathematics*, 1, 323–339.

Kennedy, A.B., Chen, Q., Kirby, J.T. and Dalrymple, R.A. (2000). Boussinesq modeling of wave transformation, breaking, and runup. I: 1D. *Journal of Waterway, Port, Coastal, and Ocean Engineering*, 126(1), 39–47.

Kim, K., Sirviente, A.I. and Beck, R.F. (2005). The complementary RANS equations for the simulation of viscous flows. *International Journal for Numerical Methods in Fluids*, 48(2), 199–229. https://doi.org/10.1002/fld.892.

Kim, C., Clément, A. and Tanizawa, K. (1999). Recent research and development of numerical wave tank — a review. *International Journal of Offshore and Polar Engineering*, 9(4), 241–256.

Kim, Y.J. (2021). Numerical improvement and validation of a naval hydrodynamics CFD solver in view of performing fast and accurate simulation of complex ship-wave interaction. Ph.D. thesis, École centrale de Nantes. https://theses.hal.science/tel-03530266.

Kobayashi, N. and Wurjanto, A. (1990). Numerical model for waves on rough permeable slopes. *Journal of Coastal Research*, 7, 149–166.

Lara, J.L., del Jesus, M. and Losada, I.J. (2012). Three-dimensional interaction of waves and porous coastal structures. *Coastal Engineering*, 64, 26–46.

Larsen, J. and Dancy, H. (1983). Open boundaries in short wave simulations a new simulation. *Coastal Engineering*, 7, 285–297.

Lee, C., Park, W.S., Cho, Y.S. and Suh, K.D. (1998). Hyperbolic mild slope equations extended to account for rapidly varying topography. *Coastal Engineering*, 34, 243–257.

Li, Y., Kelly, D.M., Li, M. and Harris, J.M. (2014). Development of a new 3D Euler-Lagrange model for the prediction of scour around offshore structures. *Coastal Engineering Proceedings*, 1(34), sediment.31.

Li, Z., Bouscasse, B., Ducrozet, G., Gentaz, L., Touzé, D.L. and Ferrant, P. (2021). Spectral wave explicit Navier-Stokes equations for wave-structure interactions using two-phase computational fluid dynamics solvers. *Ocean Engineering*, 221, 108513. https://doi.org/10.1016/j.oceaneng.2020.108513.

Liang, D., Cheng, L. and Li, F. (2005). Numerical modeling of flow and scour below a pipeline in currents: Part II. Scour simulation. *Coastal Engineering*, 52(1), 43–62.

Lin, P. and Karunarathna, S.A. (2007). A numerical study of solitary wave interaction with porous breakwaters. *Journal of Waterway, Port, Coastal, and Ocean Engineering* (ASCE), 133(5), 352–363.

Lin, P. and Liu, P.L.F. (1998). A numerical study of breaking waves in the surf zone. *Journal of Fluid Mechanics*, 359, 239–264.

Lin, P. and Liu, P.L.-F. (1999). Internal wave-maker for Navier–Stokes equations models. *Journal of Waterway, Port, Coastal and Ocean Engineering*, 125, 207–215.

Liu, Z. and Fang, K. (2016). A new two-layer Boussinesq model for coastal waves from deep to shallow water: Derivation and analysis. *Wave Motion*, 67, 1–14.

Liu, P.L.F., Lin, P., Chang, K.A. and Sakakiyama, T. (1999). Numerical modeling of wave interaction with porous structures. *Journal of Waterway, Port, Coastal and Ocean Engineering* (ASCE), 125(6), 322–330.

Liu, X., Lin, P. and Shao, S. (2015). ISPH wave simulation by using an internal wave maker. *Coastal Engineering*, 95, 160–170.

Longuet-Higgins, M.S. and Cokelet, E.D. (1976). The deformation of steep surface waves on water: I. A numerical method of computation. *Proceedings of Royal Society London Part A*, 350, 1–26.

Losada, I.J., Lara, J.L., Guanche, R. and Gonzalez-Ondina, J.M. (2008). Numerical analysis of wave overtopping of rubble mound breakwaters. *Coastal Engineering*, 55, 47–62.

Loysel, T., Sabrina, C., Eric, G., Laurent, B. and De Seze, P.-E. (2012). Results of the first sloshing model test benchmark. In *Paper Presented at the Twenty-Second International Offshore and Polar Engineering Conference*, Rhodes, Greece, June 2012.

Lynett, P.J. and Liu, P.L.F. (2004). Linear analysis of the multi-layer model. *Coastal Engineering*, 51(5–6), 439–454.

Madsen, H. and Jakobsen F. (2004). Cyclone induced storm surge and flood forecasting in the northern Bay of Bengal. *Coastal Engineering*, 51(4), 277–296.

Madsen, P.A., Murray, R. and Sørensen, O.R. (1991). A new form of the Boussinesq equations with improved linear dispersion characteristics. *Coastal Engineering*, 15(4), 371–388.

Madsen, P.A. and Sørensen, O.R. (1992). A new form of the Boussinesq equations with improved linear dispersion characteristics. Part 2: A slowly-varying bathymetry. *Coastal Engineering*, 18, 183–204.

Madsen, P.A., Bingham, H.B. and Liu, H. (2002). A new Boussinesq method for fully nonlinear waves from shallow to deep water. *Journal of Fluid Mechanics*, 462, 1–30.

Mathieu, A., Chauchat, J., Bonamy, C. and Nagel, T. (2019). Two-phase flow simulation of tunnel and lee-wake erosion of scour below a submarine pipeline. *Water*, 11(8), 1727.

McCorquodale, J.A. (1972). Wave dissipation in rockfill. In *Proceedings of 13th Coastal Engineering Conference*, ASCE, pp. 1885–1900.

Narayanaswamy, M., Crespo, A., Gómez-Gesteira, M. and Dalrymple, R.A. (2010). SPHysics-FUNWAVE hybrid model for coastal wave propagation. *Journal of Hydraulic Research*, 48(Supl), 85–93.

Nichols, B.D., Hirt, C.W. and Hotchkiss, R.S. (1980). SOLAVOF: A solution algorithm for transient fluid flow with multiple free boundaries. Report LA8355, Los Alamos Scientific Laboratory.

Nwogu, O. (1993). Alternative form of Boussinesq equations for nearshore wave propagation. *Journal of Waterway, Port, Coastal, and Ocean Engineering* (ASCE), 119, 618–638.

Osher, S. and Sethian, J.A. (1988). Fronts propagating with curvature dependent speed algorithms based on Hamilton Jacobi formulations. *Journal of Computational Physics*, 79, 12–49.

Polubarinova-Kochina, P.Y. (1962). *Theory of Ground Water Movement* (translated from the Russian by J.M.R De Wiest). Princeton University Press, United states of America.

Patankar, S.V. (1981). *Numerical Heat Transfer and Fluid Flow*. Hemisphere, New York.

Patankar, S.V. and Spalding, D.B. (1972). A calculation procedure for heat, mass and momentum transfer in three dimensional parabolic flows. *International Journal of Heat and Mass Transfer*, 15, 1787–1806.

Patankar, S.V. and Spalding, D.B. (1983). A calculation procedure for heat, mass and momentum transfer in three-dimensional parabolic flows. In *Numerical Prediction of Flow, Heat Transfer, Turbulence and Combustion*. Pergamon Press, New York, pp. 54–73.

Peregrine, D.H. (1967). Long waves on a beach. *Journal of Fluid Mechanics*, 27(4), 815–827.

Radder, A.C. (1979). On the parabolic equation method for water wave propagation. *Journal of Fluid Mechanics*, 95(1), 159–176.

Ransley, E., Yan, S., Brown, S., Mai, T., Graham, D.I., Ma, Q., Musiedlak, P., Engsig-Karup, A.P. . . . and Greaves, D. (2019). A blind comparative study of focused wave interactions with a fixed FPSO-like structure (CCP-WSI blind test series 1). *International Journal of Offshore and Polar Engineering*, 29(2), 113–127. https://doi.org/10.17736/ijope.2019.jc748.

Ransley, E.J., Scott, A.B., Hann, M., Greaves, D.M., Windt, C., Ringwood, J., Davidson, J., Schmitt, P., Yan, S., Junxian, X., Jinghua, W., Wang, H., Ma, Q.W., Xie, Z., Giorgi, G., Hughes, J., Williams, A., Masters, I., Lin, Z., Chen, H., Qian, L., Ma, Z., Chen, Q., Ding, H., Zang, J., van Rij, J., Yu, Y.-H., Li, Z., Bouscasse, B., Ducrozet, G. and Bingham, H. (2021). Focused wave interactions with floating structures: A blind comparative study. *Proceedings of the Institution of Civil Engineers — Engineering and Computational Mechanics*, 174(1), 46–6.

Ransley, E., Yan, S., Brown, S., Hann, M., Graham, D. I., Windt, C., Schmitt, P., Davidson, J. . . . and Greaves, D. (2020). A blind comparative study of focused wave interactions with floating structures (CCP-WSI blind test series 3). *International Journal of Offshore and Polar Engineering*, 30(1), 1–10. https://doi.org/10.17736/ijope.2020.jc774.

Ris, R.C., Booij, N. and Holthuijsen, L.H. (1999). A third-generation wave model for coastal regions, Part II, Verification. *Journal of Geophysical Research*, 104, 7667–7681.

Robaux, F. and Benoît, M. (2021). Development and validation of a numerical wave tank based on the harmonic polynomial cell and immersed boundary methods to model nonlinear wave-structure interaction. *Journal of Computational Physics*, 446, 110560.

Rosemurgy, W.J., Beck, R.F. and Maki, K.J. (2016). A velocity decomposition formulation for 2D steady incompressible lifting problems. *European Journal of Mechanics-B/Fluids*, 58, 70–84.

Roulund, A., Sumer, B.M., Fredsøe, J. and Michelsen, J. (2005). Numerical and experimental investigation of flow and scour around a circular pile. *Journal of Fluid Mechanics*, 534, 351–401.

Ren, B., Wen, H., Dong, P. and Wang, Y. (2014). Numerical simulation of wave interaction with porous structures using an improved smoothed particle hydrodynamic method. *Coastal Engineering*, 88, 88–100.

Saincher, S. and Banerjee, J. (2017). On wave damping occurring during source-based generation of steep waves in deep and near-shallow water. *Ocean Engineering*, 135, 98–116.

Saincher, S. and Sriram, V. (2022a). A three-dimensional hybrid fully nonlinear potential flow and Navier Stokes model for wave structure interactions. *Ocean Engineering*, 266, 112770. https://doi.org/10.1016/j.oceaneng.2022.112770.

Saincher, S, and Sriram, V. (2022b). An efficient operator-split CICSAM scheme for three-dimensional multiphase-flow problems on Cartesian grids. *Computers & Fluids*, 240. https://doi.org/10.1016/j.compfluid.2022.105440.

Saincher, S., Sriram, V., Ravindar, R., Yan, S., Stagonas, D. . . . and Wan, D. (2023). Comparative study on breaking waves interaction with vertical wall retrofitted with recurved parapet in small and large scale. *International Journal of Offshore and Polar Engineering*, 33(2), 113–122. https://doi.org/10.17736/ijope.2023.jc890.

Schäffer, H.A. and Klopman, G. (2000). Review of multidirectional active wave absorption methods. *Journal of Waterway, Port, Coastal and Ocean Engineering*, 126(2), 88–97.

Shao, S. (2010). Incompressible SPH flow model for wave interactions with porous media. *Coastal Engineering*, 57(3), 304–316.

Seiffert, B.R. and Ducrozet, G. (2018). Simulation of breaking waves using the high-order spectral method with laboratory experiments: Wave-breaking energy dissipation. *Ocean Dynamics*, 68(1), 65–89.

Seiffert, B.R., Ducrozet, G. and Bonnefoy, F. (2017). Simulation of breaking waves using the high-order spectral method with laboratory experiments: Wave-breaking onset. *Ocean Modelling*, 119, 94–104.

Shagun, A. (2022). Wave-structure interaction by coupled Boussinesq (FEM) and Navier-Stokes (particle based) models in 3D. PhD thesis, Indian Institute of Technology Madras.

Shashank, V.G., Sriram, V., Schüttrumpf, H. and Sannasiraj, S.A. (2024). A new surge index with the incorporation of cyclone track approach angle information for the Bay of Bengal. *Coastal Engineering*, 188. https://doi.org/10.1016/j.coastaleng.2023.104429.

Sommerfeld, A. (1949). *Partial Differential Equations in Physics.* Academic Press, New York, USA.

Sitanggang, K.I. (2008). Boussinesq-Equation and RANS hybrid wave model. Doctoral dissertation, Texas A&M University. https://core.ac.uk/download/pdf/147132376.pdf.

Sriram, V. (2008). Finite element simulation of nonlinear free surface waves. PhD thesis, Indian Institute of Technology Madras.

Sriram, V., Agarwal, S., Yan, S., Xie, Z., Saincher, S., Schlurmann, T., Ma, Q. . . . and Li, G. (2021). A comparative study on the nonlinear interaction between a focusing wave and cylinder using state-of-the-art solvers: Part A. *International Journal of Offshore and Polar Engineering*, 31(1), 1–10. https://doi.org/10.17736/ijope.2021.jc820.

Sriram, V. and Ma, Q.W. (2021). Review on the local weak form-based meshless method (MLPG): Developments and applications in ocean engineering.

Applied Ocean Research, 116, 102883. https://doi.org/10.1016/j.apor.2021.102883.

Sriram, V., Ma, Q.W. and Schlurmann, T. (2014). A hybrid method for modelling two dimensional non-breaking and breaking waves. *Journal of Computational Physics*, 272, 429–454. https://doi.org/10.1016/j.jcp.2014.04.030.

Stahlmann, A. and Schlurmann, T. (2012). Investigations on scour development at tripod foundations for offshore wind turbines: Modeling and application. *International Conference on Coastal Engineering* (2012), pp. 1–11.

Su, X. and Lin, P. (2005). A hydrodynamic study on flow motion with vegetation. *Modern Physics Letters B*, 19(28–29), 1659–1662.

Sun, R. and Xiao, H. (2016). Sedifoam: A general-purpose, open-source CFD-DEM solver for particle-laden flow with emphasis on sediment transport. *Computers & Geosciences*, 89, 207–219.

Sussman, M., Smereka, P. and Osher, S. (1994). A level set approach for computing solutions to incompressible two-phase flow. *Journal of Computational Physics*, 114, 146–159.

Takewaki, H. and Yabe, T. (1985). The cubic-interpolated pseudo particle (CIP) method: application to nonlinear and multi-dimensional hyperbolic equations. *Journal of Computational Physics*, 61, 261–268.

Tanizawa, K. and Clément, A.H. (2000). Report of the 2nd Workshop of ISOPE Numerical Wave Tank Group: Benchmark Test Cases of Radiation Problem, Brest, May 1999, *Proceedings of 10th International Offshore and Polar Engineering Conference*, Seattle, ISOPE.

Tian, Z., Perlin, M. and Choi, W. (2010). Energy dissipation in two-dimensional unsteady plunging breakers and an eddy viscosity model. *Journal of Fluid Mechanics*, 655(2010), 217–257.

Tolman, H.L. (1999). User manual and system documentation of WAVEWATCH-III version 1.18. NOAA / NWS / NCEP / OMB Technical Note 166.

van Gent, M.R.A. (1995). Wave interaction with permeable coastal structures. Ph.D. thesis, Delft University of Technology, Delft University Press, Delft, The Netherlands.

Wei, G. and Kirby, J.T. (1995). Time dependent numerical code for extended Boussinesq equations. *Journal of Waterway, Port, Coastal, and Ocean Engineering* (ASCE), 121(5), 251–261.

Wei, G., Kirby, J.T. and Sinha, A. (1999). Generation of waves in Boussinesq models using a source function method. *Coastal Engineering*, 36(4), 271–299.

Wellens, P. (2012). Wave simulation in truncated domains for offshore applications. PhD thesis, Delft University of Technology.

Yan, S. and Ma, Q. (2011). Improved model for air pressure due to wind on 2D freak waves in finite depth. *European Journal of Mechanics - B/Fluids*, 30(1), 1–11. DOI: 10.1016/j.euromechflu.2010.09.005.

Yang, Z. and Liu, P.L. (2022). Depth-integrated wave–current models. Part 2. Current with an arbitrary profile. *Journal of Fluid Mechanics*, 936. https://doi.org/10.1017/jfm.2022.42.

Yoshizawa, A. (1993). Bridging between eddy-viscosity-type and second-order turbulence models through a two-scale turbulence theory. *Physical Review E — Statistical Physics, Plasmas, Fluids, and Related Interdisciplinary Topics,* 48, 273–281.

Zubeldia, E.H., Fourtakas, G., Rogers, B.D. and Farias, M.M. (2018). Multiphase SPH model for simulation of erosion and scouring by means of the shields and Drucker-Prager criteria. *Advances in Water Resources*, 117, 98–114.

Chapter 9

Coastal-Riverine Confluence

9.1 Introduction

All rivers naturally drain towards the ocean, and the confluence of the river and ocean is commonly referred to as an estuary or tidal inlet. The hydrodynamics and morphodynamics at these inlets are considered to be more complex in the downstream river region as a bi-directional flow results due to interactions of the river flow with the tidal currents. The behaviour of the flow and allied sedimentation depends on the extent of tidal influence over the coast. The dynamic morphological adjustments at the river confluence make it difficult to manage land use and plan for infrastructure development (Ettema, 2008) in its vicinity. The study of river confluence needs more attention owing to the importance of flora and fauna biodiversity across estuarine inlets or tidal confluences.

Tidal inlets are ecologically important due to the flushing of the river during the high tide and the flow cleaning up the river during the low tide. This in particular is extremely important in megacities, where the population density has an impact on the quality of the river water. The regular exchange of water between shoreline openings (bays, lagoons, marsh, estuaries, etc.) and the open ocean is due to the tidal cycle. The common features observed near the confluence of a river and the sea are primarily due to changes in the morphology of the landscape. These morphological changes lead to the formation and migration of various coastal features such as sandbars, spits, and shoals. Sandbars are submerged or partially submerged ridges of sand or gravel and are formed due to the deposition of sediment carried by the river and waves. Spits are narrow landforms that extend from the coast into the water due to the longshore sediment drift. Shoals are shallow areas in a body of water and are often formed by the accumulation of sediment. These dynamic processes continuously reshape the coastline, creating a diverse and ever-changing environment at the confluence of river and sea. These formations are significant along a coast exposed to dominant longshore sediment motion and act as a barrier and prevent tidal influx. Such formations pose challenges to the smooth flow of vessels when inlets are used as navigable channels. To mitigate sandbar formation, measures such as constructing a pair of training walls or implementing sand bypassing techniques are often employed. Sandy shoal depositions are characteristic of the tidal channels on the landward and seaward sides of the inlet. In the event of flooding from the riverine basin, tidal inlet openings are crucial to avoid flooding all along the riverbanks. During extreme weather, the drastic increase in the flow will cause upland flooding, which is also governed by the characteristics of the sandbar which, in turn, depends on the littoral drift and the sediment characteristics at the confluence of the river and sea.

River confluences are known to create sediment sinks where the longshore currents are most likely to deposit sediment transported alongshore. The interactions of the flow parameters, sediment transport and morphological change as conceptualized by Leeder (1983) and Ashworth and Ferguson (1986) are unified and projected in Fig. 9.1. Bolla Pittaluga *et al.* (2015) employed an one-dimensional model to investigate the equilibrium morphological profile under various tidal parameters. Xie *et al.* (2018) reported results obtained through field investigations and numerical simulations. It is to be noted that most of the parameters are both spatially and temporally varying, and therefore a refined understanding of the complex

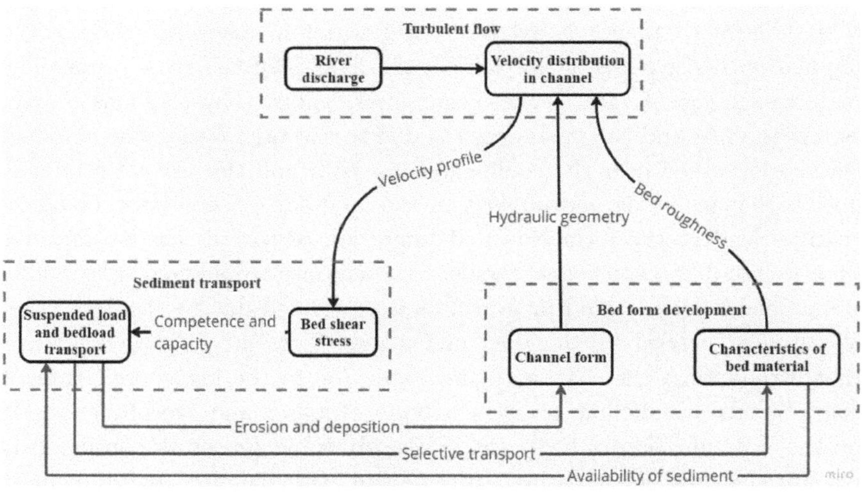

Fig. 9.1. Conceptualized unified models proposed by Leeder (1983) and Ashworth and Ferguson (1986).
Source: Bilal *et al.* (2020).

hydrodynamics is mandatory to comprehend the physics of near-confluence sedimentation.

9.2 Impacts of Tidal Inlets

Tidal inlets are the portion of the water body between the sea and a river/backwater that is subjected to flooding during high tide and ebbing during low tide. The exchange of water in an inlet is either maintained by the cyclic flushing of tides or closed due to heavy longshore drift. Tidal inlets take shape through the gradual accumulation of sediment along barrier beaches flanking the entrance. When tidal currents surge toward the shore, it is termed a flood current or flood tide. This influx of sediment brought in by the flood tide contributes to the formation of a shoal along the riverside or backwater, known as a flood tidal delta or flood shoal. Conversely, when the tidal current retreats seaward, it is referred to as ebb current or ebb tide.

The sediment carried out by the ebb tide is called the ebb shoal or ebb-tidal delta. The formation of ebb/flood shoals will grow larger and result in the closure of the inlet mouth. Wave-induced sediment motion is either along the coast or normal to the shoreline, referred to as cross-shore mode of sediment transport, whereas tide motion is only normal to the shoreline. Tidal range governs the duration of the waves that act on the shore, meaning that the effect of waves on sediment motion decreases with an increase in the tidal range. Longshore currents dominate in locations of lesser tidal range, and tide-induced currents dominate in locations exposed to high tidal range.

Micro-tidal inlets, locations of tidal range of less than 2 m, promote siltation, which usually forms sandbars that disconnect the inlet from the sea, posing challenges to the free flow of river water, resulting in a stagnant pool of water and leading to the deterioration of water quality; at locations where sewage water is discharged into the river, this causes hazards to health. This has been a problem at several inlets around the globe. There is a growing interest in keeping these inlets permanently open across all seasons which can enable potential land use through the development of harbours or recreational facilities.

9.3 Tidal Inlet Classification

The hydrodynamics of an inlet is hugely influenced by the mean sea level, tides, wave climate and sediment supply (Hayes, 1980). Besides coasts with predominant longshore currents, inlets can act as potential sediment sinks, trapping large amounts of coastal sediments (Stive *et al.*, 1991). This sinkage of sediments will in turn impact the adjoining coasts (Fitzgerald, 1984; FitzGerald and Montello, 1993). These inlets serve as an entry point to sheltered bays and lagoons. It is significant to study the complex hydrodynamics at an inlet in detail since it serves as a major sediment sink, capturing huge volumes of littoral sediment. It is crucial to stabilize the inlet and maintain a balanced ecosystem, which requires an in-depth understanding of the sediment circulation as well as the relationship between the inlet throat dimensions, tidal prism and back-barrier area.

Hayes (1979) updated the classification of shorelines of Davis (1964) specifically focusing on those with medium wave energy. In his redefinition, the study area was identified as a micro-tidal inlet, characterized by minimal tidal range and limited wave energy. This classification helps in understanding the unique morphological and sedimentary processes occurring in such environments, where the influence of tides is relatively small compared to other coastal forces. Bruun (1986) considered several inlets, among which it was reported that the minimum and maximum depths in the gorge were 4.5 m and 18m, respectively. Such inlets, in the USA and Europe, are found in semi-diurnal tidal regimes, with a spring tidal range of 3 m. It was concluded that if the ratio of the tidal prism volume to the annual gross volume of sediment transported past the channel entrance exceeds 150, the probability of formation of shoals near the confluence is less. If this ratio reduces to around 20–50, it could result in significant shoaling, thus forcing maintenance dredging or planning for permanent mitigation measures to combat the formation of sandbars. These criteria developed by Bruun (1986) characterize the nature of inlets for sandbar formation.

9.4 Mechanism of Inlet Closure

The nature of the confluence of the estuary and the ocean is dictated by the interaction of wave-induced longshore currents and the flow characteristics in the estuary, which is controlled by the tidal range. The tidal inlet disrupts the uninterrupted movement of longshore sediment transport caused by wave-induced currents. This interruption leads to sediment trapping and hence, the formation of shoals. The growth of these shoals depends on various factors, including the tide and wave-driven flow fields and the sediment characteristics. Shoals are developed both on the ebb and flood tidal deltas; smaller shoals may be developed on the downward drift side of the inlet. Sand deposition due to the impedance of ebb currents being diverted by longshore currents gets lifted away from the inlet entrance towards the downdrift direction (Hayes, 1975; Komar, 1976; FitzGerald, 1988, 1996; Komar, 1996). The existence of these shoals is attributed to the formation of eddies on the downdrift side (Oertel, 1972). Continued sediment supply to the updrift shoal will lead to its growth and eventually result in spits and sandbars. When the current flow from the river is strong enough to take away the deposited sediment at the inlet mouth, spit and sandbar formation will be eradicated. However, for inlets with a relatively weaker flow, spits will continue to grow until the mouth is completely blocked.

Initially, the shoal would form from the updrift side of the inlet and would continue to grow due to the weak ebb current against the longshore current (Hayes, 1975; Komar, 1976; Fitzgerald, 1988, 1996; Komar, 1996). The downdrift shoal has also been attributed to the development of a separation eddy on the downdrift side of the inlet (Oertel, 1972). The continuous supply of sediment alongshore will lead to the growth of the shoal and the eventual emergence of a spit across the inlet entrance along the downdrift side of the coast. The phenomenon is illustrated in Fig. 9.2.

A spit or sand spit is a deposition bar or beach landform off coasts or lakeshores. A spit is developed at the entrance of an estuary, usually propagating towards the direction of littoral drift as long as the sediment in motion continues due to the retardation of flow alongshore. A spit thus formed, probably over decades, may tend to become too stable to get washed away, even permitting inhabitants. A long spit can also serve as a natural attenuator, reducing the incident wave energy and thus promoting a stable beach on its leeward side.

As an example, the growth of a spit, known as Hope Island (16.97°N 82.35°E), near the confluence of the Godavari River and the Bay of Bengal has provided the advantage of a natural harbour off Kakinada, India, which now is a deep-water port. Nageswara Rao *et al.* (2005) have

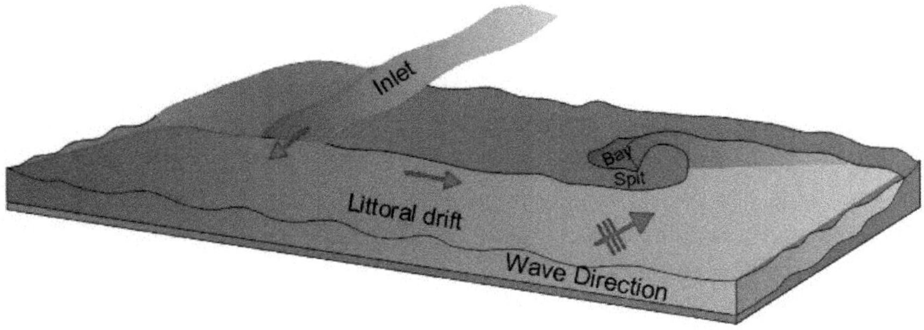

Fig. 9.2. Phenomenon of formation of spits.

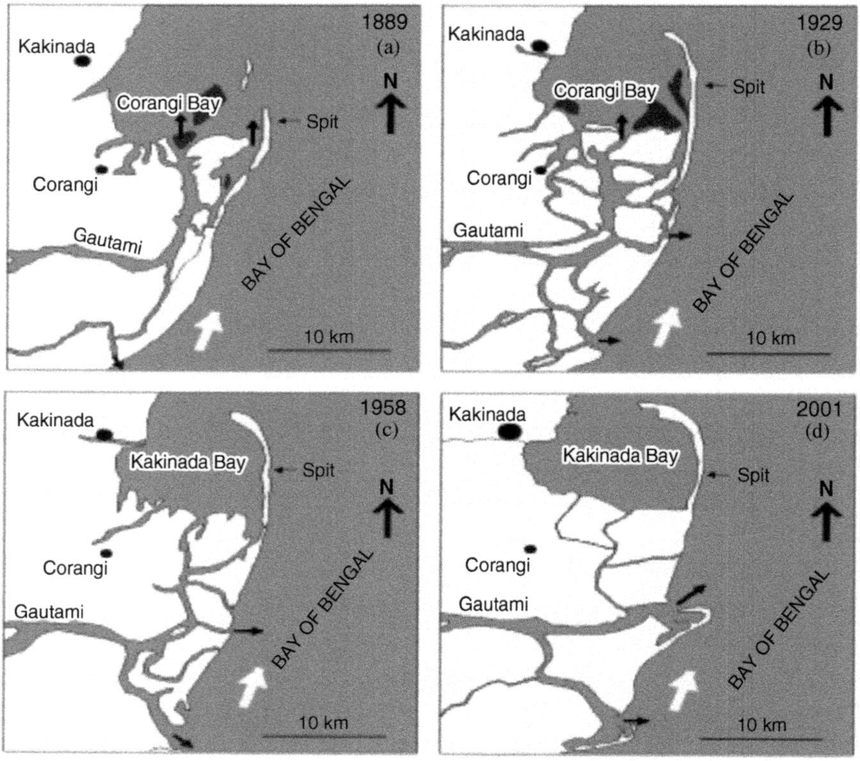

Fig. 9.3. Stages of growth of the spit, Hope Island (white arrow: discharge points from River Godavari; black arrow: direction of littoral drift) (Nageswara Rao (2005)).

discussed in detail the formation and the growth of the spit, which is shown in Fig. 9.3.

If the flow in the inlet is strong enough to clear the sediment bars near the mouth, then the growth of the spit will reduce or cease, in which case sandbars might form on either side of the inlet, which again depends on the seasonal variation of wave climate, which is the main driving parameter of the flow field in the nearshore. If the current in the inlet is weak, the sediments will gradually settle and promote the growth of the sandbar in the confluence of the estuary and the sea.

A standard model for flood tidal deltas formed in meso-tidal inlets was developed by Hayes (1980). The components of the flood delta include a flood ramp, a flood channel, an ebb shield, ebb spits and spillover lobes as shown in Fig. 9.4. Although, the aforesaid model was primarily derived

Fig. 9.4. Morphodynamic processes at a wave-induced outlet, as defined by Hayes (1980).
Source: Schüttrumpf.

for a meso-tidal inlet, studies by Davis and Fox (1981), Gallivan and Davis (1981), Nummedal *et al.* (1977), and Hubbard *et al.* (1979) suggest that the model can be extended to be applicable to more wave-dominated micro-tidal inlets.

In general, the sediment size will tend to decrease while in motion from the estuary towards the major channels to large creeks and finally into small tidal creeks. The elevation of intertidal flats also plays a role as the grain size of the sediment tends towards a finer dimension with an increase in its elevation. Although the transition of grain size of sediments from coarse to fine is induced by the increase in tidal elevation, it is also a function of two factors (DaBoll, 1969; Hartwell, 1970) as stated in the following:

- The sediments under ebb flow are coarser as stated by Daboll (1969).
- In an estuary, the variation in the tidal-induced currents results in an unsteady and non-uniform flow both in the direction of the tidal flow and across the estuary (Daboll, 1969; Hartwell, 1970).

9.4.1 *Shoals and sandbars*

Shoals of different sizes are formed due to sediment along the bed and suspended load driven by ocean waves merging together to form sandbars. While the waves break in the nearshore, the currents are induced. If the waves approach the shore at an angle from the shore normal, the induced current can be divided into two components: one parallel to the shoreline, called longshore current and another perpendicular to the coast, called cross shore current. The cross shore flow towards the shoreward would return back as seaward flow. The residual flow field will cause the sediment to settle down or move in suspended form, which will depend on its characteristics.

A river bar or harbour bar is a sedimentary bar present at a river mouth or the entrance of a harbour, which forms landforms and can result from the action of waves on up-current beaches or the seafloor, or by the deposition of freshwater sediment. If the bed and suspended loads of a river are big enough, or if the beaches are highly mobile, deposition can create a sandbar that blocks the river mouth entirely and dams the river.

Wave-induced longshore and cross-shore currents generated due to the breaking of waves in the nearshore area carry sediment set in motion and at locations of lesser velocity. The sediment settles down, which depends on the grain size and the fall velocity. The aforementioned two modes of flow will depend on the obliqueness of the wave attack along the coast. At locations where the waves break on the coast, particularly during a storm, sediment tends to get carried towards the sea and settles to form an offshore bar. The bars thus formed may again drift back towards the coast when the storm recedes as the waves may also break over the bar. The ratio of the rate of alongshore littoral drift (M) and the volume of water discharged from the inlet (Q) can be used to predict the bar- or tidal flow-bypassing conditions (Fig. 9.5). If $M/Q > 200$, it is a bar-bypassing condition, where the littoral drift is more pronounced than the discharge from the river, and thereby the sediments will bypass the inlet opening and result in spit and sandbar formations. If $M/Q < 7$, it is a tidal flow-bypassing condition, where the strong discharge from the river flushes out the spits and sandbars from the inlet.

9.5 Stabilization of Tidal Inlets

A tidal inlet may be improved by just dredging, including the installation of sand traps, which may be located on the updrift side where they are most practical and convenient. This is where sand traps are most efficient for trapping sediment and at the same time presenting minimum nuisance to navigation. A suitable measure to stabilize an inlet can be identified

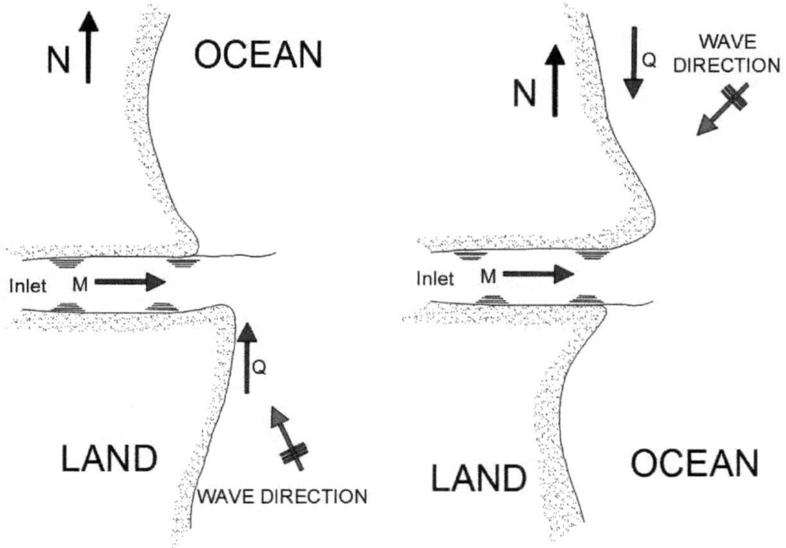

Fig. 9.5. Littoral drift and inlet discharge.

Table 9.1. Check for stability using the formulae from the inlet stability criteria established by Brunn (1986).

Stability factor	Navigability
$P/M_{\text{tot}} > 150$	Little or no ocean bar outside inlet.
$100 < P/M_{\text{tot}} < 150$	Low ocean bar, minor navigation problems.
$50 < P/M_{\text{tot}} < 100$	Wider, high ocean bar, increasing navigation problems.
$20 < P/M_{\text{tot}} < 50$	Wide shallow ocean bar, navigation difficult.
$P/M_{\text{tot}} < 20$	Very shallow ocean bar, navigation very difficult.

Source: Bruun (1986).

based on the work by Carr and Krauss (2001). The stability of the inlet is maintained by installing structural measures like training walls and jetties (Table 9.1).

Structures for improvements include the following:

- Jetties or training walls with/without spurs if the banks of the inlet need to be stabilized.
- Breakwaters built parallel to the shore.

- Weirs in breakwaters with traps inside for periodic dredging.
- Dredging of the navigation channel.

Hicks and Hume (1996) and Carr and Kraus (2001) infer that the tidal prism is expected to control the size and location of the ebb shoal. The values from different studies equating tidal prism to tidal stability are listed in Tables 9.2 and 9.3. Table 9.4 reports the relationship between inlet bypassing and the classification of channel area cross-sectional stability as per Bruun and Gerritsen (1959).

$$A_c = CP^n, \tag{9.1}$$

where A_c = minimum cross-sectional area (m^2), P = tidal prism (m^3), C and n = correlation coefficients (jetties/training walls).

A stable inlet should possess equilibrium, where if the cross-sectional area of the inlet fluctuates about a mean, it is in stable equilibrium, which is a function of the currents and the shear stresses over a spring tidal cycle as stated by Escoffier (1940). The relationship between the spring tidal prism "P" and the closure curve area "A" established by O'Brien (1931, 1969) does not apply to all tidal inlets. It was claimed by Styles *et al.* (2016) that the similar characteristics of the tidal variation over a much larger cross-sectional area will generate lesser velocities so as to preserve continuity. In the case of weaker flows, the sediment transport-carrying capacity will reduce, leading to difficulties in maintaining a larger inlet cross-section. Although the relatively weak initial forcing would be sufficient to deepen the channel by the removal of the sediment, it may still be not good enough to establish equilibrium over a stipulated time interval. In an estuarine tidal inlet, numerous parameters contribute towards the changes in the local morphodynamics. A tidal inlet is governed by a number of parameters and is characterized by the changes at the confluence of the inlet and the

Table 9.2. Cross-sectional area and tidal prism relationship.

Regression Values found by (Jarrett, 1976) for (m^2, tidal inlets on U.S. coasts)						
	All Inlets		Unjettied, Single jettied		Dual Jettied	
Location	C	n	C	n	C	n
All inlets	$1.576*10^{-4}$	0.95	$3.797*10^{-5}$	1.03	$7.490*10^{-4}$	0.86
Atlantic Coast	$3.039*10^{-5}$	1.05	$2.261*10^{-5}$	1.07	$1.584*10^{-4}$	0.95
Gulf Coast	$9.311*10^{-4}$	0.84	$6.992*10^{-4}$	0.86	Insufficient data	
Pacific Coast	$2.833*10^{-4}$	0.91	$8.950*10^{-6}$	1.10	$1.015*10^{-3}$	0.85

Source: Jarrett (1976).

Table 9.3. Equilibrium relationships for tidal inlet morphology.

Authors	Morphological feature	Relation
LeConte (1905), O'Brien (1931, 1969), Riedel and Gourlay (1980), and Hume and Herdendorf (1992)	Channel cross-sectional area Ac (considers the longshore transport rate magnitude).	$Ac = C_1P$
Escoffier (1940)	Inlet cross-sectional area stability.	Close curve
Floyd (1968)	Channel depth and minimum depth over the ebb shoal.	–
Jarrett (1976)	Channel cross-sectional area, with and without jetties.	$Ac = C_2P^n$
Bruun and Gerritsen (1959)	Inlet stability and bypassing type.	P/M
Walton and Adams (1976), Marino and Mehta (1988), and Hicks and Hume (1996)	Ebb shoal volume V_E(note: relationships differ according to wave climate).	$V_E = C_3P^m$
Shigemura (1981)	Throat width W_e("e" denoting equilibrium).	$W_e = C_7P$
Gibeatu and Davis (1993)	Ebb shoal area A_E.	$A_E = C_4P^k$
Kraus (1998)	Channel cross-sectional area relation including longshore sediment transport rate in C_2.	$A_c = C_2P^n$
Carr de Betts (1999) and Carr de Betts and Mehta (2001)	Flood shoal area A_F and volume V_F.	$A_F = C_5P^P$ $V_F = C_6P^q$
Buonaiuto and Kraus (2003)	Limiting depth h_{cr} over the crest of ebb shoal; limiting slopes around the ebb shoal.	$H_{\mathrm{cr}} - P^j$

Note: P = tidal prism; A = area; V = volume; subscripts C, E and F = channel, ebb shoal and flood shoal, respectively; C = empirical or derived coefficient; j, m, n, p, q = empirical or derived power; W = width of inlet throat; M = gross longshore transport in a year.

ocean as well as by the characteristics of the sediment bars and shoals as discussed by several investigators (Oertel, 1977; Fitzgerald, 1984; Hume and Herdendorf, 1992; Fitzgerald *et al.*, 2002; Siegle *et al.*, 2004; Elias and van der Spek, 2006; Cheung *et al.*, 2007; Cooper *et al.*, 2007).

The dimensions of tidal inlets keep varying due to the variations in the flow in space and time along the channel. In case the inlet has potential for navigation, probably leading to a harbour or berthing facility along the estuary, a great deal of attention needs to be given to dredging in order to maintain the required depth for vessels moving in and out of the estuary. Construction of training walls is one of the most commonly adopted measures for preventing or reducing shoal formation and thus bringing down

Table 9.4. Inlet bypassing and classification of channel area cross-sectional stability.

r-value	Channel stability	Dominant bypassing mode
$r < 20$	Unstable. Inlet may be closed by deposition of sediment during a storm. Not typically a navigable channel.	Bar bypassing
$r = 20–50$	Highly variable channel in location and area, with multiple channels possible. Dredging and jetties are typically required to maintain navigable depths.	Bar bypassing; may have several bars
$r = 50–150$	Clear main channel and well-developed ebb shoal.	Bar bypassing and tidal bypassing
$r > 150$	Reasonable stable channel.	Episodic bypassing and tidal bypassing

Note: $r = P/M$; P = Spring tidal prism and M = Total transport rate per annum.

the dredging quantity to ensure sufficient draft required for vessels. The structural measures as well as the dredging in the channel followed by the disposal of dredged spoil will govern the hydrodynamic as well as morphodynamic equilibrium (Komar, 1996; Kraus, 2009; Oost *et al.*, 2012). Several studies have discussed the effects of training walls across an estuarine inlet. The training walls tend to modify the local littoral sediment transport rate, the evolution of the shoreline, geometry and hydrodynamics of inlet channels (Tomlinson, 1991; French, 2001; Hanson and Kraus, 2001; Castelle *et al.*, 2007; Fontolan *et al.*, 2007; Buonaiuto and Bokuniewicz, 2008; Kraus, 2009; Pacheco *et al.*, 2010; Wang and Beck, 2012; Sundar and Sannasiraj, 2016).

In the case of a trained inlet, if equilibrium in the volume of the water exchanged over a tidal cycle is achieved, the sediments will be transported continuously from updrift to downdrift side across the inlet as discussed by Dean (1988).

The grain size of the sediment in general decreases as it moves upstream due to an increase in the elevation over intertidal flats. Daboll (1969) reported that the higher sandy portions (ebb shields) are usually coarser than the lower flood ramp areas on the flood tidal delta. Within an estuary, the flow is unsteady and non-uniform owing to the tidal current fluctuations along and in the lateral directions. The magnitude and the direction of the flow and the boundary shear stress govern the rate and location of siltation or erosion. Herein, the flow is more complex with a no-diffusion upper zone or jet, where the turbulence beneath this layer does not affect

Table 9.5. Parameters governing morphological changes at the inlet.

Dynamic parameters	Static parameters
Net longshore sediment transport	Offshore and nearshore bathymetry
Tidal prism	Sediment grain size
Riverine sediment supply	Properties of structures
Dredging quantity of the channel	Geometry of structures
Flood shoal evolution	Size and shape of the back bay
Wave refraction and diffraction over shoals	

Source: Carr and Kraus (2001).

the no-diffusion zone. In the central zone, the flow is turbulent, which enables mixing from above the jet with the almost-static/motionless water beneath. The lower zone has no flow or backflow dynamics, induced by the separation turbulence above. As the flow in the aforesaid three zones progresses, the mixing zone flow reattaches to the bed further downstream, where intense turbulence and scour may occur, the magnitude of which depends on the original flow velocity. This turbulence in flow near the bed carries finer grains downstream as suspended load and the coarser grains slide down the leeward side as bed load sediments. This can also be in the lateral direction across the width of the estuary, which will depend on the width. The larger grains in the suspended load controlled by the fall velocity settle through the zone of mixing to become deposited either in front of or on the slip face. The smaller grains are then carried to the reattachment zone and thus to the next bed further downstream, while simultaneous erosion at the point of flow reattachment removes material from the channel bed (Davis, 1978).

An ideally stable inlet would retain its cross-section by settling sediments either further away from flood and ebb delta or within the sediment traps in the bed (Bruun, 1986, 1989). The current vortices are generated when tides move over the irregular topography within the inlet basin. These currents are advected by the tidal flow, resulting in a complex flow (Zimmerman, 1980; Ridderinkhof, 1989; Stanev *et al.*, 2003, 2007). Carr and Kraus (2001) listed the various dynamic and static parameters responsible for morphological changes at the vicinity of an inlet, which are mentioned in Table 9.5.

The existing analytical models are not capable of accurately assessing the phase lag between the confluence and the farthest point on long inlets due to nonlinearities caused by friction, which is responsible for modulating the volume flux and the water surface, an aspect detailed by Styles *et al.* (2016).

9.6 Importance of Inlets

For several reasons, inlets are quite important but complex; the water inside a bay contains less dissolved salts in it unlike ocean water. This kind of water is known as brackish water, which falls between fresh and saltwater. Many kinds of marine animals depend on such brackish water inlets to stay alive. The intensity of salt content in a bay is controlled by the inlets. If an inlet closes, the salt content in the water decreases, resulting in inhabitable conditions for many organisms living in the bay, and vice versa if a new inlet opens up. As the areas behind barrier islands are protected from swells and huge storm waves, most harbours are located behind barrier islands. Inlets act as a highway between the ocean and the bay.

Tidal inlets are breaks in barrier island systems that allow water, organisms, ships and people easy access and exchange between the high-energy open ocean and the low-energy back-barrier environment consisting of bays, lagoons, tidal marshes and creeks. Most tidal inlets are within barrier island systems, but others may separate barrier islands from rocky or glacial headlands. Tidal inlets are extremely important for navigation between sheltered ports on the back-barrier bays and the open ocean; thus, they are the sites of many coastal modifications such as jetties, breakwaters and dredged channels to keep the channels stable and open. The strong tidal currents move water in and out of tidal inlets as the tides wax and wane and also carry out of the channel large amounts of sediment brought in by waves and longshore transport.

As tides rise on the ocean side of tidal inlets, the propagation of seawater takes place up to a distance along the river, and this region of the river up to which the salt water penetrates is termed an estuary. In tidal inlets with large tidal ranges, during flood tide (rising tide), water flows into the inlet. During ebb tide (falling tide), the water flows out towards the ocean due to high water level within the inlet. It facilitates the removal of floating debris from the inlet into the broader water body, the ocean. The mangroves, which serve as natural barriers in reducing inundation during an extreme event like a tsunami, can survive only in a zone where an exchange of seawater and fresh river water takes place. The strongest currents in tidal inlets are found where the inlet is narrow, which is a place with the deepest water, called the inlet throat. Water rushes at high velocity into and out of this throat, carrying sand in and out of the back-barrier environment. Since the velocity of the water decreases after it passes through the throat, large lobes and sheets of sand are typically deposited as tidal deltas on both the inside and the outside of tidal inlets. The delta deposited by the incoming (flood) tide on the inside or landward side of the inlet is known as a flood-tidal delta and the delta deposited on the outside of the inlet by the ebb tide is known as an ebb-tidal delta. Tidal inlets may also form through longshore currents, building a spit across a bay or drowned river valley. As the spit grows across an open bay, the area open to the sea gradually becomes narrower until it begins to host strong tidal currents when it becomes deep and reaches equilibrium between the amount of sediment transported to the inlet by longshore drift and the amount of sand moved out of the inlet by tidal currents and waves.

9.7 Training of Inlets

9.7.1 *General*

The most common coastal protection measure for inlet stabilization is the construction of training walls. These are shore-connected structures protruding from either end of the river bank. Entrance training is achieved using rock walls (or breakwaters) to stabilize inlets and improve navigation at river entrances. The basic design consists of two parallel walls of approximately equal length, mostly oriented perpendicular to the shoreline. More sophisticated training wall designs which include non-parallel walls, e.g. Colombia River (Gelfenbaum and Kaminsky, 2010), or overlapping walls (Nguyen and Luong, 2015) have been constructed. In some cases, the entrance training is combined with coastal port development, often incorporating novel breakwater shapes (Mangor *et al.*, 2017). In all of these cases, design caveats have focused on longshore sediment transport and direct blockage due to the training walls at the shoreline (CERC, 1984; Mangor *et al.*, 2017; van Rijn, 2013). Training walls should not be aligned to coincide with the internal river orientation. Instead, the alignment relative to the shoreline should cause exiting currents to orient with the direction of net longshore sediment transport, obtained after initial analysis of the wave climate and sediment supply sources. The downdrift wall will be shorter than the updrift wall to maximize bypassing. With these simple additions, downdrift erosion will be reduced while the ebb delta is expected to be deeper for navigational safety (Black *et al.*, 2021).

9.7.2 *Riverine inlets – Case studies*

The discussion herein pertains to the southeast and southwest coasts of the Indian peninsula, with various coastal features including lagoons, bays, barrier islands, river mouths and backwaters. There also exist a number of harbours and ports across the coastline that promote fisheries and trade.

A number of perineal rivers and their tributaries also drain into the ocean along these coasts, which leads to the formation of riverine or estuarine inlets with tidal flushing. In general, the tides across the South Indian maritime states of Kerala and Tamil Nadu are in the range of 0.7–1.2 m, and the width of the inlets is usually in the order of 100–150 m with a few exceptions.

A majority of the inlets in this region can be classified as micro-tidal inlets. Both the southeast and the southwest coasts are longshore predominant and the rate of longshore sediment transport across the different stretches is governed by a number of local parameters including sediment grain size, bathymetry, orientation of the coast, angle of wave approach and offshore wave climate. The inlets along the coast act as sediment sinks where the longshore sediments are likely to be trapped and deposited. This leads to the formation of shoals, sandbars and spits at the inlet mouths and eventually leads to the closure of the inlet, provided that the discharge from the river source is insufficient to keep the mouth open across all seasons. The dynamic nature of the coasts leads to shifting of the shoreline across seasons and over the years, owing to the complexity of the hydrodynamics and morphodynamics at the inlet and the adjoining shoreline. The shoreline fluctuations in the vicinity of inlets need to be monitored. A few case studies of riverine inlets at different locations are discussed in the following sections.

9.7.2.1 *Karaikal*

The complex involved in modelling hydrodynamics and morphodynamics in the vicinity of a tidal inlet has been described in previous sections. Sundar *et al.* (2022) and Saichenthur *et al.* (2022) studied the dynamics and shoal formation processes at the Karaikal inlet (10°54′52″N; 79°51′09″E) through a finite volume solver using field-measured data to calibrate and validate the model. Sample hydrodynamic and morphodynamic results for a typical month at Karaikal inlet are projected in Fig. 9.6. In view of finding a long-term solution to the siltation problem and erosion of the river banks, a revetment structure was proposed to be constructed along the southern bank of the river, along the concave eroded region. This proposal was subjected to the aforementioned modelling scheme and the results (see Fig. 9.7) indicate reduced siltation owing to the presence of the revetment structure (marked in red line on the southern bank of the inlet).

9.7.2.2 *Thottappally pozhy*

Thottappally pozhy is located at latitude 9°18′35.76″N and longitude, 76°22′55.52″E in Alappuzha district, Kerala. The Thottappally River

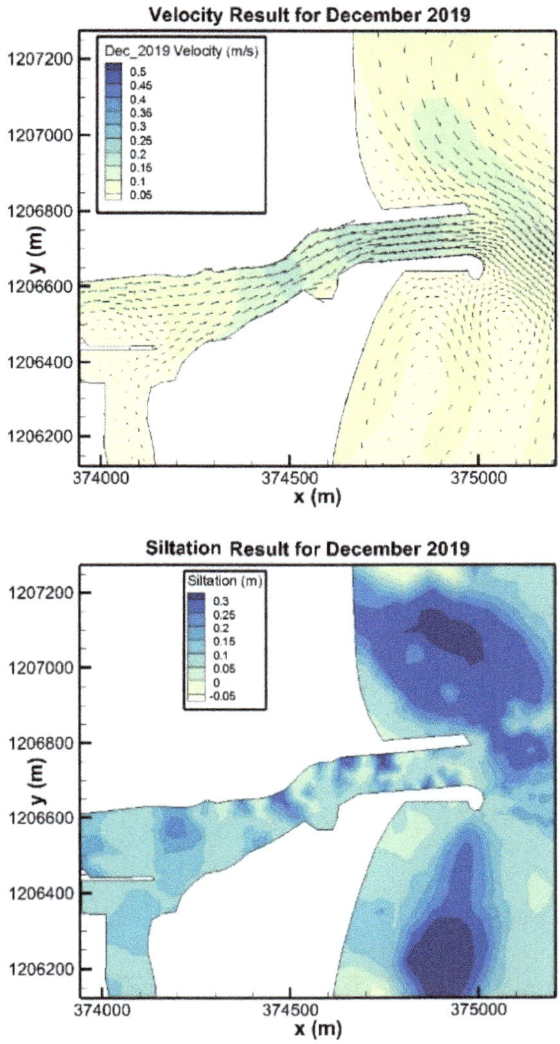

Fig. 9.6. Hydrodynamic and morphodynamic modelling results for Karaikal inlet.

drains into the Arabian Sea and the inlet opening is about 350 m wide, which is often blocked due to sandbar and spit formations at its mouth. Based on field-measured data and numerical modelling (as seen in Fig. 9.8), a proposal was made to construct a pair of training walls (see Fig. 9.9) at Thottappally inlet.

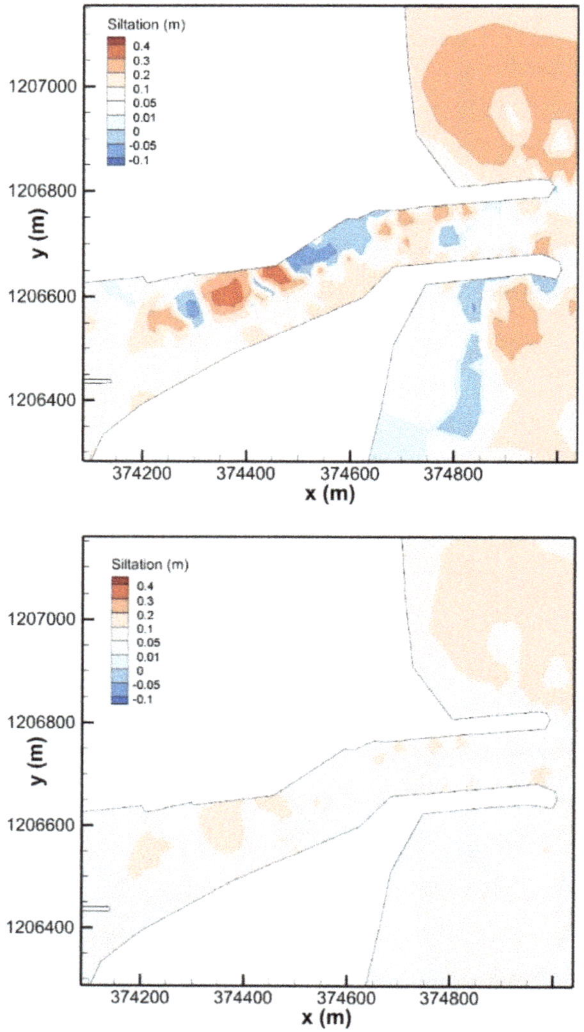

Fig. 9.7. Morphodynamic modelling with proposed revetment structure.

9.7.2.3 *Vanagiri*

Vanagiri is located at latitude 11°8′11.18″N and longitude 79°51′27.00″E, in Nagapattinam district, Tamil Nadu. Here, the Kaveri River drains into the Bay of Bengal and this site lies in the vicinity of an active fishing community. The existing training walls at the site are insufficient, and excessive shoal formation can be observed on both banks of the river, which

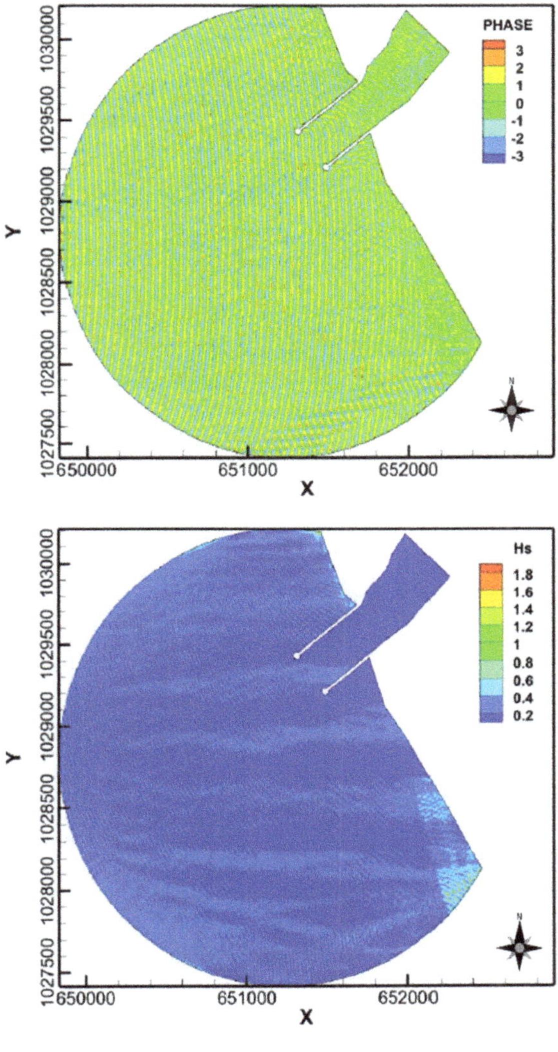

Fig. 9.8. Variation of phase and significant wave heights for an offshore wave approach angle of 290°.

has intensified over the years and almost closed the inlet. Based on field-measured data and numerical modelling (as seen in Fig. 9.10), a proposal was made to extend the training walls up to the surf zone width and plan for additional groynes to combat erosion on the downstream side (see Fig. 9.11) at the Vanagiri inlet.

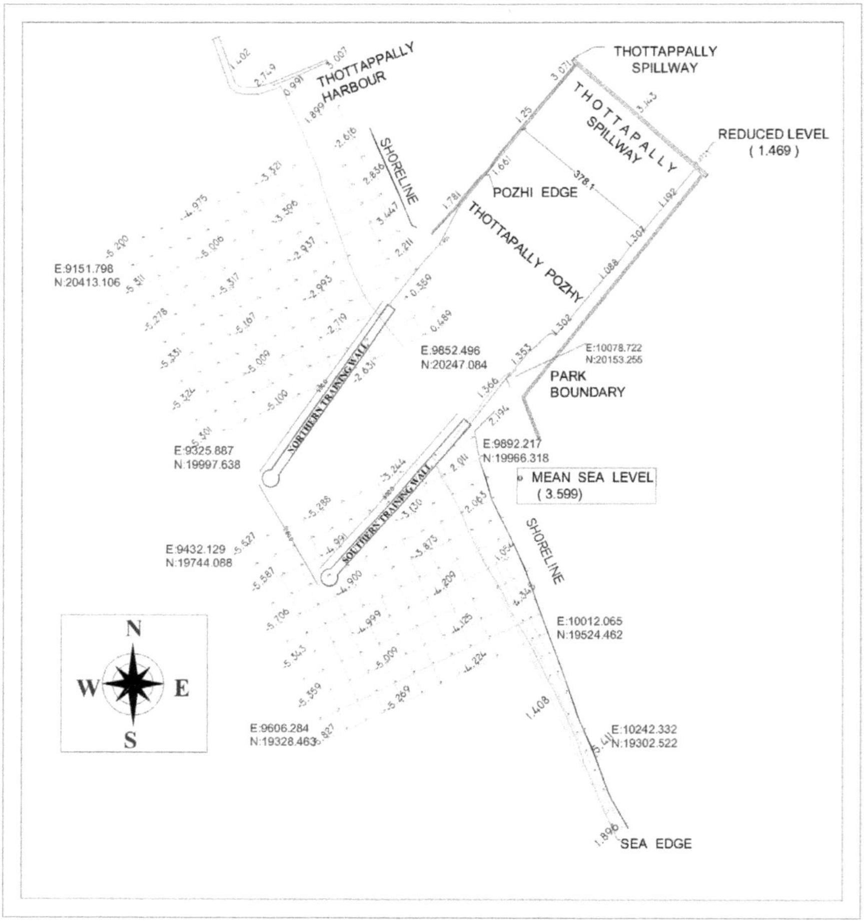

Fig. 9.9. Proposed construction of training walls at Thottappally pozhy inlet.

9.7.2.4 *Shoreline evolution*

Numerical models offer the capability to study the effect of wave characteristics, structure dimensions and other associated parameters in providing reasonable estimates of shoreline response. As the ocean waves approach the shore, they undergo transformations like shoaling, refraction, diffraction and breaking. The phenomena of wave breaking throw sediments to the surface due to the turbulence generated. The sediments in suspension are then driven by wave-induced currents. When a structure normal to the shoreline is constructed, it will intercept the free passage of longshore sediment

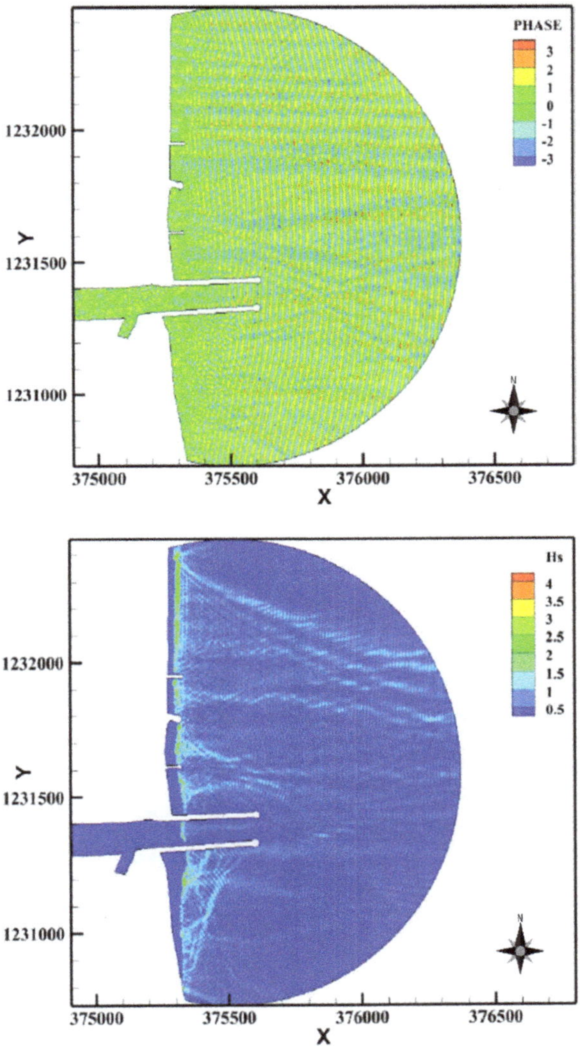

Fig. 9.10. Variation of phase and significant wave heights for an offshore wave approach angle of 90°.

transport, which results in an imbalance in the quantity of sediment in the nearshore region, especially near the structure. This leads to accretion on the updrift side and erosion on the downdrift side of the structure. Kraus and Harikai (1983) proposed a numerical scheme to solve the one-line model using the Crank Nicholson implicit finite difference method. Then, shoreline

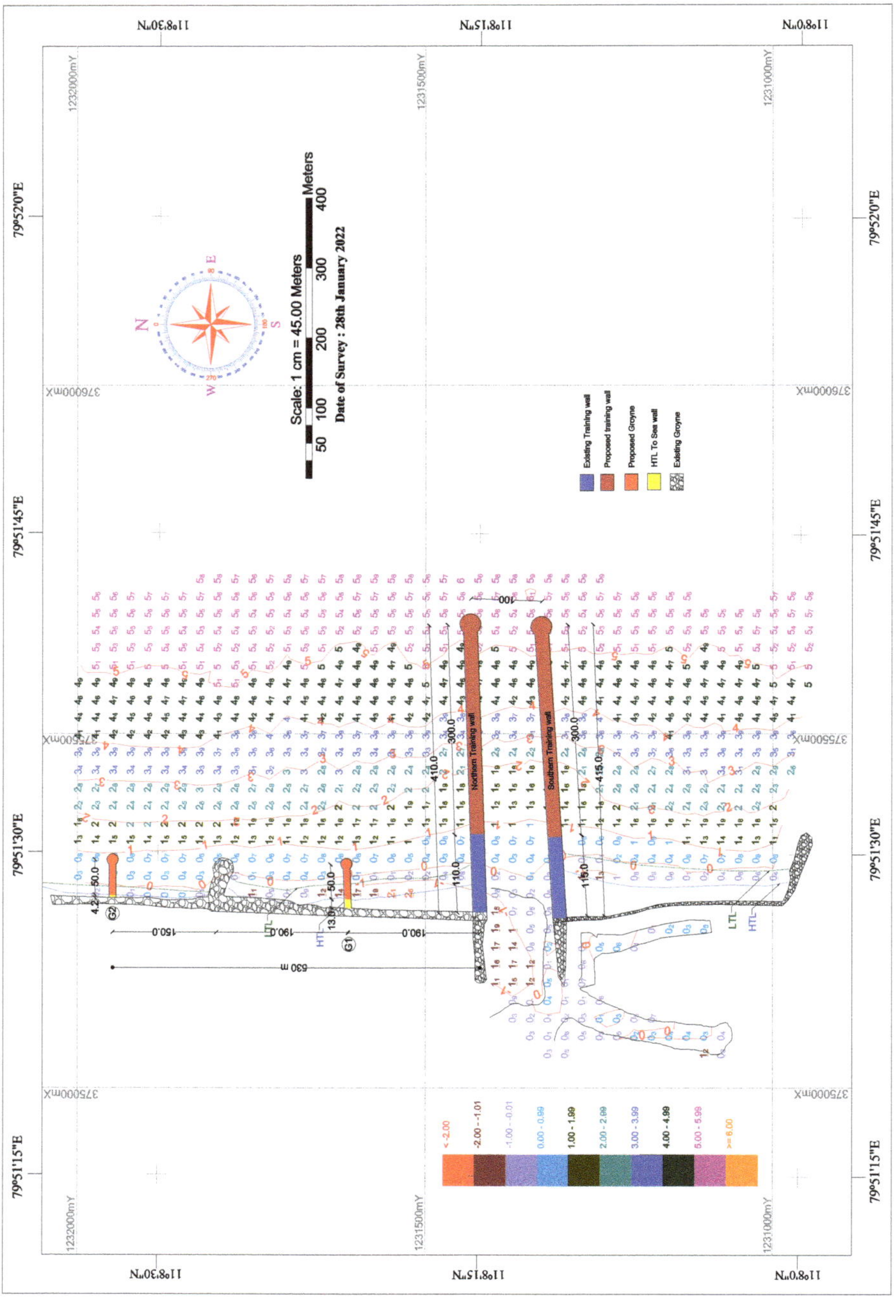

Fig. 9.11. Proposed extension of training walls at Vanagiri.

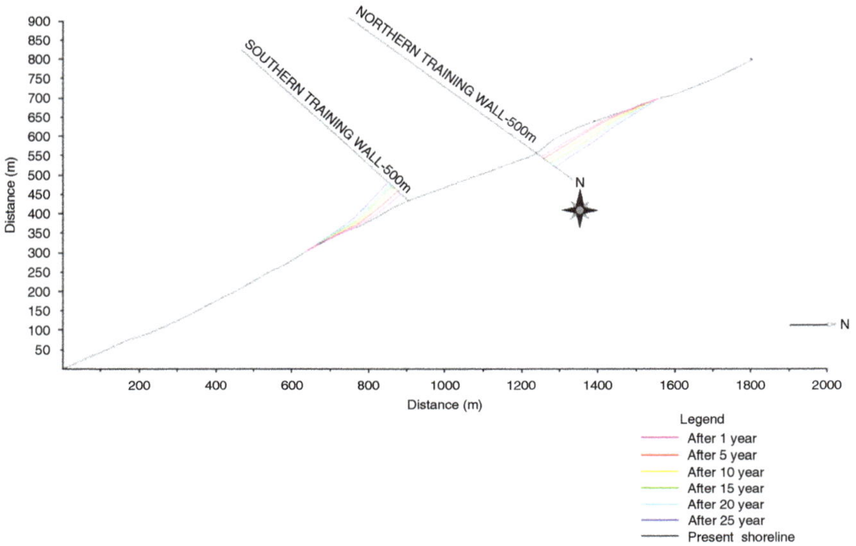

Fig. 9.12. Shoreline evolution changers at Thottappally inlet.

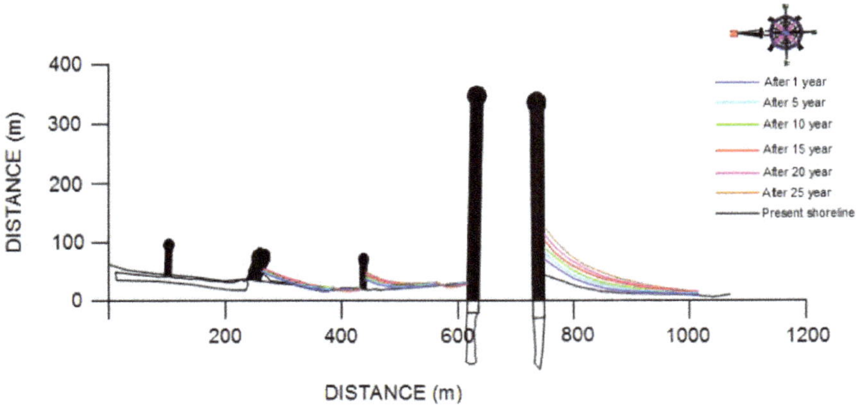

Fig. 9.13. Shoreline evolution changes at Vanagiri.

changes over a non-dimensional time δt across the space δx are calculated. In this method, Q^* (longshore sediment transport rate) at the time interval (t^*+1) is expressed in terms of the shoreline coordinate of y^*, first isolating the term involving $\alpha_{\rm sp}$ (angle of shoreline normal to x-axis) using trigonometric identities. One of the terms involving $\alpha_{\rm sp}$ is then expressed as first-order quantities in y^* at time step (t^*+1). The numerical model was executed for the most frequently occurring wave characteristics for a typical calendar year. The results of the predicted shoreline variations over the years for Thottappally inlet and Vanagiri are projected in Fig. 9.12 and Fig. 9.13, respectively. The shoreline prediction has been made at the end of 1 year, 5 years, 10 years, 15 years, 20 years and 25 years after the construction of the groynes and has been presented by superimposing the shoreline patterns.

9.7.2.5 *Shoreline evolution in between a pair of inlets*

The shoreline between any two training walls is influenced by factors such as the distance between them, the discharge of water (both in quantity and quality) from the inlets, and the magnitude and direction of longshore sediment transport. A case study considering two inlets along the South of Chennai port are herein discussed in brief. The Cooum and Adyar rivers run within the urban area of the city of Chennai and drain into the Bay of Bengal adjoining Marina Beach (world's second-longest beach), which is an important tourist destination. The coastal stretch between the South of Cooum (13°4′N and 80°17′20″E) and Zone II: Adyar (13°0′49″N and 80°16′38″E) has been analyzed from year 2000 to year 2019. Based on the findings by Gracy Margret Mary *et al.* (2022), sizeable accretion is observed at the Cooum inlet and considerable erosion is observed at the Adyar inlet (see Fig. 9.14). The Adyar River zone was classified as highly eroded with an average erosion rate of about $-4.15\,\mathrm{m/yr}$ through statistical regression. Furthermore, as the net drift is towards the north, the Cooum inlet being in the south traps a significant quantity of sediment, leading to significant sandbar formation near its mouth.

9.8 Stratification in Estuaries

9.8.1 *General*

Salt water intrusion into estuaries is commonly observed across all tidal inlets. In the case of a riverine estuary, there exists a continuum along a fresh–brackish–saltwater gradient, which influences density stratification based on local geophysical, geochemical and geomorphological conditions at the site under study. A mixing of salt water and fresh water occurs at estuarine inlets; the extent of the mixing depends on the riverine flow rate and tidal prism. The occurrence of strong tides, cyclones, storm surges,

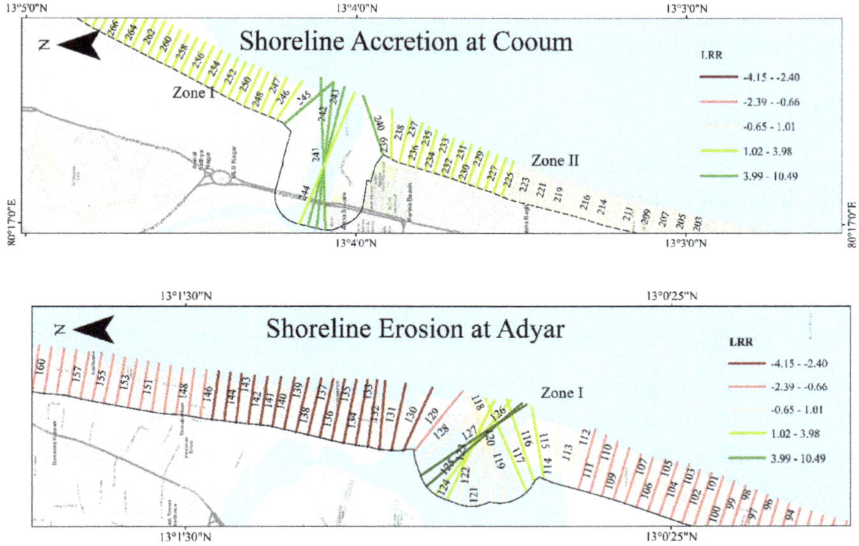

Fig. 9.14. Eroding/accreting transects at river mouths.

tidal bores, flash floods or other extreme events could also trigger sudden changes in the physicochemical conditions in the vicinity of estuaries.

Geologically, stratified estuaries can be broadly classified as coastal plain estuaries, bar-built estuaries, fjords and tectonic estuaries. The dimensions, stratification densities and the rate of water exchange are all governed primarily by the geomorphological features. A graphical representation of the different types of stratified estuaries is shown in Fig. 9.15.

Sea level rise is commonly witnessed across most coastlines worldwide; the increased water levels in the vicinity of an inlet flood the river basin with seawater, where the mixing of fresh and seawater leads to brackish conditions at the inlet. The water depths of coastal plain estuaries are shallow and the rivers carry a heavy sediment-laden flow, which results in the formation of a number of depositional features such as shoals, spits, sandbars and barrier islands.

9.8.1.1 *Bar-built estuaries*

The existence of geomorphological features such as sandbars, spits and barrier islands leads to the formation of bar-built estuaries. In such estuaries, a barrier island is created between the mainland and the ocean, and the water contained within the sandbars is cut off from the mixing and receives fresh water from the land-side riverine source.

9.8.1.2 *Fjords*

These are very deep basins, carved out by advancing glaciers. As glaciers melt down and disintegrate, the increased sea level fills in the carved-out glaciers, creating deep, steep-walled fjords. Fjords can be observed along mountainous coastlines once covered by glaciers.

9.8.1.3 *Tectonic estuaries*

Owing to the movement of tectonic plates across fault lines, certain sections subside from their original level. These sections of lower elevations then get flooded with seawater, resulting in the formation of tectonics estuaries.

9.8.2 *Stratification index*

The stratification index (S_p) is widely used to indicate the intensity of stratification within the water column. It is the ratio of difference in salinity between the surface and bottom waters (δS) and the vertically averaged salinity (S_0). Based on the intensity of stratification, it can be classified as well mixed ($S_p \leq 0.1$), partially mixed ($0.1 < S_p \leq 1.0$) or highly stratified ($S_p > 1.0$).

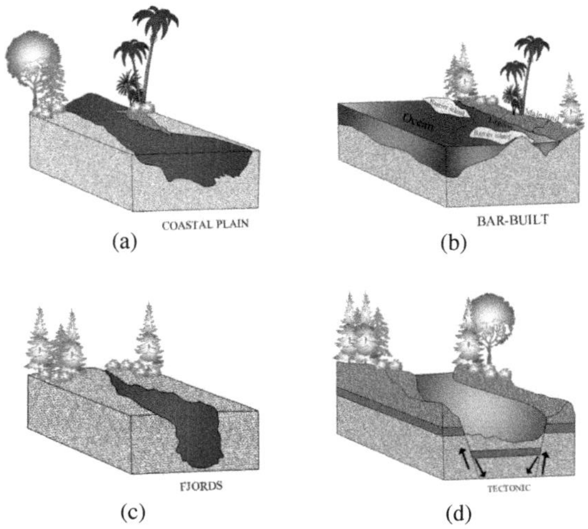

Fig. 9.15. Types of stratified estuaries.

In a *well-mixed* estuary, the fresh and salt water are completely mixed from the surface to the bottom. The observed salinity at a fixed coordinate (i.e. x, y) will be uniform throughout the depth (z), whereas across the estuary, lower salinity can be witnessed near the entrance of freshwater riverine source and higher salinity can be observed at the mouth, where seawater enters the estuary. This is often the characteristic of shallow water estuaries, which enables complete vertical mixing.

Partially mixed estuaries have a salinity profile similar to that of a well-mixed estuary, with an exception in the distribution of the salinity along the depth. Partial mixing or slightly stratified estuaries are characteristic of deeper estuaries where the surface waves and currents enable mixing on the top layer, which may not be extended all the way to the bottom.

Highly stratified estuaries are characteristic of very deep estuaries, such as fjords. Owing to the steep vertical profiles, mixing is witnessed along the

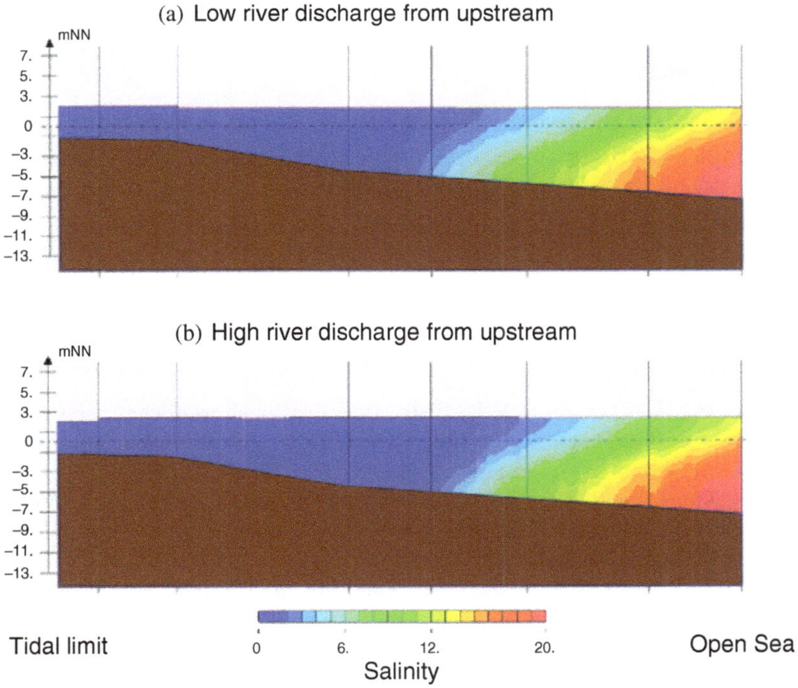

Fig. 9.16. Density-induced circulation.
Source: Schüttrumpf.

surface alone and the bottom layer is unaffected under the morphological conditions.

9.8.3 *Density-induced estuarine circulation*

Density-induced estuarine circulation is primarily governed by the depth of the estuary, eddy viscosity in the mixing layers and the density gradient distribution along the channel (Officer, 1976). A number of simplifying assumptions have been made to understand the fundaments of the driving parameters for density stratification, which is the strength of the estuarine exchange flow associated with gravitational circulation. The river discharge from upstream exerts first-order control over vertical stratification, which in turn governs the eddy viscosity, which further influences the strength of the exchange flow (see Fig. 9.16).

Extensive literature is available for interested readers to pursue, in which various methods and techniques have been studied to control and influence stratification density across various types of inlets, particularly wind speed. This falls under the purview of physical oceanography.

9.9 Summary

The confluence of the river and ocean is a coastal geomorphological feature with complex hydrodynamics and morphodynamics with the potential to alter the status of adjoining coasts. The training of riverine inlets is a vital measure to ensure the navigability of the inlet against the blocking of the inlet mouth. This chapter extensively covers the fundamentals of river-coastal confluence, supported by field examples.

References

Ashworth, P.J. and Ferguson, R.I. (1986). Interrelationships of channel processes, changes and sediments in a proglacial braided river. *Geografiska Annaler: Series A, Physical Geography*, 68(4), 361–371.

Bilal, A., Xie, Q. and Zhai, Y. (2020). Flow, sediment, and morpho-dynamics of river confluence in tidal and non-tidal environments. *Journal of Marine Science and Engineering*, 8(8), 591.

Black, K.P., Kulkarni, K.B., Naik, G.B., Naik, M. and Mathew, J. (2021). Improving training wall orientation at river and estuary entrances. *Journal of Coastal Research*, 37(3), 484–493.

Bolla Pittaluga, M., Tambroni, N., Canestrelli, A., Slingerland, R., Lanzoni, S. and Seminara, G. (2015). Where river and tide meet: The morphodynamic equilibrium of alluvial estuaries. *Journal of Geophysical Research: Earth Surface*, 120(1), 75–94.

Bruun, P. (1986). Morphological and navigational aspects of tidal inlets on littoral drift shores. *Journal of Coastal Research*, (2), 123–146.

Bruun, P. (1989). *Port Engineering*, Vols. III and IV. Gulf Publishing Company, Houston, Texas, 1,700 p. (1981), p. 800.

Bruun, P. and Gerritsen, F. (1959). Natural by-passing of sand at coastal inlets. *Journal of the Waterways and Harbors Division*, 85(4), 75–107.

Buonaiuto, F.S. and Kraus, N.C. (2003). Limiting slopes and depths at ebb-tidal shoals. *Coastal Engineering*, 48(1), 51–65.

Buonaiuto, F.S. and Bokuniewicz, H.J. (2008). Hydrodynamic partitioning of a mixed energy tidal inlet. *Journal of Coastal Research*, 24(5), 1339–1348.

Carr de Betts, E.E. (1999). An examination of flood deltas in Florida's tidal inlets. Doctoral dissertation, Coastal & Oceanographic Engineering Department, University of Florida.

Carr de Betts, E.E. and Mehta, A.J. (2001). An assessment of inlet flood deltas in Florida. In *Coastal Dynamics'01*, pp. 252–262.

Carr, E.E and Kraus, N.C. (2001). Morphologic asymmetries at entrances to tidal inlets. Coastal and Hydraulic Engineering Technical Note ERDC/CHL CHETN-IV-33. U.S. Army Engineer Research and Development Center, Vicksburg, MS. http://chl.wes.army.mil/library/publications/chetn.

Castelle, B., Bourget, J., Molnar, N., Strauss, D., Deschamps, S. and Tomlinson, R. (2007). Dynamics of a wave-dominated tidal inlet and influence on adjacent beaches, Currubin Creek, Gold Coast, Australia. *Coastal Engineering*, 54(1), 77–90.

CERC (Coastal Engineering Research Center). (1984). Shore Protection Manual, Volume I and II. Vicksburg. US Army Corps of Engineers, Waterways Experiment Station, Mississippi, 337 p. and 222 p.

Cheung, K.F., Gerritsen, F. and Cleveringa, J. (2007). Morphodynamics and sand bypassing at Ameland inlet, the Netherlands. *Journal of Coastal Research*, 23(1), 106–118.

Cooper, J.A.G., McKenna, J., Jackson, D.W.T. and O'connor, M. (2007). Mesoscale coastal behavior related to morphological self-adjustment. *Geology*, 35(2), 187–190.

Daboll, J.M. (1969). Holocene sediments of the Parker River estuary, Massachusetts: Cont. No. 3, Coastal Research Group, Department of Geology, University of Massachusetts, 137.

Davies, J.L. (1980). *Geographical Variation in Coastal Development*, 2nd edn. Longman, New York, 212 p.

Davis, R.A. (1964). Foraminiferal assemblages of Alacran Reef, Campeche Bank, Mexico. *Journal of Paleontology*, 38(2), 417–421. http://www.jstor.org/stable/1301569.

Davis, R.A. (1978). *Coastal Sedimentary Environments.* Springer, New York. https://doi.org/10.1007/978-1-4684-0056-4.

Davis Jr, R.A. and Fox, W.T. (1981). Interaction between wave-and tide-generated processes at the mouth of a microtidal estuary: Matanzas River, Florida (USA). *Marine Geology*, 40(1–2), 49–68.

Dean, R.G. (1988). Sediment interaction at modified coastal inlets: Processes and policies. In *Hydrodynamics and sediment Dynamics of Tidal Inlets.* Springer Verlag, New York, pp. 412–439.

Elias, E.P. and van der Spek, A.J. (2006). Long-term morphodynamic evolution of Texel Inlet and its ebb-tidal delta (The Netherlands). *Marine Geology,* 225(1–4), 5–21.

Escoffier, F. (1940). The stability of tidal inlets. *Shore and Beach,* 8(4), 114–115.

Ettema, R. (2008). Management of confluences. In *River Confluences, Tributaries and the Fluvial Network.* John Wiley & Sons, Chichester, pp. 93–118.

Fitzgerald, D.M. (1984). Interactions between the ebb-tidal delta and landward shoreline; Price Inlet, South Carolina. *Journal of Sedimentary Research,* 54(4), 1303–1318.

FitzGerald, D.M. and Montello, T.M. (1993). Backbarrier and inlet sediment response to the breaching of Nauset Spit and formation of new inlet, Cape Cod, Massachusetts. In *Formation and Evolution of Multiple Tidal Inlets,* 44, 158–185.

Fitzgerald, D.M. (1988). Shoreline erosional-depositional processes associated with tidal inlets. In *Hydrodynamics and Sediment Dynamics of Tidal Inlets.* Springer, New York, pp. 186–225.

Fitzgerald, D.M. (1996). Geomorphic variability and morphologic and sedimentologic controls on tidal inlets. *Journal of Coastal Research,* 47–71.

Fitzgerald, D.M., Buynevich, I.V., Davis Jr, R.A. and Fenster, M.S. (2002). New England tidal inlets with special reference to riverine-associated inlet systems. *Geomorphology,* 48(1–3), 179–208.

Floyd, C.D. (1968). River mouth training in New South Wales, Aust. In *Coastal Engineering 1968,* pp. 1267–1281.

Fontolan, G., Pillon, S., Quadri, F.D. and Bezzi, A. (2007). Sediment storage at tidal inlets in northern Adriatic lagoons: Ebb-tidal delta morphodynamics, conservation and sand use strategies. *Estuarine, Coastal and Shelf Science,* 75(1–2), 261–277.

French, P.W. (2001). Coastal defences: Processes, problems and solutions. *Earth Surface Processes and Landforms* (John Wiley & Sons, Ltd., London.), 28(8), 923–924.

Gallivan, L.B. and Davis Jr, R.A. (1981). Sediment transport in a microtidal estuary: Matanzas River, Florida (USA). *Marine Geology,* 40(1–2), 69–83.

Gelfenbaum, G. and Kaminsky, G.M. (2010). Large-scale coastal change in the Columbia River littoral cell: An overview. *Marine Geology,* 273(1–4), 1–10.

Gibeatu, J.C. and Davis Jr, R.A. (1993). Statistical geomorphic classification of ebb-tidal deltas along the west-central Florida coast. *Journal of Coastal Research,* 165–184.

Gracy Margret Mary, R., Sundar, V. and Sannasiraj, S.A. (2022). Analysis of shoreline change between inlets along the coast of Chennai, India. *Marine Georesources & Geotechnology,* 40(1), 26–35.

Hanson, H. and Kraus, N.C. (2001). Chronic beach erosion adjacent to inlets and remediation by composite (T-head) groins. Engineer Research and Development Center Vicksburg MS Coastal and Hydraulics Lab.

Hartwell, A.D. (1970). Hydrography and Holocene sedimentation of the Merrimack River estuary. Massachusetts, Department of Geology, University of Massachusetts Contribution, (5-CRG).

Hayes, M.O. (1975). Morphology of sand accumulation in estuaries: An introduction to the symposium. In *Geology and Engineering.* Academic Press, New York, New York, USA, pp. 3–22.

Hayes, M.O. (1979). Barrier island morphology as a function of wave and tide regime. In Leatherman, S.P. (ed.) *Barrier Islands from the Gulf of St. Lawrence to the Gulf of Mexico*, Academic Press, New York, New York, USA, pp. 1–29.

Hayes, M.O. (1980). General morphology and sediment patterns in tidal inlets. *Sedimentary Geology*, 26(1–3), 139–156.

Hicks, D.M. and Hume, T.M. (1996). Morphology and size of ebb tidal deltas at natural inlets on open-sea and pocket-bay coasts, North Island, New Zealand. *Journal of Coastal Research*, 12, 47–63.

Hubbard, D.K., Oertel, G. and Nummedal, D. (1979). The role of waves and tidal currents in the development of tidal-inlet sedimentary structures and sand body geometry; examples from North Carolina, South Carolina, and Georgia. *Journal of Sedimentary Research*, 49(4), 1073–1091.

Hume.T. M. and Herdendorf. C.E. 1990. Morphologic and hydrologic characteristics of tidal inlets on a headland dominated, low littoral drift coast, Northeastern New Zealand, https://www.researchgate.net/journal/Journal-of-Coastal-Research-1551-5036?_tp=eyJjb250ZXh0Ijp7ImZpcnN0UGFnZSI6InNjaWVudGlmaWNDb250cmlidXRpb25zIiwicGFnZSI6InB1YmxpY2F0aW9uIiwicHJldmlvdXNQYWdlIjoic2NpZW50aWZpY0NvbnRyaWJ1dGlvbnMiLCJwb3NpdGlvbiI6InBhZ2VIZWFkZXIifX0 *Journal of Coastal Research* 9, 527–563, Coastal Education and Research Foundation (CERF).

Hume, T.M. and Herdendorf, C.E. (1992). Factors controlling tidal inlet characteristics on low drift coasts. *Journal of Coastal Research*, 8, 355–375.

Jarrett, J.T. (1976). *Tidal Prism: Inlet Area Relationships*, Vol. 3. U.S. Army Corps of Engineers, Waterways Experiment Station, Vicksburg, Mississippi, 54 p.

Komar, P.D. (1976). *Beach Processes and Sedimentation.* Prentice-Hall, Englewood Cliffs, NJ, 429 p.

Komar, P.D. (1996). Tidal-inlet processes and morphology related to the transport of sediments. *Journal of Coastal Research*, 23, 23–45.

Kraus, N.C. (1998). Inlet cross-sectional area calculated by process-based model. In *Coastal Engineering 1998,* pp. 3265–3278.

Kraus, N.C. (2009). Engineering of tidal inlets and morphologic consequences. In Kim, Y.C. (ed.), *Handbook of Coastal and Ocean Engineering.* World Scientific, Singapore, pp. 867–901.

Kraus, N.C. and Harikai, S. (1983). Numerical model of the shoreline change at Oarai Beach. *Coastal Engineering*, 7(1), 1–28.

LeConte, L.J. (1905). Discussion on the paper, "Notes on the improvement of river and harbor outlets in the United States" by D. A. Watt, paper no. 1009. *Transactions of ASCE,* 55(December), 306–308.

Leeder, M.R. (1983). On the interactions between turbulent flow, sediment transport and bedform mechanics in channelized flows. *Modern and Ancient Fluvial Systems*, Wiley, John Lewin, pp. 3–18.

Mangor, K., Dronen, N.K., Kaergaard, K.H. and Kristensen, S.E. (2017). Shoreline management guidelines. COADAPT Project of the DHI, Danish Coastal Authority and the University of Copenhagen, Horsholm, Denmark, 452 p. http://www.dhigroup.com.

Marino, J.N. and Mehta, A.J. (1988). Sediment trapping at Florida's east coast inlets. *Hydrodynamics and Sediment Dynamics of Tidal Inlets*, Vol. 29, Springer, USA, pp. 284–296.

Nageswara Rao, K., Sadakata, N., Hema Malini, B. and Takayasu. K. (2005). Historical treatise of shore dynamics and spit growth near Kakinada coast, Godavari delta, Andhra Pradesh, 1–14. Research Gate. https://www.researchgate.net/.

Nguyen, Q.D.A. and Luong, P.H. (2015). Efficiency analysis of river entrance construction works in Central Vietnam. In *Proceedings of the Vietnam-Japan Workshop on Estuaries, Coasts and Rivers*, Hˆoi An, Vietnam, pp. 99–105.

Nummedal, D., Oertel, G.F., Hubbard, D.K. and Hine, A.C. (1977). Tidal inlet variability—Cape Hatteras to Cape Canaveral. In *Coastal Sediments. 5th Symposium of the Waterway.*

O'Brien, M.P. (1931). Estuary tidal prisms related to entrance areas. *Civil Engineering, ASCE*, 1(8), 738–739.

O'Brien, M.P. (1969). Equilibrium flow areas of inlets on sandy coasts. *Journal of the Waterways and Harbors Division*, 95(1), 43–52.

Oertel, G.F. (1972). Sediment transport of estuary entrance shoals and the formation of swash platforms. *Journal of Sedimentary Research*, 42(4), 858–863.

Oertel, G. F. (1977). Geomorphic Cycles in Ebb Deltas and Related Patterns of Shore Erosion and Accretion. *Journal of Sedimentary Research*, 47(3), 1121–1131. Society for Sedimentary Geology (SEPM).

Officer, C.B. (1976). *Physical Oceanography of Estuaries and Associated Coast Waters.* John Wiley and Sons, New York.

Oost, A.P., Hoekstra, P., Bartholdy, J., van der Berg, M.W., Vos, P.C., de vries, S. and Wang, Z.B. (2012). Barrier island management: Lessons from the past and directions for the future. *Ocean Coastal Management*, 68, 18–38.

Pacheco, A., Ferreira, Ó., Williams, J.J., Garel, E., Vila-Concejo, A. and Dias, J.A. (2010). Hydrodynamics and equilibrium of a multiple-inlet system. *Marine Geology*, 274(1–4), 32–42.

Ridderinkhof, H. (1989). Tidal and residual flows in the western Dutch Wadden sea III: Vorticity balances. *Netherlands Journal of Sea Research*, 22(1), 9–26.

Riedel, H.P. and Gourlay, M.R. (1980). Inlets/estuaries discharging into sheltered waters. In *Coastal Engineering 1980,* pp. 2550–2564.

Saichenthur, N., Babu, S.R., Sundar, V., Murali, K. and Rajasekar, A. (2022). Numerical prediction of siltation patterns at a tidal inlet in Karaikal, India. *International Journal of Ocean and Coastal Engineering*, 4(03n04), 2250003.

Shigemura, T. (1981). Tidal prism-throat width relationships of the bays of Japan. *Shore and Beach*, 49, 34–39.

Siegle, E., Huntley, D.A. and Davidson, M.A. (2004). Physical controls on the dynamics of inlet sandbar systems. *Ocean Dynamics*, 54 (3–4), 360–373.

Stanev, E.V., Flöser, G. and Wolff, J.O. (2003). First-and higher-order dynamical controls on water exchanges between tidal basins and the open ocean. A case study for the East Frisian Wadden Sea. *Ocean Dynamics*, 53(3), 146–165.

Stanev, E.V., Flemming, B.W., Bartholomä, A., Staneva, J.V. and Wolff, J.O. (2007). Vertical circulation in shallow tidal inlets and back-barrier basins. *Continental Shelf Research*, 27(6), 798–831.

Stive, M.J., Roelvink, D.J.A. and de Vriend, H.J. (1991). Large-scale coastal evolution concept: The Dutch coast: Paper no. 9. In *Coastal Engineering 1990*, pp. 1962–1974.

Styles, R., Brown, M.E., Brutsché, K.E., Li, H., Beck, T.M. and Sánchez, A. (2016). Long-term morphological modeling of barrier island tidal inlets. *Journal of Marine Science and Engineering*, 4(4), 65.

Sundar, V. and Sannasiraj, S.A. (2016). Training of river mouths of the Kerala Coast in India. In *6th International Symposium on Hydraulic Structures: Hydraulic Structures and Water System Management, ISHS 2016*, Portland, USA, June 27–30, 2016, pp. 161–170. https://doi.org/10.15142/t3570628160853.

Sundar, V., Murali, K., Ramesh Babu, S. and Arun Rajasekar, A. (2022). Tidal inlet morphodynamics through numerical prediction and measurements. *Marine Geo-resources & Geo-technology*, 40(11), 1316–1327.

Tomlinson, R.B. (1991). Processes of sediment transport and ebb tidal delta development at a jettied inlet. In *Proceedings of Coastal Sediments'91*, pp. 1404–1418.

Van Rijn, L.C. (2013). Design of Hard Coastal Structures against Erosion. Training document, 33 p. http://www.leovanrijn-sediment.com.

Walton, T.L. Jr. and Adams, W.D. (1976). Capacity of inlet outer bars to store sand. In *Proceedings, 15th International Conference on Coastal Engineering*. American Society of Civil Engineers, New York, pp. 1919–1937.

Wang, P. and Beck, T.M. (2012). Morphodynamics of an anthropogenically altered dual-inlet system: John's Pass and Blind Pass, west-central Florida, USA. *Marine Geology*, 291, 162–175.

Xie, Q., Yang, J., Lundström, S. and Dai, W. (2018). Understanding morphodynamic changes of a tidal river confluence through field measurements and numerical modeling. *Water*, 10(10), 1424.

Zimmerman, J.T.F. (1980). Vorticity transfer by tidal currents over an irregular topography. *Journal of Marine Research*, 38(4), 601–631.

Chapter 10

Overview of Ecosystem-Based Coastal Structures

10.1 Background

There are two major aspects concerning coastal communities: extreme coastal hazards like storm surges and tsunamis and coastal erosion. The causes of the extreme events may be due to cyclonic activities, underwater landslides, earthquakes, volcanic eruptions, etc. The intensity of such seafront events that threaten coastal communities is escalating due to climate change activities on a global scale. Furthermore, natural processes (such as extreme waves, currents and floods) and coastal activities like dredging, sand mining, and construction of structures and dams across rivers that trap sediment from reaching the coast are the possible causes of coastal erosion.

Conventional structures, although serving as an immediate solution, are expensive and need regular rehabilitation, often considered eyesores and classified as non-eco-friendly solutions. Hence, researchers and engineers have been forced to look for alternate eco-friendly solutions for sustainable coastal zone development and management. The term "ecosystem-based coastal structure" refers to two major aspects, namely the structure in the marine environment built to improve the ecosystem and the existing ecological system present in that coastal region contributing to the betterment of that locality. Several researches have proved that the former concept mainly focuses on the field of biodiversity improvement and restoration of coastal features, while the latter concept has been discussed in terms of its significance towards the aspects of coastal protection through mangrove swamps, coral reefs, etc. Eco-based structures comprise soft and hard engineering measures permitting the growth of greenery over them and serving as a promising adaptation measure to cope with the anticipated climate change. The adaptation of such strategies is a welcome change in several engineering domains, provided they protect the coast but do not contribute much towards increased global warming. Sea level rise is yet another major threat to coastal communities, for which conventional hard measures may not offer a solution.

The presence of hard measures like dikes or embankments along wetlands can contribute to the degradation of the ecosystem. Similarly, reinforcing the banks of an estuary mouth with such hard solutions often exacerbates storm effects by funnelling the flow, resulting in increased flooding upstream. An effective approach to mitigating this issue involves integrating dikes, revetments or embankments with wetlands. Removing hard solutions in specific areas aids in stabilizing wetlands, enabling excess floodwaters to permeate the land and thereby reduce surge levels. During storms, sandy beaches protected by hard measures like groynes and seawalls often experience erosion of sediment due to the intense wave action. The return flow carries away most of the sediment, leading to erosion in front of the protected coast. This erosion can contribute to the failure of the aforesaid hard structures during extreme events. Conversely, sandy coasts with submerged reefs facilitate the premature breaking of waves, leading to reduced storm intensity effects on the coast, thus promoting natural beach nourishment with the sand trapped within the swash zone due to the presence of the reef. Hence, combined hard and soft measures could be a much better mitigation measure as detailed by Temmerman *et al.* (2013).

The following are the major merits of ecosystem-based adaptation measures:

- Better conservation and management of water resources,
- Enhancement of water quality due to better circulation,
- Better control over air quality and coastal erosion, landslides and subsidence,
- Better public health,
- Higher aesthetic value,
- Lesser damage to coastal infrastructure,
- CO_2 sequestration.

The coastal ecosystem comprises all the features along coasts and estuaries. The major aspects that are being looked into by the coastal engineers are the behaviour of waves in the ocean, sediment transport on a regular basis, coastal inundation and forces on coastal building during storm surges and tsunamis. Figure 10.1 shows a schematic flowchart considered for modelling the above-mentioned aspects. Coastal structures are built to reduce adverse impacts. These structures generally act as energy dissipaters and are modelled based on experimental studies/field data. Ecosystem-based measures are now slowly replacing conventional measures to attain the

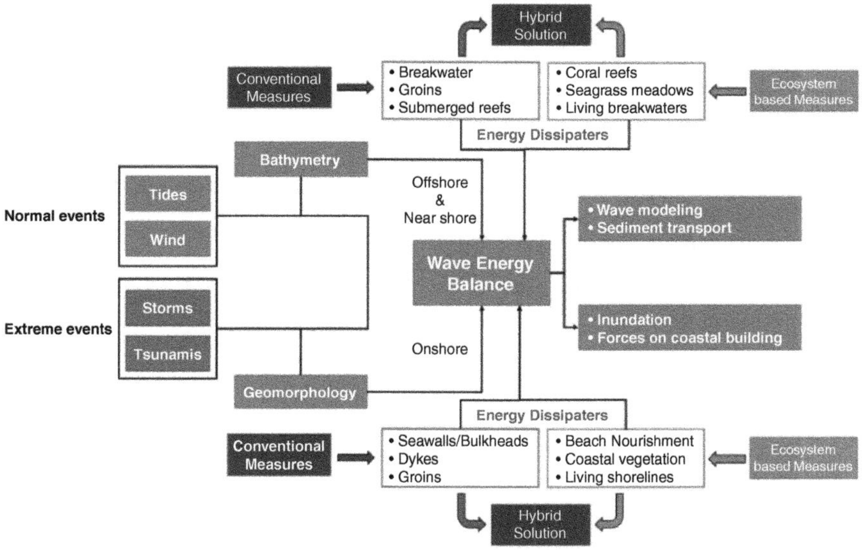

Fig. 10.1. Schematic flowchart for wave modelling studies.

required safe environment without compromising environmental quality. In a few cases, a combination of both soft and hard measures has been considered, popularly known as "hybrid coastal measures," which accommodates the merits of both coastal protection measures.

Drag force is a key parameter considered in the design of the eco-based coastal measures. Mesoscale models consider details from either experimental studies of a prototype/model or microscopic numerical studies. Figure 10.2 gives a representative idea of the modelling studies that are carried out.

Erduran and Kutija (2003) considered a quasi 3D model that incorporated cantilever beam theory to account for the effects of deflection in vegetation due to the wave loads and handled the turbulence using a parabolic eddy viscosity approach in combination with mixing length theories. Maza *et al.* (2015) studied solitary wave evolution over vegetation using Hi FOAM, considering both macroscopic analyses using drag coefficient terms and direct numerical modelling. Both analysis and modelling highlighted the strengths and weaknesses concerning computational efficiency and the sensitivity to coefficients. These findings provided an insight into the impact due to varying setups and the reliability of results. However, both methods predicted wave transmission over the vegetation for the considered flow. Divya *et al.* (2020) modelled the wave–vegetation

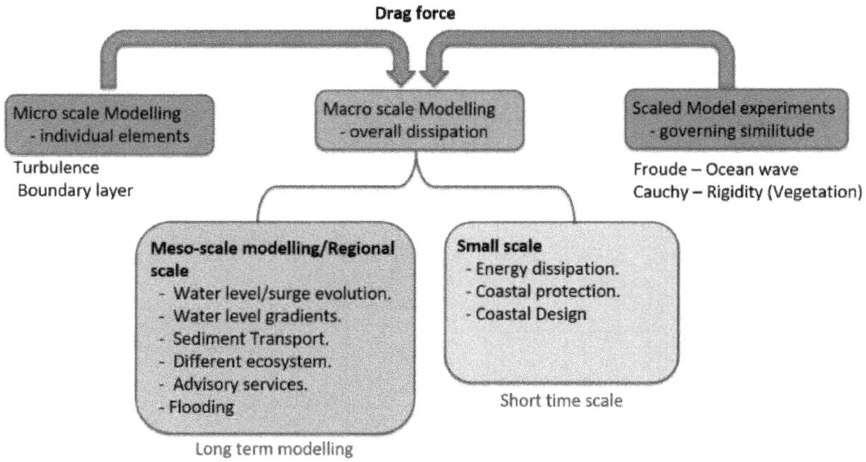

Fig. 10.2. Methodology being applied in modelling the eco-based coastal structures.

interaction using a meshless scheme, considering the vegetation patch to be a porous structure by incorporating Darcy's law. The study was extended by considering the vegetation model to be a cylinder and square strip for emergent and submerged condition. It was observed that the influence in emergent conditions due to the different shapes was insignificant when compared with the submerged condition in which the cylindrical model was more effective in wave attenuation. The wave attenuation characteristics of vegetation as a single patch and two patches were also studied, and the reduction patterns were reported.

10.2 Living Structures and Shorelines

"Living breakwaters" are engineered structures that are partially submerged in the water that act to reduce the risk imposed by high-energy waves during storm events and also provide habitat for local marine life. These are also known as "submerged artificial reefs" and are classified under hybrid solutions as they are combined with structural units. The structural units are generally made of concrete, but can be made more eco-friendly by considering naturally available materials without degrading the nearby biodiversity. Several studies have reported the effectiveness of structural units in wave attenuation (Dhinakaran *et al.*, 2012; Lokesha *et al.*, 2019). They are also implemented in sites of high erosion to prevent further loss followed by accretion and stabilization of the land. The region in the Gulf of Mannar suffered severe erosion due to excessive coral mining and fishing

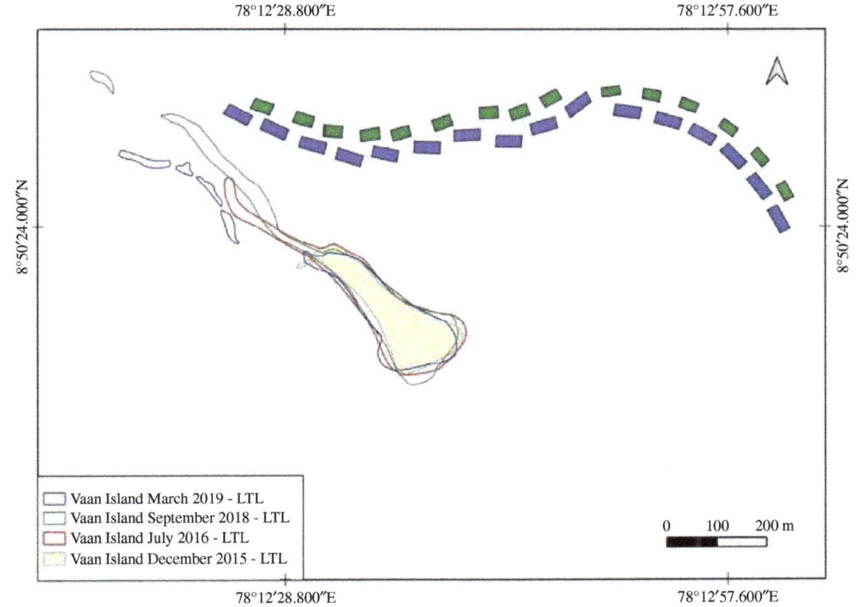

Fig. 10.3. Morphological variation of Vaan Island after deployment of artificial reef units for restoration.
Source: Jayanthi *et al.* (2020).

that resulted in the sinking of the island. The implementation of a trapezoidal artificial reef seized soil erosion and resulted in the accretion of the lost island areas (see Fig. 10.3) as detailed by Jayanthi *et al.* (2020). Further, the artificial reef promoted the growth of coral reefs over it and formed a suitable habitat for indigenous marine life. Vegetated floating breakwaters, known as BioHaven® floating breakwaters (BFBs), survived 90-MPH category-1 hurricanes in South Louisiana marshes and protected the coast. Wang *et al.* (2019) investigated this type of platform numerically using the OlaFlow solver and showed effective attenuation by vegetation canopy above the water level and by roots below the water level. Another soft measure, called a living seawall, has been introduced in the Sydney Harbour where the walls are integrated with reef model tiles that can improve water quality by attracting marine life that feeds on pollutants.

Several types of precast concrete units have been developed and deployed to promote ecological enhancement of marine structures while also serving as wave energy dissipaters (Coastrack, 2010). The reef ball proposed by Harris (2003) is a popular structure that has been installed

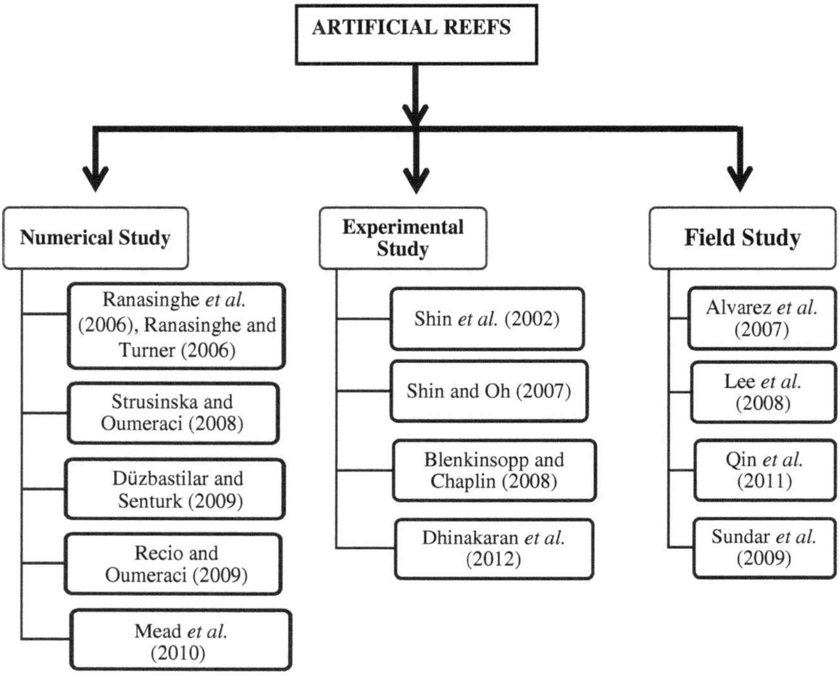

Fig. 10.4. Summary of literature review on artificial reefs.
Source: Lokesha *et al.* (2018).

at a few locations; however, off the Dominican coast, the occurrence of a hurricane swept the reef balls away (Coastrack, 2010).

A few other concrete units for armouring of multiple habitat breakwaters have been discussed by Firth *et al.* (2016); it has been claimed that biotic-origin materials can be adopted in place of conventional concrete or rocks.

Lokesha *et al.* (2013) conducted a comprehensive literature review of studies on artificial reefs, a summary of which is shown in Fig. 10.4.

10.3 Seagrass and Meadows

"Seagrass meadow" ecosystems are capable of attenuating incident waves and are thereby effective in reducing the risk of erosion. The root system of species like the Seagrass meadow has naturally evolved to thrive in wavy environment. It enhances its ability to anchor firmly in the soil. This adaptation plays a crucial role in preventing soil erosion in its vicinity. By securely binding with the soil, the root system maintains stability, effectively mitigating the impact of environmental forces such as waves or currents while safeguarding the surrounding area from erosion. The effect of the seagrass in Derawan Island, Indonesia, during storm events has been studied by Christianen *et al.* (2013). Bouma *et al.* (2007) and Van Zelst *et al.* (2021) have discussed the trapping of sediments in the marsh due to the dissipation of flows that promote the vertical growth of the bed, a phenomenon that is important to understand flow–vegetation interaction. John *et al.* (2015) through experimental investigation have reported that the relative plant height and relative meadow width influence the energy dissipation rate of seagrass meadows. A run-up reduction of 41% was observed on a 1:12 slope fronted by a 2-m seagrass meadow. Stratigaki *et al.* (2011) conducted large-scale experimental studies exhibiting higher attenuation for shallower waters with denser vegetation fields. The dependence of stiffness of the seagrass in attenuating waves was reported by Bouma *et al.* (2005), Fonseca and Koehl (2006) and Paul *et al.* (2012). It should also be noted that seagrass meadows are not suitable for protecting all shorelines. They are more effective in shallow waters and low-energy environments (sheltered shorelines with low-period waves may be considered optimal conditions in which seagrass can provide enhanced coastal protection) (Bradley and Houser, 2009). Olanrewaju and Lazzaro's (2023) experimental studies on seaweeds reported 15% wave damping and also stated that this can increase with an increase in the width of the patch.

10.4 Coral Reefs, Mangroves and Vegetation

"Coral reefs" are yet another important feature of coastal ecosystems that are the habitat of several marine creatures. They are called the "rainforests of the seas." These features are mainly formed by calcium carbonate. The morphodynamics of a coast fronted by coral reef is generally stable when compared to that of a beach without a reef (Sanderson, 2000; Sheppard *et al.*, 2005; Gallop *et al.*, 2012). De Alegria-Arzaburu *et al.* (2013) reported that vertical morphological changes of 0.5 m and 2 m were measured in Puerto Morelos and Cancun, respectively, for the same offshore wave conditions. Field investigations by Jeanson *et al.* (2013) on coral reefs in a lagoon off the Mayotte Islands in the Indian Ocean reported that attenuation characteristics are inversely proportional to the relative wave height, and the influence of reef width was not significant. Baba *et al.* (2008) investigated the Great Barrier Reef's role in protecting Australia's coasts during the 2007 Solomon Islands Tsunami through the numerical method. Based on their model study, it was reported that the coral reefs provided a buffer time of 5–10 min for the time of impact and also reduced the amplitude of the first pulse of the tsunami wave by half.

"Mangroves" are generally found in coastal regions of brackish water, intertidal regions and swamps. Coastal vegetation cover made a significant contribution during the 2004 Great Indian Ocean Tsunami (Danielsen *et al.*, 2005; Kathiresan and Rajendran, 2005; Sundar *et al.*, 2007; Mascarenhas and Jayakumar, 2008). Chang *et al.* (2006), through remote sensing analysis of the 2004 tsunami event along the North Andaman coast off Thailand, stated that communities established behind mangroves sustained minimal damage. The assessments were reported based on the tsunami intensity scale established by Papadopoulos and Imamura (2001). Enough evidence was reported by Noarayanan *et al.* (2012a) on the role of vegetation in reducing the damage and protecting the hinterland from the 2004 tsunami.

The death toll during a supercyclone that struck Odisha in 1999 was found to be significantly less in the areas behind a wider mangrove patch compared to area with narrow mangrove patches or no mangroves (Das and Vincent, 2009). Krauss *et al.* (2009) described the protective role of mangroves during the 2004 and 2005 hurricane events along the coast of Florida, which resulted in a reduction of about 4.2 cm/km and 9.4 cm/km in the surge in the riverine mangroves and interior mangrove marshes, respectively.

Experimental investigations on wave–vegetation interaction, considering the different characteristics of vegetation, have given detailed insights into the attenuation behaviours of vegetation patches. Noarayanan *et al.* (2012a) reported that the reduction in run-up was influenced more by the width of the vegetation belt than its density through experimental studies considering regular and cnoidal waves. Hari Ram *et al.* (2022) investigated the interaction of solitary and elongated solitary waves with vegetation patches and observed that the percentage of energy reduction remains the same for both models for a given vegetation width. Wang *et al.* (2022) described the significant role of the complex submerged root morphology in mangrove forests in wave dissipation characteristics. Ram *et al.* (2022) described the energy attenuation process occurring through vegetation belts by considering monochromatic waves with higher harmonics. The incoming wave energy was observed to dissipate primarily from higher harmonics. Table 10.1 provides details on ecosystems and their functionality in coastal protection, based on which, the implementation of the ecosystem can be planned considering the type of protection required in the coastal region.

It is essential for planners and designers to have knowledge of the reduction of forces on seaside structures due to the presence of vegetation or a green belt; this knowledge is necessary both in planning for the green belt

Table 10.1. Functional aspects of different ecosystems towards coastal protection.

Ecosystem type	Characteristics of ecosystem type	Support provided
Seagrass meadow	Shoot – above sea bead	Wave attenuation through turbulence, CO_2 sink,
	Root system – below sea bed	Soil stabilization.
Mangroves	Trunk and Canopy complex root system – above ground	Strom surges and tsunami energy dissipation.
	Root system – below sea bed	Wave attenuation and sediment deposition.
		Soil stabilization.
Beach nourishment	Increased beach width	Wave dissipation gradually by bed friction.
Coral reefs	Complex perforated rough surface	Wave energy dissipation by surface friction.
		CO_2 sink.
Salt marshes	Promotes bed evolution and vegetation	Attenuates storm surges.

and in designing structures that will face excessive wave loading during extreme coastal hazards. Noarayanan *et al.* (2012b) analyzed an area with a constant green belt width and measured the forces on a structure model of width B for relative spacing $G/B = 0$, 0.5, 1.0 and 1.5, where G is the distance between the green belt and the structure. The tests were carried out with the model over a beach slope of 1:30 in the presence and absence of the green belt. The variation of dimensionless force F^* (through a percentile analysis) with G/B, where F^* is defined by Equation (10.1), is reported as projected in Fig. 10.5:

$$F^* = \left(\frac{F_{\max}}{0.5 * \rho g H^2 B_s}\right), \tag{10.1}$$

wherein $F_{\max}$ = maximum force on the structure, ρ = mass density of water, g = gravitational acceleration and H = wave height.

According to Fig. 10.6(a), the % of occurrence of F^* being less than equal to unity for $G/B = 0.0$, 0.5, 1.0 and 1.5 is found to be about 92%, 90%, 55% and 96%, respectively. The results show that in the absence of vegetation, the F^* is highest for $G/B = 1.0$. Similar results in Fig. 10.6(b) for the structure fronted by a green belt with G/B ratios of 0.0, 0.5, 1.0, and 1.5 show protection levels of 100%, 30%, 75%, and 90% respectively. These results prove that $G/B = 0$ and $G/B > 1$–5 will certainly be beneficial.

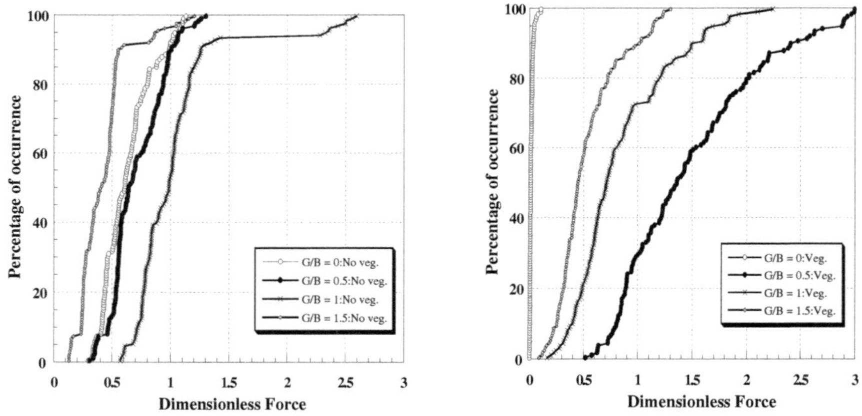

(a) (b)

Fig. 10.5. Probability distribution of wave forces in the absence and presence of vegetation.
Source: Noarayanan *et al.* (2012b).

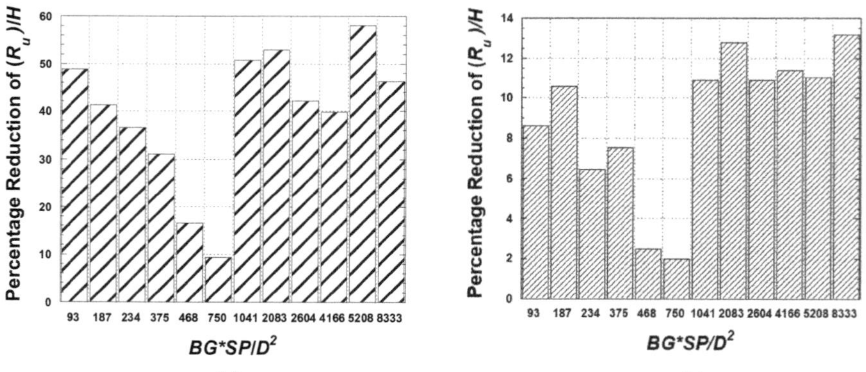

(a) (b)

Fig. 10.6. Variation of percentage reduction of $(R_u)/H$ with BG^*SP/D^2: (a) cnoidal waves and (b) regular waves.
Source: Noarayanan *et al.* (2012c).

In summary, the forces seemed to be the least only for $G/B = 0$ and $G/B = 1.5$.

In addition, Noarayanan *et al.* (2012c) reported (Fig. 10.6) that the percentage reduction in wave run-up varies across different configurations of BG*SP/D2 (where G represents the distance between green belts, B

denotes the width of the structure, SP indicates the spacing between stems, and D represents stem diameter). Despite the maximum run-up observed for Cnoidal waves (long waves), the study found a higher percentage reduction in run-up for these configurations. This suggests that the effectiveness of wave reduction measures, as indicated by the percentage reduction in wave run-up, remains prominent across varying structural parameters. Hence, this reveals that vegetation along coasts frequented by long waves is one of the best soft solutions.

10.5 Adaptation of Nature-Based Solution to Climate Change

The coastal wetland habitat employs two primary mechanisms to mitigate the threat posed by rising sea level: vertical accretion and horizontal transgression onto higher grounds (Kirwan *et al.*, 2016). Vertical accretion involves the accumulation of sediment and organic matter, which raises the marsh surface elevation, keeping pace with or exceeding the rate of sea level rise. This process helps to maintain the wetland's critical elevation relative to mean sea level, and thereby preserving its ecological functions and habitat suitability.

In addition to vertical accretion, wetlands also exhibit horizontal transgression, where they expand inland or onto higher areas adjacent to existing wetlands. This expansion allows wetlands to migrate and occupy new areas as sea levels rise. It further facilitates to create new habitats and potentially buffering coastal communities against flooding and erosion. Together, these adaptive strategies enable coastal wetlands to dynamically respond to changing environmental conditions, ensuring their resilience and persistence in the face of sea level rise. The former is dependent on tidal inundation, sediment supply and biomass in that locality, whereas the latter depends on the topography and anthropogenic modification of the shoreline (Brinson *et al.*, 1995; Titus *et al.*, 2009; Kirwan *et al.*, 2016). Kirwan *et al.* (2016) reported that the rate of landward transgression of marshes was proportional to the historic rates of sea level rise. Nature-based solutions adapt to climate change, whereas adaptation is limited in the case of conventional hard measures. Similarly, oyster reefs tend to grow vertically by attracting oyster larvae that drift through the water and attach with the existing reefs, contributing to their growth (Rodriguez *et al.*, 2014). The success of mangrove restoration at West Lake, situated near Fort Lauderdale, Florida, without the need for any plantation of mangroves has been reported by Lewis (1990).

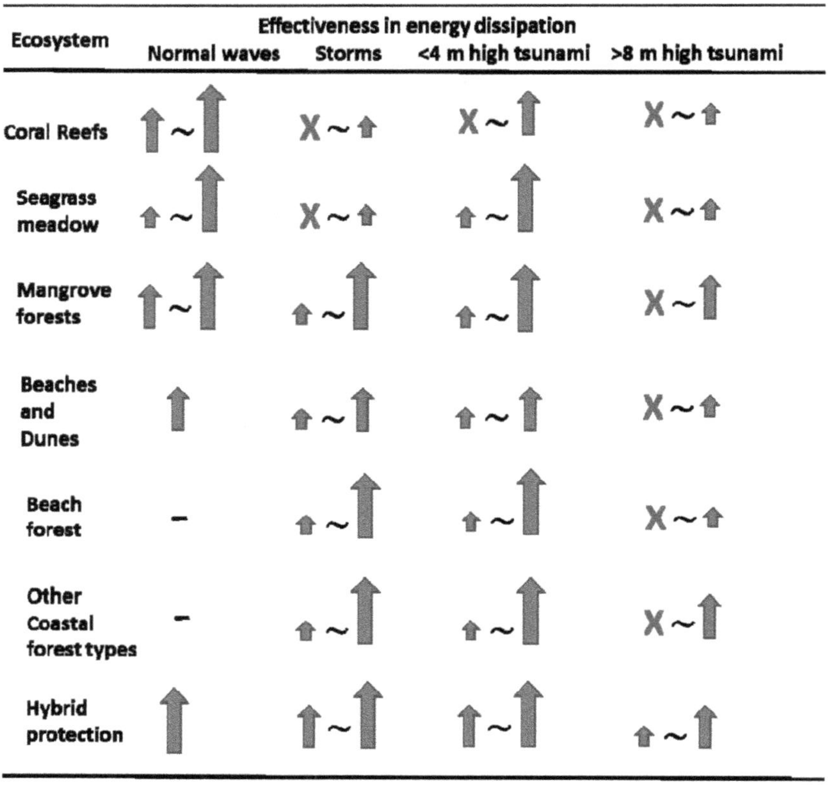

Fig. 10.7. Wave attenuation characteristics for different ecological measures against tsunamis.
Source: Hari Ram and Sundar (2024).

A quantitative level of protection that different features of the ecosystem provide for different wave conditions is shown in Fig. 10.7. It can be seen that most of the features do not provide the desired protection level for extremely high tsunamis, which would be similar to the protection provided by traditional hard measures too.

Ecosystem-based solutions also reduce the rate of expansion of climate change, especially serving as sinks for global warming gases. Nature-based solutions play a key role in blue carbon sequestration. Several studies have proven that coastal vegetation aids in reducing carbon dioxide (Bianchi *et al.*, 2013; Ward, 2020). The sediments of mangroves, salt marshes and seagrass meadows are found to adapt to the rising sea level as they accrete vertically, thus never reaching a saturation for carbon sequestration (McKee *et al.*, 2007).

10.6 Hybrid Protection with Vegetation

Protection against coastal erosion has been widely ensured through the construction of hard and soft measures, the salient details of which are discussed in Chapter 2. Both hard and soft measures have inherent advantages and disadvantages as listed in Table 10.2.

Hence, a combination of both the hard and soft measures, termed a hybrid solution, can accommodate the merits of both measures for coastal protection. Schoonees *et al.* (2019) proposed the following nature-based solutions to minimize negative impacts as discussed in Table 10.2: (i) applying soft solutions complementary to hard solutions, (ii) mitigating morphological and hydrodynamic changes, and (iii) ecologically enhancing hard coastal structures. The solution for coastal protection is classified into grey and green infrastructures. Under this broad classification, grey infrastructure will refer to conventional hard structures. Green infrastructure can subcategorized as (i) environmentally friendly grey infrastructure, (ii) hybrid infrastructure and (iii) soft infrastructure. The broad classification is illustrated in Fig. 10.8. Among the soft measures, vegetation along the anticipated eroding stretches of the coast has gained tremendous momentum. The role of coastal vegetation in protecting the coast had gained special attention soon after the Great Indian Ocean Tsunami of

Table 10.2. A comparison between soft and hard measures for coastal protection.

Factors	Hard measures	Soft measures
Aesthetics	Not appealing	Appealing
Duration for restoration	Immediately effective.	Not suitable for immediate enforcement.
Cost-effectiveness	Huge capital is involved, and maintenance is also required in some cases.	Capital cost is high for installing slurry pumps; geo-synthetic materials are also expensive.
Environmental impact	Leaves behind a huge footprint and suitable for almost all climatic and environmental conditions.	Leaves behind minimum footprint and their implementation is limited to certain climatic and environmental conditions.
Implementation	Speedy and easier.	Time-consuming and skilled.
Design criteria	Shore protection manuals and coastal engineering manuals available.	No standard published materials for the design of structures using geo-synthetics.
Design life	30 years to 100 years.	Not defined.

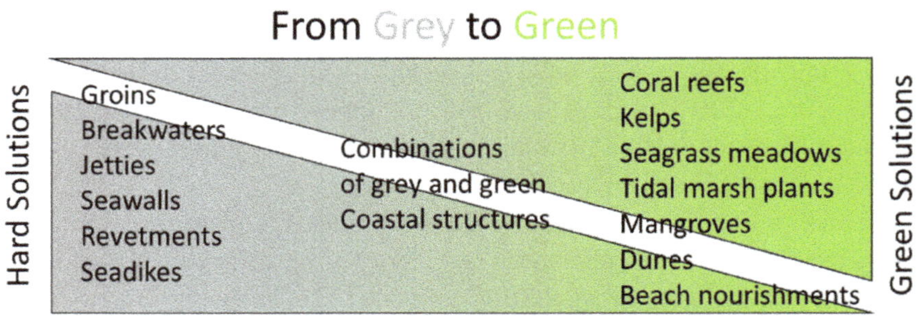

Fig. 10.8. From Grey to Green.

Table 10.3. Merits and demerits of hybrid coastal structures.

Merits	Demerits
Adopts natural materials for shore protection.	More studies on its performance characteristics are essential.
The changes in the shoreline dynamics are expected to be a minimum.	Coastal stretches exposed to high waves may not be suitable for implementation.
Helps in regaining lost beaches by trapping the sand in motion.	It may not be suitable for a shoreline that is already protected using hard measures like the seawalls formed by armour stones.
Promotes habitat along the shoreline.	Guidelines for design, implementation and maintenance are not well established compared to conventional hard structural measures.
Attenuation of incident wave energy serves as a possible mitigation measure against natural coastal hazards like storm surges.	The performance characteristics of living shorelines for normal and extreme conditions are not well documented.
Reclaimed beaches serve as a good location for walking, jogging and relaxation for the public.	
The capital cost may be less if carefully planned and implemented.	
Can act as a carbon sink and mitigate climate change effects.	
The space requirement is less.	

2004, the post-mortem of which proved that mangroves, vegetation and reefs behaved as natural wave attenuators. This conclusion follows field investigations conducted after the tsunami by Danielsen *et al.* (2005) and Sundar *et al.* (2007).

The merits and demerits of hybrid infrastructure are reported in Table 10.3.

Through case studies along the U.S. coast, Sutton-Grier *et al.* (2015) discussed the strengths and limitations of hard, soft and hybrid measures for coastal protection. Likewise, Nakamura (2022) studied the concepts of integrating green infrastructure (soft measures) and grey infrastructure (hard measures) by considering climate change through several events and adaptation strategies in Japan and other countries. Cunniff and Schwartz (2015) have discussed the strengths of different nature-based solutions and the favourable conditions in which they provide the best returns in terms of protection, based on the knowledge gained from different strategies that were adopted across the globe. The adaptive mangrove hybrid platform proposed by Takagi (2019) is yet another solution for establishing green infrastructure along urban coasts with deep water columns and high waves. Moraes *et al.* (2022) have discussed the aspects of different nature-based solutions adopted across Europe. Most of these solutions are hybrid solutions created to address coastal challenges like flood reduction, erosion control and biodiversity restoration/conservation.

Palinkas *et al.* (2022) discussed three case studies of hybrid coastal protection measures that included the associated challenges in their implementation. The following case studies were considered: (i) Living Breakwaters in New York Harbour attempted to address the aspects of wave attenuation, beach restoration and shoreline erosion during short-terms events like storms as well as in the long term considering aspects like climate change; (ii) The Coastal Texas Protection and Restoration Study focused on problems related to habitat loss, erosion, sea level rise and storm surges. (iii) The South Bay Salt Pond Restoration Project in Francisco Bay focused on habitat restoration, protection against tidal flows and providing public access to connect people with the Bay without disturbing the wildlife.

Chim *et al.* (2022) conducted comprehensive physical modelling and examined the performance characteristics of a hybrid coastal protection system consisting of a natural system (i.e. mangrove) and a hard structure armoured with XblocPlus and a berm. The hard structure was reported to effectively attenuate incident waves, thus enhancing the opportunity for mangrove colonization. The presence of mangroves (of prototype width 18 m) was reported to reduce wave overtopping by 18–45% compared to the standalone hard structure. It was found that the mangroves provide more effective wave dissipation with higher root emergence at a lower water level, whereas the revetment berm alone was not effective at a high water level and submerged mangrove roots induced more interaction for wave dissipation. It was claimed that hybrid structure could be upgraded as a system with multiple lines of defence to ensure resiliency and adaptability.

Singhvi *et al.* (2022) reviewed 105 coastal protection case studies and examined external and internal factors of resilience, which showed that external factors are typically used to characterize the grey–green spectrum. The review showed that dikes, reefs, saltmarshes, sand nourishment and dunes span a wider segment of the grey–green spectrum than generally categorized. The study suggests that coastal planners will benefit from a more diverse range of options when considering the incorporation of grey and green interventions in the context of each spatial scale. It was stressed that internal resilience should be accounted for when infrastructure options are comparatively evaluated.

Morris *et al.* (2023) have proved that using rock fillets as a nature-based method for coastal protection is successful in trapping sediment, thereby serving as a habitat for mangrove colonization. It was claimed that structures formed by rock fillets are more environmentally and economically sustainable nature-based solutions for coastal protection compared to natural rocks.

10.7 Summary

A large fraction of the world's population (41%) and many megacities (60%) are situated near the coastal shoreline (Martínez *et al.*, 2007). Considering the present scenario of global climate change, natural/eco-friendly coastal protection measures are discussed in this chapter. Ecological solutions are thus encouraged in several countries due to the multi-functional support and adaptive response they offer. They support fisheries, biodiversity, water quality improvement, and recreational and cultural benefits (Barbier *et al.*, 2011).

Apart from this, the implementation of nature-based solutions on a local scale is affordable and sustainable. These solutions can be region specific and can be adapted to incorporate native ecosystem features. Such features can be combined with engineering ideas, creating a prefect countermeasure. Similar to the quality assessments carried out for hard structures, risk assessment and management need to be conducted for the ecosystem as it is considered a component of coastal protection. Cormier *et al.* (2013) have given details on ecosystem-based risk assessment and management. Ecosystem-based adaptation measures are the most effective ways of adapting to climate change by integrating the use of biodiversity and ecosystems so as to derive maximum co-benefits. When carefully planned, designed and implemented, with effective management, ecosystem-based adaptation can be a sustainable solution. Even though the literature on such eco-based solutions for coastal problems is scarce, such solutions are certain to gain favour in the field of coastal protection in the future. While numerous researchers have endeavoured to explore nature-based solutions for living structures, there is still a considerable distance to traverse.

References

Alvarez, I.E., Rubio, R. and Ricadle, H. (2007). Beach restoration with geotextile tubes as submerged breakwaters in Yucatan, Mexico. *Journal of Geotextiles and Geomembranes*, 25, 233–241.

Baba, T., Mleczko, R., Burbidge, D., Cummins, P.R. and Thio, H.K. (2008). The effect of the Great Barrier Reef on the propagation of the 2007 Solomon Islands tsunami recorded in Northeastern Australia. *Pure and Applied Geophysics*, 165(11–12), 2003–2018.

Barbier, E.B., Hacker, S.D., Kennedy, C., Koch, E.W., Stier, A.C. and Silliman, B.R. (2011). The value of estuarine and coastal ecosystem services. *Ecological Monographs*, 81(2), 169–193.

Bianchi, T.S., Allison, M.A., Zhao, J., Li, X., Comeaux, R.S., Feagin, R.A. and Kulawardhana, R.W. (2013). Historical reconstruction of mangrove expansion in the Gulf of Mexico: Linking climate change with carbon sequestration in coastal wetlands. *Estuarine, Coastal and Shelf Science*, 119, 7–16.

Blenkinsopp, C.E. and Chaplin, J.R. (2008). The effect of relative crest submergence on wave breaking over submerged slopes. *Coastal Engineering*, 55(12), 967–974.

Bouma, T.J., De Vries, M.B., Low, E., Peralta, G., Tánczos, I.V., van de Koppel, J. and Herman, P.M.J. (2005). Trade-offs related to ecosystem engineering: A case study on stiffness of emerging macrophytes. *Ecology*, 86(8), 2187–2199.

Bouma, T.J., Van Duren, L.A., Temmerman, S., Claverie, T., Blanco-Garcia, A., Ysebaert, T. and Herman, P.M.J. (2007). Spatial flow and sedimentation patterns within patches of epibenthic structures: Combining field, flume and modelling experiments. *Continental Shelf Research*, 27(8), 1020–1045.

Bradley, K. and Houser, C. (2009). Relative velocity of seagrass blades: Implications for wave attenuation in low-energy environments. *Journal of Geophysical Research: Earth Surface*, 114, 1–13. https://doi.org/10.1029/2007JF000951.

Brinson, M.M., Christian, R.R. and Blum, L.K. (1995). Multiple states in the sea-level induced transition from terrestrial forest to estuary. *Estuaries*, 18, 648–659.

Chang, S.E., Adams, B.J., Alder, J., Berke, P.R., Chuenpagdee, R., Ghosh, S. and Wabnitz, C. (2006). Coastal ecosystems and tsunami protection after the December 2004 Indian Ocean tsunami. *Earthquake Spectra*, 22(3_suppl), 863–887.

Chim, Y.Q., Muilwijk, M., Ruwiel, T., Reedijk, J.S. and Yang, Z.Q. (2022). Physical Model Investigation of Hybrid Coastal Protection, Combining Nature-Based and Hard Infrastructure, Singapore International Water Week, April.

Christianen, M.J., van Belzen, J., Herman, P.M., van Katwijk, M.M., Lamers, L.P., van Leent, P.J. and Bouma, T.J. (2013). Low-canopy seagrass beds still provide important coastal protection services. *PloS One*, 8(5), e62413.

Coastrack. (2010). Artificial reefs concept & Indian experience. *ENVIS - MoEF Newsletter on Coastal Zone Management*, 9(2), July - September 2010.

Cormier, R., Kannen, A., Elliott, M., Hall, P. and Davies, I.M. (2013). Marine and Coastal Ecosystem-Based Risk Management Handbook.

Cunniff, S. and Schwartz, A. (2015). Performance of natural infrastructure and nature-based measures as coastal risk reduction features. Environmental Defense Fund.

Danielsen, F., Sørensen, M.K., Olwig, M.F., Selvam, V., Parish, F., Burgess, N.D., ... and Suryadiputra, N. (2005). The Asian tsunami: A protective role for coastal vegetation. *Science*, 310(5748), 643–643.

Das, S. and Vincent, J.R. (2009). Mangroves protected villages and reduced death toll during Indian super cyclone. *Proceedings of the National Academy of Sciences*, 106(18), 7357–7360.

De Alegria-Arzaburu, A.R., Mariño-Tapia, I., Enriquez, C., Silva, R. and González-Leija, M. (2013). The role of fringing coral reefs on beach morphodynamics. *Geomorphology*, 198, 69–83.

Dhinakaran, G., Sundar, V., Sundaravadivelu, R. and Graw, K.U. (2012). Performance of a perforated submerged semicircular breakwater due to non-breaking waves. *Proceedings of the Institution of Mechanical Engineers, Part M: Journal of Engineering for the Maritime Environment*, 226(1), 36–50.

Divya, R., Sriram, V. and Murali, K. (2020). Wave-vegetation interaction using improved meshless local Petrov Galerkin method. *Applied Ocean Research*, 101, 102116.

Düzbastılar, F.O. and Sentürk, U. (2009). Determining the weights of two types of artificial reefs required to resist wave action in different water depths and bottom slopes. *Ocean Engineering*, 36(12–13), 900–913.

Erduran, K.S. and Kutija, V. (2003). Quasi-three-dimensional numerical model for flow through flexible, rigid, submerged and non-submerged vegetation. *Journal of Hydroinformatics*, 5(3), 189–202.

Firth, L.B., Knights, A.M., Bridger, D., Evans, A.J., Mieszkowska, N., Moore, P.J., O'Connor, N.E., Sheenan, E.V., Thompson, R.C. and Hawkins, S.J. (2016). Ocean sprawl: Challenges and opportunities for biodiversity management in a changing world. *Oceanography and Marine Biology: An Annual Review*, 2016, 193–269.

Fonseca, M.S. and Koehl, M.A.R. (2006). Flow in seagrass canopies: The influence of patch width. *Estuarine, Coastal and Shelf Science*, 67(1–2), 1–9.

Gallop, S.L., Bosserelle, C., Eliot, I. and Pattiaratchi, C.B. (2012). The influence of limestone reefs on storm erosion and recovery of a perched beach. *Continental Shelf Research*, 47, 16–27.

Hari Ram, N. and Sundar, V. (2024). A comprehensive review on wave–vegetation interaction. In *Hydrodynamics of Wave-Vegetation Interactions*, Wworld Scientific, Singapore, 203–238. https://doi.org/10.1142/13619.

Hari Ram, N., Sriram, V. and Murali, K. (2022). Experimental investigation on the characteristics of solitary and elongated solitary waves passing over vegetation belt. *Journal of Ocean Engineering and Marine Energy*, 8(3), 305–318.

Harris, L.E. (2003). Artificial reef structures for shoreline stabilization and habitat enhancement. In *Proceedings of the 3rd International Surfing Reef Symposium*, Raglan, New Zealand, June 22–25, 2003, pp. 176–178.

Jayanthi, M., Patterson Edward, J.K., Malleshappa, H., Gladwin Gnana Asir, N., Mathews, G., Diraviya Raj, K. and Sannasiraj, S.A. (2020). Perforated trapezoidal artificial reefs can augment the benefits of restoration of an island and its marine ecosystem. *Restoration Ecology*, 28(1), 233–243.

Jeanson, M., Anthony, E.J., Dolique, F. and Aubry, A. (2013). Wave characteristics and morphological variations of pocket beaches in a coral reef–lagoon setting, Mayotte Island, Indian Ocean. *Geomorphology*, 182, 190–209.

John, B.M., Shirlal, K.G. and Rao, S. (2015). Effect of artificial vegetation on wave attenuation–an experimental investigation. *Procedia Engineering*, 116, 600–606.

Kathiresan, K. and Rajendran, N. (2005). Coastal mangrove forests mitigated tsunami. *Estuarine, Coastal and Shelf Science*, 65(3), 601–606.

Kirwan, M.L., Temmerman, S., Skeehan, E.E., Guntenspergen, G.R. and Fagherazzi, S. (2016). Overestimation of marsh vulnerability to sea level rise. *Nature Climate Change*, 6(3), 253–260.

Krauss, K.W., Doyle, T.W., Doyle, T.J., Swarzenski, C.M., From, A.S., Day, R.H. and Conner, W.H. (2009). Water level observations in mangrove swamps during two hurricanes in Florida. *Wetlands*, 29, 142–149.

Lee, H.S., Sidhartham, M., Shim, C.S., Kim, Y.D., Lim, C.Y., Ko, J.W., Han, M.D., Rang. M.J., Bim, L.S., Cho, H.S. and Shin, H.W. (2008). Screening and formulation of chemoattractant coatings for artificial reef structures. *Journal of Environmental Biology*, 29(4), 605–612.

Lewis III, R.R. (1990). Creation and restoration of coastal plain wetlands in Florida. In Kusler, J.A. and Kentula, M.E. (eds.), *Wetland Creation and Restoration: The Status of the Science*, Island Press, Washington D.C, USA.

Lokesha, Sundar, V. and Sannasiraj, S.A. (2013). Artificial reefs: A review. *International Journal of Ocean and Climate Systems, IJOS*, 4(2), 117–124. https://doi.org/10.1260/1759-3131.4.2.117.

Lokesha, Sannasiraj, S.A. and Sundar, V. (2018). Hydrodynamic characteristics of a submerged trapezoidal artificial reef unit. *Journal of Engineering for Maritime Environment*. DOI: 10.1177/1475090218825178.

Lokesha, Sannasiraj, S.A. and Sundar, V. (2019). Hydrodynamic characteristics of a submerged trapezoidal artificial reef unit. *Proceedings of the Institution of Mechanical Engineers, Part M: Journal of Engineering for the Maritime Environment*, 233(4), 1226–1239.

Martínez, M.L., Intralawan, A., Vázquez, G., Pérez-Maqueo, O., Sutton, P. and Landgrave, R. (2007). The coasts of our world: Ecological, economic and social importance. *Ecological Economics*, 63(2–3), 254–272.

Mascarenhas, A. and Jayakumar, S. (2008). An environmental perspective of the post-tsunami scenario along the coast of Tamil Nadu, India: Role of sand dunes and forests. *Journal of Environmental Management*, 89(1), 24–34.

Maza, M., Lara, J.L. and Losada, I.J. (2015). Tsunami wave interaction with mangrove forests: A 3-D numerical approach. *Coastal Engineering*, 98, 33–54.

McKee, K.L., Cahoon, D.R. and Feller, I.C. (2007). Caribbean mangroves adjust to rising sea level through biotic controls on change in soil elevation. *Global Ecology and Biogeography*, 16(5), 545–556.

Mead, S.T., Blenkinsopp, C., Moores, N. and Borrero, J. (2010). Design and construction of the Boscombe multi-purpose reef. In *Proceedings of 32nd Conference on Coastal Engineering*, Shanghai, China, pp. 1–9.

Moraes, R.P., Reguero, B.G., Mazarrasa, I., Ricker, M. and Juanes, J.A. (2022). Nature-based solutions in coastal and estuarine areas of Europe. *Frontiers in Environmental Science*, 10, 829526.

Morris, R.L., Fest, B., Stokes, D., Jenkins, C. and Swearer, S.E. (2023). The coastal protection and blue carbon benefits of hybrid mangrove living shorelines, *Journal of Environmental Management*, 331, 117310.

Nakamura, F. (2022). *Green Infrastructure and Climate Change Adaptation: Function, Implementation and Governance.* Springer Nature, Fukuoka, Japan, p. 506.

Noarayanan, L., Murali, K. and Sundar, V. (2012a). Role of vegetation on beach run-up due to regular and Cnoidal waves. *Journal of Coastal Research*, 28(1A), 123–130.

Noarayanan, L., Murali, K. and Sundar, V. (2012b). Effect of flexible vegetation on the forces with Keulegan carpenter number on structures due to long waves. *Journal of Marine Science and Application*, 11, 24–33. https://doi.org/10.1007/s11804-012-1102-9.

Noarayanan, L., Murali, K. and Sundar, V. (2012c). Role of vegetation on beach run-up due to regular and Cnoidal waves. *Journal of Coastal Research*, 28, 123–130. https://doi.org/10.2112/jcoastres-d-10-00078.1.

Olanrewaju, O.S. and Lazzaro, U. (2023). Review of Nature-Based Echo-Hydraulic Aqua-Forest Technology for Coastal Resilience and Sea Level Rise Climate-Induced Adaptation.

Palinkas, C.M., Orton, P., Hummel, M.A., Nardin, W., Sutton-Grier, A.E., Harris, L., Gray, M., Li, M., Ball, D., Burks-Copes, K., Davlasheridze, M., De Schipper, M., George, D.A., Halsing, D., Maglio, C., Marrone, J., McKay, S.K., Nutters, H., Orff, K., Taal, M., Van Oudenhoven, A.P.E., Veatch, W. and Williams, T. (2022). Innovations in coastline management with natural and nature-based features (NNBF): lessons learned from three case studies. *Frontiers in Built Environment.* DOI: 10.3389/fbuil.2022.814180.

Papadopoulos, G.A. and Imamura, F. (2001). A proposal for a new tsunami intensity scale. In ITS 2001 Proceedings (Vol. 5, pp. 569–577). This paper introduces a 12-point tsunami intensity scale that describes the effects on humans, objects, and buildings? (IGN) (SpringerLink).

Paul, M., Bouma, T.J. and Amos, C.L. (2012). Wave attenuation by submerged vegetation: Combining the effect of organism traits and tidal current. *Marine Ecology Progress Series*, 444, 31–41.

Qin, C.-X., Chen, P.-M. and Jia, X.-P. (2011). Effects of artificial reef construction to marine ecosystem services value: A case of Yang-Meikeng artificial reef region in Shenzhen. *Chinese Journal of Applied Ecology*, 22(8), 2160–2166.

Ram, N.H., Sriram, V. and Murali, K. (2022). Investigation of spectral energy distribution in wave groups due to presence of vegetation. *Coastal Engineering Proceedings*, (37), 53.

Ranasinghe, R., Turner, I. and Symonds, G. (2006). Shoreline response to multi-functional artificial surfing reefs: A numerical and physical modeling study. *Coastal Engineering*, 53, 589–611.

Ranasinghe, R. and Turner, I. (2006). Shoreline response to submerged structures: A review. *Coastal Engineering*, 53, 65–79.

Recio, J. and Oumeraci, H. (2009). Process based stability formulae for coastal structures made of geotextile sand containers. *Coastal Engineering*, 56(5–6), 632–658.

Rodriguez, A.B., Fodrie, F.J., Ridge, J.T., Lindquist, N.L., Theuerkauf, E.J., Coleman, S.E., ... and Kenworthy, M.D. (2014). Oyster reefs can outpace sea-level rise. *Nature Climate Change*, 4(6), 493–497.

Sanderson, P.G. (2000). A comparison of reef-protected environments in Western Australia: The central west and Ningaloo coasts. *Earth Surface Processes and Landforms: The Journal of the British Geomorphological Research Group*, 25(4), 397–419.

Schoonees, T., Gijón Mancheño, A., Scheres, B., Bouma, T.J., Silva, R., Schlurmann, T. and Schüttrumpf, H. (2019). Hard structures for coastal protection, towards greener designs. *Estuaries Coasts*, 42, 1709–1729.

Sheppard, C., Dixon, D.J., Gourlay, M., Sheppard, A. and Payet, R. (2005). Coral mortality increases wave energy reaching shores protected by reef flats: Examples from the Seychelles. *Estuarine, Coastal and Shelf Science*, 64(2–3), 223–234.

Shin, E.C. and Oh, Y.I. (2007). Coastal erosion prevention by geotextile tube technology. *Journal of Geotextiles and Geomembranes*, 25, 264–277.

Shin, E.C., Ahn, K.S., Oh, Y.I. and Das, B.M. (2002). Construction and monitoring of geotubes. In *Proceedings of The Twelfth International Offshore and Polar Engineering Conference*, Kitakyushu, Japan, May 26–31.

Singhvi, A., Luijendijk, A.P. and van Oudenhoven, A.P.E. (2022). The grey – green spectrum: A review of coastal protection interventions. *Journal of Environmental Management*, 311, 114824.

Stratigaki, V., Manca, E., Prinos, P., Losada, I.J., Lara, J.L., Sclavo, M., ... and Sánchez-Arcilla, A. (2011). Large-scale experiments on wave propagation over Posidonia oceanica. *Journal of Hydraulic Research*, 49(sup1), 31–43.

Strusinska, A. and Oumeraci, H. (2008). Geotextile reef as a coastal protection against tsunami. *Geotechnical Engineering for Disaster Mitigation and Rehabilitation - Proceedings of the 2nd International Conference GEDMAR08*, pp. 742–747.

Sundar, V., Maiti, D.K., Sannasiraj, S.A. and Venkatraman, M. (2009). Geosynthetic application for coastal protection at Shankarpur, West Bengal, India. In *Proceedings of 5th International Conference on Asian and Pacific Coasts (APAC)*, 23–25 September, Singapore, pp. 58–64.

Sundar, V., Sannasiraj, S.A., Murali, K. and Sundaravadivelu, R. (2007). Runup and inundation along the Indian peninsula, including the Andaman Islands, due to Great Indian Ocean Tsunami. *Journal of Waterway, Port, Coastal, and Ocean Engineering*, 133(6), 401–413.

Sutton-Grier, A.E., Wowk, K. and Bamford, H. (2015). Future of our coasts: The potential for natural and hybrid infrastructure to enhance the resilience of our coastal communities, economies and ecosystems. *Environmental Science & Policy*, 51, 137–148.

Takagi, H. (2019). "Adapted mangrove on hybrid platform"–coupling of ecological and engineering principles against coastal hazards. *Results in Engineering*, 4, 100067.

Temmerman, S., Meire, P., Bouma, T.J., Herman, P.M., Ysebaert, T. and De Vriend, H.J. (2013). Ecosystem-based coastal defence in the face of global change. *Nature*, 504(7478), 79–83.

Titus, J.G., Hudgens, D.E., Trescott, D.L., Craghan, M., Nuckols, W.H., Hershner, C.H., ... and Wang, J. (2009). State and local governments plan for development of most land vulnerable to rising sea level along the US Atlantic coast. *Environmental Research Letters*, 4(4), 044008.

Van Zelst, V.T.M., Dijkstra, J.T., Van Wesenbeeck, B.K., Eilander, D., Morris, E.P., Winsemius, H.C., Ward, P.J. and de Vries, M.B. (2021). Cutting the costs of coastal protection by integrating vegetation in flood defences. *Nature Communications*, 12. Article number: 6533.

Wang, Y., Yin, Z. and Liu, Y. (2019). Numerical study of solitary wave interaction with a vegetated platform. *Ocean Engineering*, 192, 106561.

Wang, Y., Yin, Z. and Liu, Y. (2022). Experimental investigation of wave attenuation and bulk drag coefficient in mangrove forest with complex root morphology. *Applied Ocean Research*, 118, 102974.

Ward, R.D. (2020). Carbon sequestration and storage in Norwegian Arctic coastal wetlands: Impacts of climate change. *Science of the Total Environment*, 748, 141343.

Chapter 11

Flood Control Structures

11.1 Introduction

Extreme weather events, such as coastal flooding, flash floods, urban floods and riverine floods, have become increasingly prevalent in recent years due to various factors including climate change and urbanization. These events include a coastal area inundated by rising sea levels, a flash flood in steep mountain regions, heavy precipitation in an urban area, and a river overflowing its banks. Under certain circumstances, different types of floods can even occur at the same time and exacerbate each other, resulting in serious disasters. These events pose significant threats to human life, infrastructure and the environment, necessitating effective flood control measures. Therefore, protection against severe flood events is a permanent task for society, especially in densely populated areas.

Flood control structures are essential components of flood management systems, playing a crucial role in mitigating the impacts of flooding events. These structures include levees, dams, floodwalls, detention basins and other engineered systems designed to regulate water flow, reduce flood risk, and protect communities and assets. In addition to manmade structures, nature-based solutions supplement or replace technical solutions under certain circumstances. Some examples are dunes or mangrove forests in coastal areas, bogs, floodplains and wetlands along riverine areas, and parks and other infiltration or storage volumes in urban areas. A generalized method flowchart is shown in Fig. 11.1 for a flood control strategy that facilitates decision-making.

This chapter on flood control explores the probability aspects involved in estimating the occurrence of flooding events, considering site-specific conditions such as topography, land use, soil conditions, rainfall patterns and hydrological characteristics. Understanding the probability and magnitude of flooding helps authorities and planners anticipate and prepare for potential disasters, allowing for the implementation of proactive mitigation measures.

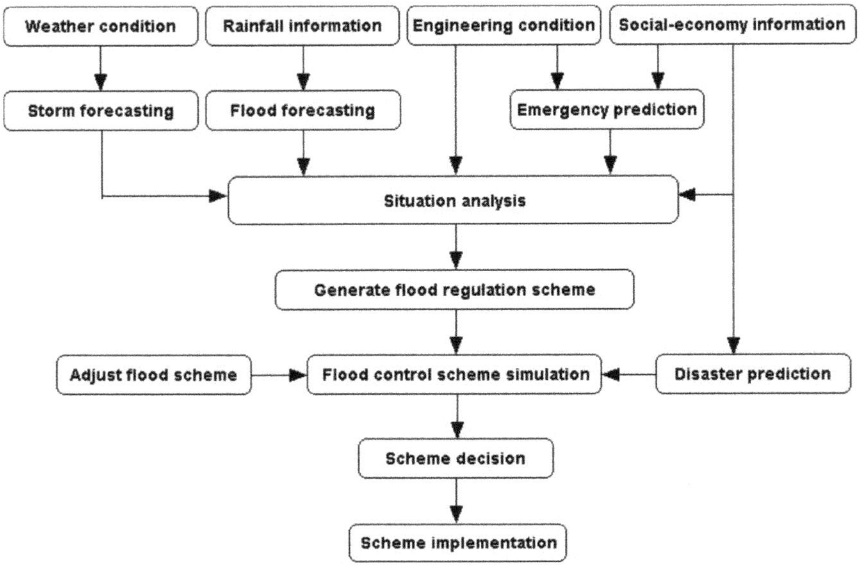

Fig. 11.1. Flow in decision-making for flood control.

Mitigation measures for flood control encompass a range of strategies aimed at reducing the likelihood and severity of flooding impacts. These may include structural interventions, such as building and maintaining levees and floodwalls, and non-structural approaches, such as land use planning, zoning regulations and floodplain management.

Moreover, this chapter delves into the importance of comprehending the interaction between incoming flows, flood control structures and protected buildings in the hinterland – the area beyond the immediate flood zone but still susceptible to indirect impacts. This understanding is critical for optimizing the design and operation of flood control systems to minimize risks and maximize effectiveness.

Experimental and computational models play a significant role in addressing the complex interactions between water flow, flood control infrastructure and built environments. These models allow researchers and engineers to simulate various scenarios, assess the performance of different flood control strategies and refine designs to enhance resilience and adaptability in the face of changing environmental conditions. In addition, extreme or future scenarios can be tested and verified which cannot be tested in the field. Therefore, both experimental and computational models are important tools in hydraulic and coastal engineering to assess the stability and performance of flood protection structures under (extreme) design conditions.

In summary, this chapter emphasizes the importance of proactive measures to mitigate the impacts of extreme weather events, discusses the role of flood control structures in this endeavour, explores probability estimation techniques and highlights the significance of understanding the intricate dynamics between incoming flows, infrastructure and vulnerable assets.

11.2 Probability Aspects in Estimating Flood Events

Estimating extreme flood events is a complex process that requires a comprehensive analysis of multiple factors to assess the probability of occurrence and potential severity. This estimation involves integrating various data sources, models and technologies to account for historical trends, hydrological dynamics and future climate projections.

Studying the historical records of past flood occurrences yields invaluable insights into the recurrence, severity and repercussions of flooding within a specific locality. Through a meticulous analysis of historical data pertaining to river or storm surge levels, precipitation trends and flood extents, researchers can discern recurring patterns and tendencies that serve as foundational knowledge for probabilistic assessments of forthcoming flood hazards. The utilization of historical flood event data is paramount in gauging the potential risk posed by floods in coastal or riverine regions. However, limitations in the availability and accessibility of water level data over extended durations necessitate supplementation from historical chronicles, which furnish details regarding the timing, magnitude and scope of past flood incidents. This comprehensive information aids planners and designers in more accurately evaluating the likelihood of future extreme flood events.

Furthermore, climate projections are indispensable for forecasting future flood risks, given that alterations in temperature, precipitation patterns and sea levels can profoundly impact the frequency and intensity of extreme weather phenomena. Climate models furnish forecasts of forthcoming climatic conditions across various emission scenarios, enabling researchers to gauge the potential ramifications of these changes on regional hydrology and flood risk in the years ahead.

Hydrological models play a crucial role in simulating the behaviour of river systems and predicting how they respond to different meteorological conditions, land use changes and human interventions. These models utilize mathematical equations to simulate the movement of water through the landscape, incorporating factors such as rainfall, evaporation, infiltration and runoff to forecast river discharges and flood levels. Hydrological model results can be used to govern hydraulic models and assess water levels, inundation areas, flow velocities and other parameters. Both hydrological and hydraulic models can be utilized to simulate various scenarios and variations, providing valuable insights into the interactions and dynamics within a water body. These models help to assess the impact of flow behaviour due to different engineering structures or management solutions and, thus a more comprehensive understanding of potential outcomes can be made. Through parametric study using these models, it is possible to predict and analyze the effects of changes in water flow, sediment transport, and other factors, that facilitating more informed decision-making and effective planning in water resource management.

Site-specific conditions, including topography, land use and drainage systems, play a critical role in shaping the likelihood and severity of flooding in a particular area. Areas with steep slopes, impermeable surfaces or poorly designed drainage infrastructure are more susceptible to flash floods and urban flooding. Assessing these local characteristics is essential for accurately estimating flood risks and designing effective mitigation measures tailored to specific vulnerabilities.

Advances in technology, such as remote sensing and geographic information systems (GIS), have revolutionized our ability to collect, analyze and visualize spatial data relevant to flood risk assessment. Remote sensing techniques, including satellite imagery and LiDAR (Light Detection and Ranging) data, provide high-resolution elevation models, land cover maps and flood extent datasets that enhance the accuracy of flood hazard mapping and modelling.

Despite these advancements, estimating extreme flood events still involves significant uncertainties, particularly regarding the impacts of climate change and human interventions on hydrological systems. Climate projections inherently carry uncertainties due to the complexity of Earth's climate system and concerns regarding future greenhouse gas emissions. Additionally, human interventions such as urbanization, deforestation and land use changes can alter local hydrology and exacerbate flood risks, further complicating the estimation process.

In summary, estimating extreme flood events requires a multidisciplinary approach that integrates historical data, hydrological and hydraulic models, climate projections and site-specific information. While advances in technology have improved our ability to predict and model flood events, uncertainties associated with climate change and human interventions underscore the importance of ongoing research and adaptive management strategies to mitigate future flood risks.

It is important to acknowledge that flood risk modelling revolves around hypothetical hydrological scenarios and does not constitute a forecast for real-world situations. Real scenarios are contingent upon current weather conditions, topographical features, soil characteristics and various other factors, which may result in outcomes that differ from the modelled scenarios, potentially being less or more severe than the design projections.

11.3 Flood Risk Management

Flood risk is determined by both the likelihood of an event occurring and the water level reached during that event, which in turn dictates the resultant damage. Situations involving exceptionally high water levels typically have a very low probability of occurrence but pose substantial stress on protective infrastructure like levees and flood defence walls. During such extreme events, these protective structures may become overwhelmed, leading to structural failures such as levee breaches, consequently inundating previously protected areas. This breach allows for significant volumes of water to flow from the river into the hinterland, resulting in extensive damage. The severity of the damage is influenced by several factors including the extent of the flooded areas, the duration and height of flooding, as well as considerations of land use, encompassing economic, ecological and psycho-social aspects.

Flood risk management has many facets with flood prevention as a critical component. This involves appropriate land use and building practices in flood-prone areas, as well as enhancing water retention at the source by preserving and restoring natural retention areas, such as floodplains. This also includes information provision and behavioural precautions by warning, informing and educating the affected population. Another key component of flood risk management is the management of floods through disaster control, reconstruction aid and reconstruction. All elements of flood risk management must be considered, optimized and coordinated to ensure optimum flood protection in order to minimize the impact of floods on the people and assets.

An important element of flood protection in densely populated areas is still traditional technical flood protection through structural measures. Past experience has led to a rethink in the design of levees or (mobile) flood defence walls. Whereas levees were initially dimensioned on the basis of empirical values, i.e. the highest flood levels ever measured at a location, deterministic dimensioning has become standard practice in recent decades. Based on the available water level data, a water level is statistically determined that occurs with a fixed probability, for example, once every 100 years. Nowadays, a probabilistic and risk-orientated design is used. All failure mechanisms of a flood protection structure are considered in their temporal and spatial sequence in order to determine the probability of failure of the structure. The advantages of this approach are the consideration of all relevant failure mechanisms, the identification of weak points and risk-reducing measures in the levee cross-section and in the levee line, the consideration of uncertainties and the possibility of cost optimisation.

Against the backdrop of expected global climate change and its effects on the regional climate, the consequences and risks for water management – in particular flood protection and flood risk management – are of great economic, ecological and social importance. Due to trend-related extreme precipitation and runoff events as a result of climate change, the design and the planning of flood protection measures based on past observations are only possible to a limited extent. However, the functionality and stability of flood defences must be guaranteed over a service life of several decades. Therefore, the long-term changing hydrological boundary conditions must be considered in the planning and design of flood defences. The aim of current research is to identify the effects and consequences of global climate change on flood protection and the necessary research requirements in order to develop climate-adapted flood protection systems and facilities on this basis. This is the only way to avoid or reduce extreme damage such as those described in Section 11.7.

In the following, an example of climate adaptation for sea dikes is explained to demonstrate how climate adaptation can be implemented in flood risk management.

Sea dikes are used in Denmark, Germany and the Netherlands for storm surge protection to avoid flooding of the hinterland. A sea dike is a very simple earthen structure with a smooth seaward slope (often in between 1:3 and 1:10) and a smooth landward slope (often in between 1:3 and 1:4). Several seadikes have failed in the past due to wave overtopping, which is considered as the primary failure mechanism for these structures. Current design aims to limit wave overtopping by applying critical overtopping rates.

Wave run-up and wave overtopping are the most important design parameters for the design of sea dikes and levees. The present design guidelines have been provided by the EurOtop (Eurotop, 2018) and were prepared for different types of structures from smooth embankments to vertical seawalls to help designers, coastal authorities and coastal engineers assess the amount of wave overtopping. This requires designing with parameters that anticipate conditions at the end of the structure's design lifespan, rather than the present. Guidance is given for coastal sea dikes and embankments, armoured rubble slopes and mounds as well as vertical and steep walls. In the following, the focus is on the use of the Eurotop manual to provide recommendations to adapt sea dikes to climate change.

Climate change is expected to lead to rising sea levels, alongside heightened storm surge water levels and wave loads. Consequently, seawalls must be engineered to withstand future hydrodynamic loads. This requires designing with parameters that anticipate conditions at the end of the structure's design lifespan, rather than the present.

A case study is discussed to show the effect of a variation of geometric parameters for sea dikes. The main parameters are typical for German and Dutch sea dikes. The sea dike is assumed to have a crest height of 8 m a. The still water level, SWL and the design water level are at 5m apart. This results in a freeboard of 3 m. The seaward slope is 1:6 and no overtopping reduction factors are considered ($\gamma_f = \gamma_\theta = \gamma_b = \gamma_v = 1.0$). The incoming wave height at the toe of the structure is assumed to be $H_S =$ 2.0 m and the associated wave period is $T_{m-1,0} = 7.0$ s. This structure is used as a reference structure with an estimated mean overtopping rate of $q =$ 0,0027 m^3/(s*m).

The mean overtopping rate q is calculated according to the Eurotop manual; the following formula is used as recommended for design and safety assessments:

$$\frac{q}{\sqrt{g \cdot H_{m0}^3}} = \frac{0.026}{\sqrt{\tan\alpha}} \cdot \gamma_b \cdot \xi_{m-1,0} \cdot \exp \times \left[-\left(2,5 \frac{R_C}{\xi_{m-1,0} \cdot H_{m0} \cdot \gamma_b \cdot \gamma_f \cdot \gamma_\beta \cdot \gamma_V} \right)^{1,3} \right], \quad (11.1)$$

with a maximum of

$$\frac{q}{\sqrt{g \cdot H_{m0}^3}} = 0,1035 \cdot \exp \left[-\left(1,35 \frac{R_C}{H_{m0} \cdot \gamma_f \cdot \gamma_\beta \cdot \gamma^*} \right)^{1,3} \right],$$

wherein q = mean wave overtopping rate [m^3/(s·m)], R_c = freeboard height [m], H_{m0} = design wave height [m], g = acceleration of gravity [m/s^2], γ_b, γ_f, γ_β, γ_V = coefficients to include the effect of berm, roughness, wave direction and a crest wall [−], respectively, α = seaward slope of structure [−], $\xi_{m-1,0}$ = surf similarity parameter = $\tan\alpha/(H_{m0}/L_0)^{1/2}$ and L_0 = deep-water wavelength calculated with the wave period $T_{m-1,0}$.

The crest height, slope, presence of roughness elements, incorporation of berms, utilization of crest walls and integration of foreshore structures all influence the degree of wave overtopping. In the following, we explore variations in these parameters to illustrate various options for adapting to climate change.

The height of the foreland, summer dikes or other constructions seaward of the dike has an impact on the incoming wave's energy and the wave parameters at the toe of the structure. This is a very effective but also expensive solution. A possible reduction of the incoming wave height by 50% results in a significant reduction of the mean overtopping rate. This is due to the exponential relationship between wave height H_S and mean overtopping rate q. The relationship is shown in Fig. 11.2.

In addition, the wave run-up and wave overtopping process can be influenced by a variation of geometrical parameters of the sea dike itself. In this context, the crest height is the most important factor. The crest height and therefore the freeboard are considered in wave overtopping formulas with

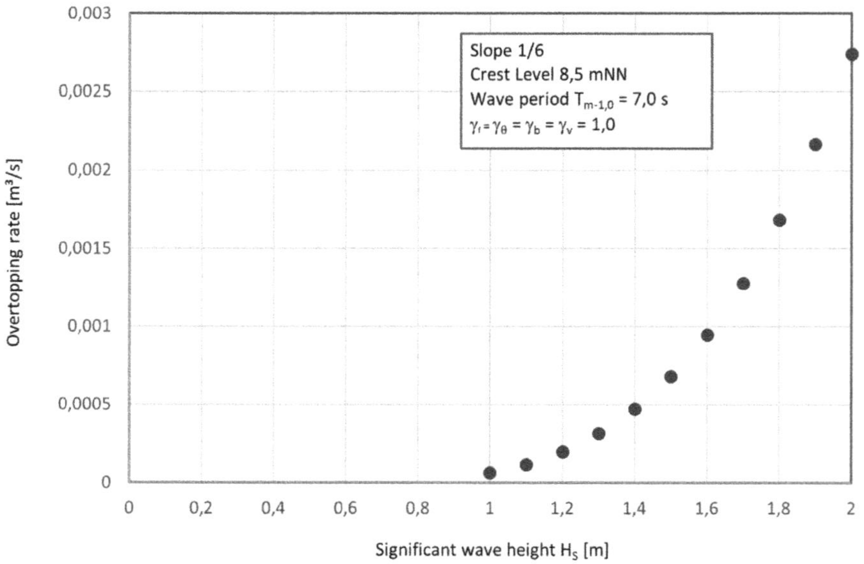

Fig. 11.2. Effect of significant wave height on the mean wave overtopping rate.

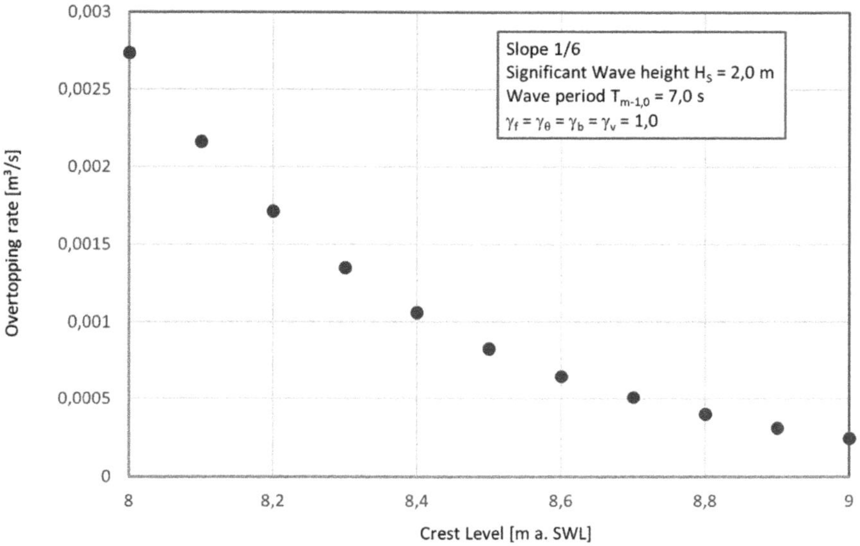

Fig. 11.3. Effect of crest level height on the mean overtopping rate.

an exponential relationship. The higher the freeboard R_C, the lower the amount of wave overtopping. The influence of the crest level on the mean overtopping rate is shown in Fig. 11.3. The reduction of the mean overtopping rate while increasing the crest level from 8 to 9 m above the SWL is obvious. Therefore, an increase in the crest level is a solution that is often considered in practice.

In a similar way, the slope of the seaward side influences wave overtopping. The smoother the slope, the more the mean overtopping rate decreases. In this, we can see two effects. Incoming waves do not break on the seaward slope of very steep structures. Therefore, overtopping behaviour is higher as can be seen in Fig. 11.4. If the slope is smoother, the overtopping rate decreases significantly, especially for very smooth slopes. However, it is important to note that smoother slopes typically require a larger base for the sea dike.

Lastly, roughness elements are integrated into the sea dike design. These elements, such as blocks, play a crucial role in influencing the wave run-up process by facilitating energy dissipation on the seaward slope. As the slope becomes rougher and more permeable, the mean overtopping rate is notably reduced, as illustrated in Fig. 11.5.

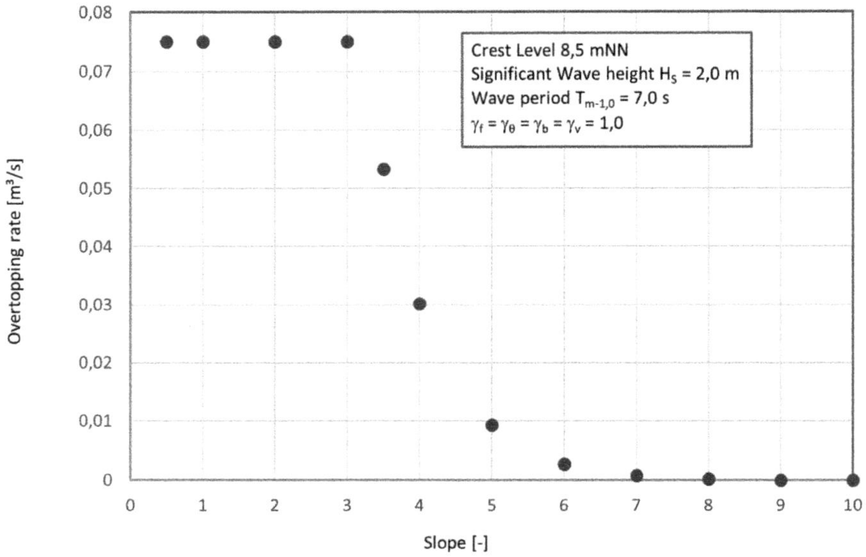

Fig. 11.4. Influence of slope on the mean overtopping rate.

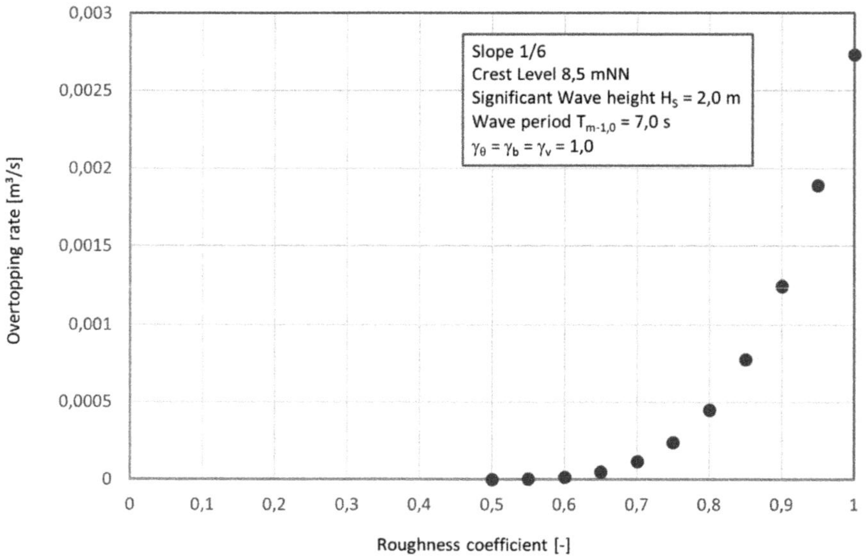

Fig. 11.5. Influence of roughness on the mean overtopping rate.

11.4 Mitigation Measures for Flood Control

Mitigating the impact of flooding necessitates a comprehensive strategy that encompasses both structural and non-structural measures aimed at reducing vulnerability and enhancing resilience to flood hazards. Therefore, effective flood prevention is composed of a number of different elements:

- Waterproof construction,
- Room for the river,
- Organization,
- Technical flood protection,
- Information and awareness,
- Assurance,
- Water retention,
- Property protection,
- Early warning and flood forecasting.

Structural measures are tangible interventions that physically modify the landscape to control the flow of water and mitigate flood impacts. These measures typically involve the construction of engineered structures designed to manage floodwaters and protect communities, infrastructure and ecosystems. Examples of structural measures include the following:

Dams: Dams are large structures built across rivers and streams to regulate water flow, store excess water during periods of high precipitation and release it gradually to prevent downstream flooding.

Levees: Levees or dikes are earthen embankments or walls constructed along riverbanks or coastlines to contain floodwaters within designated areas and prevent them from inundating adjacent lands and the hinterland.

Floodwalls: Floodwalls are vertical barriers made of concrete, steel or other materials erected along riverbanks or in urban areas to block floodwaters and protect properties from flooding.

Storm surge barriers: Storm surge barriers are large gates or movable barriers installed at the mouths of rivers or estuaries to prevent storm surges and tidal inundation from entering inland areas during coastal flooding events.

Polders: Polders are regulated or non-regulated areas along rivers that store water during flood events and mitigate the impacts of flood events downstream.

Illustrations of dams, levees, floodwalls, storm surge barriers and retention basin are shown in Fig. 11.6.

These structural measures are designed to provide physical protection against flooding by controlling water levels, redirecting flow paths and reducing the extent of inundation. Structural measures are engineered based on design flood events typically associated with a specific return interval, such as a 100-year event. These measures ensure adequate safety against the designated event if properly designed, planned, constructed and maintained. Consequently, structural flood protection measures are widely implemented globally to safeguard communities from severe flood occurrences.

Non-structural measures complement structural interventions by focusing on land use planning, policy development and community preparedness to reduce exposure and vulnerability to flood hazards. Examples of non-structural measures include the following:

Land use planning: Land use planning involves zoning regulations, building codes and development restrictions to steer development away from flood-prone areas and limit exposure to flood risk. Construction of buildings and infrastructures should be avoided in flood-prone areas.

Floodplain management: Floodplain management strategies aim to preserve natural floodplains, restore wetlands and implement measures such as setback requirements and floodplain buyouts to reduce property damage and enhance ecosystem resilience.

Early warning systems: Early warning systems are set up on a technical and a non-technical component. The technical component utilizes real-time monitoring data, weather forecasts and communication networks to provide advance notice of impending floods, allowing communities to evacuate residents, deploy emergency response resources and implement flood preparedness measures. The non-technical component comprises the behaviour of public institutions, stakeholders and the population as a response to the technical flood alert information.

By integrating both structural and non-structural measures, communities can adopt a holistic approach to flood risk management that addresses the root causes of vulnerability, enhances adaptive capacity and builds resilience to future flood events. Effective flood risk management requires collaboration among government agencies, policymakers, engineers, urban planners, community stakeholders and the public to develop and implement strategies tailored to local conditions and priorities. Effective flood risk management is tailor-made to adequately consider local geographic, hydrological, hydraulic, land use and social conditions.

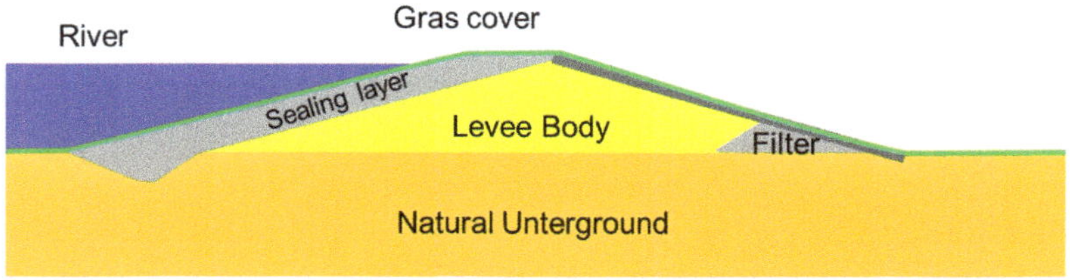

(a) Three-Zone-Levee

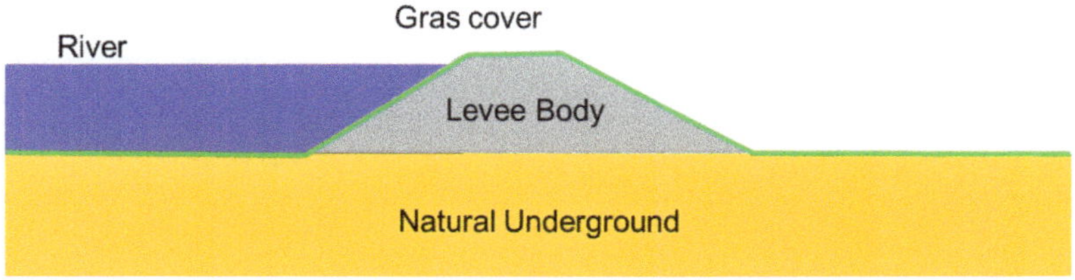

(b) Homogenous Levee

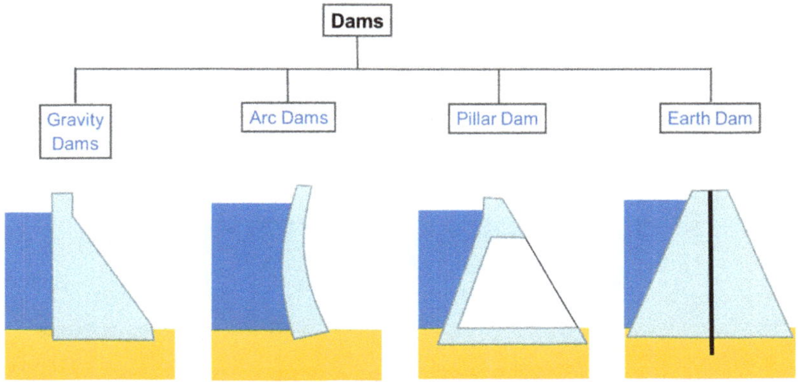

(c)

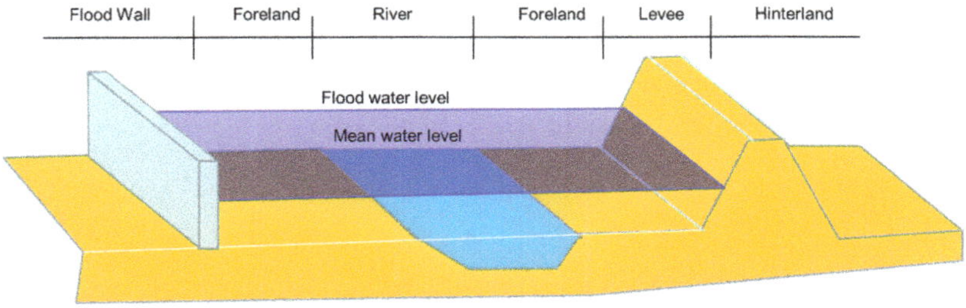

(d)

Fig. 11.6. (a) Pictorial representation of levees: (a) Three-zone levee, (b) homogenous levee, (c) different types of dams, (d) view of flood control measure with a levee and floodwall, (e) view of a storm barrier and (f) view of a retention basin.

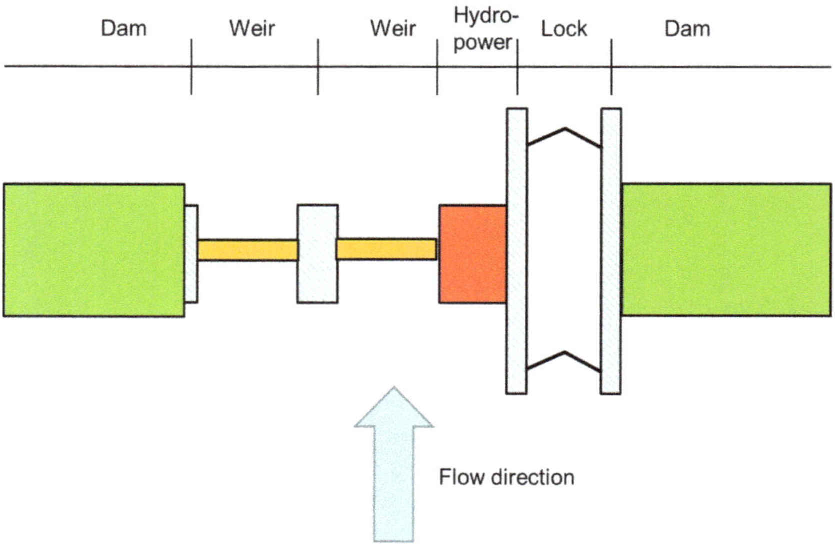

(e)

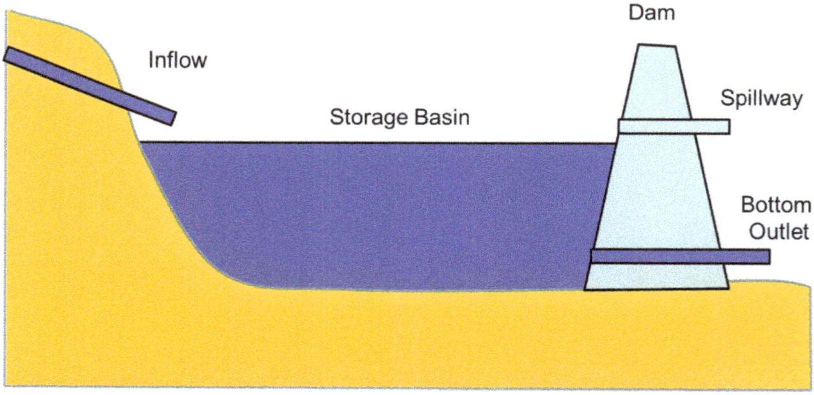

(f)

Fig. 11.6. (*Continued*)

11.5 Understanding Interaction between Flows and Structures

The effectiveness of flood control structures hinges on their capacity to interact with incoming flows and safeguard surrounding areas from inundation. Different types of flooding, such as coastal flooding, flash floods, urban floods and riverine floods, present distinct challenges, demanding tailored engineering solutions to mitigate their impact effectively.

11.5.1 *Coastal flooding*

Coastal areas are vulnerable to flooding caused by storm surges, high tides and rising sea levels. To combat these threats, engineers employ structures such as seawalls, dikes and tidal barriers. Seawalls act as barriers along coastlines, absorbing and deflecting wave energy to prevent the inundation of coastal communities. Dikes or levees, often built parallel to the shoreline, provide additional protection by confining floodwaters within designated areas. Tidal barriers, including sluice gates, are strategically positioned at estuaries or river mouths to regulate tidal flows and prevent seawater intrusion during storm events.

11.5.2 *Flash floods*

Flash floods are characterized by rapid onset and intense rainfall, posing significant challenges to flood control. In response, engineers implement rapid response measures such as detention basins, channel improvements and urban drainage systems. Detention basins temporarily store excess runoff, attenuating peak flows and reducing downstream flood risk. Channel improvements, such as widening and deepening channels or installing culverts, aim to enhance conveyance capacity and reduce the likelihood of blockages during flash flood events. Urban drainage systems, including stormwater sewers and retention ponds, help manage surface runoff and prevent localized flooding in urban areas.

11.5.3 *Riverine floods*

Riverine floods result from prolonged rainfall or snowmelt, causing rivers and streams to overflow their banks. Managing riverine floods requires a combination of upstream reservoirs, levees and floodplain restoration efforts. Upstream reservoirs regulate river flows by storing excess water during periods of high precipitation and releasing it gradually to mitigate downstream flooding. Levees, similar to those used in coastal areas, are constructed along riverbanks to contain floodwaters within designated channels and protect adjacent lands. Floodplain restoration initiatives aim to reconnect rivers with their natural floodplains, allowing floodwaters to spread out and dissipate energy, reducing the risk of catastrophic flooding downstream.

11.5.4 *Urban floods*

Urban floods can arise from intense precipitation events and are not solely reliant on river systems. Consequently, any building or house can be susceptible to urban flooding during periods of heavy rainfall. These floods can manifest in various locations, often leading to the inundation of tunnels, underground areas, basements, ground floors and other subterranean spaces.

By tailoring engineering solutions to the specific characteristics of each flood type, such as coastal, flash, urban or riverine flood, communities can effectively mitigate flood risk and protect lives, property and critical infrastructure from the devastating impacts of inundation. These measures not only enhance resilience to current flood threats but also help communities to adapt to future changes in climate and land use patterns. In addition, self-protection is required for urban or riverine flood events. Houses and buildings can be protected by stop logs and waterproof doors or windows.

11.6 Experimental Models for Understanding Interaction

Experimental models serve as invaluable tools for simulating the intricate interactions between incoming flows, flood control structures and protected buildings in the hinterland. These models, which encompass both physical scale models and computational simulations, enable engineers to study complex hydraulic phenomena, evaluate flood control strategies and refine design parameters to enhance flood resilience. Two distinct categories of models emerge: (a) floodscape models, which encompass expansive flood-prone regions to analyze the dynamic interplay among rivers, topography, infrastructure and flood defence systems, and (b) flood defence structure models, specifically designed to scrutinize the resilience of flood protection structures against elevated water levels, strong currents and intense wave impact.

11.6.1 *Physical scale models*

Physical scale models involve the construction of miniature replicas of real-world flood control systems, including rivers, levees, dams and buildings, in laboratory settings. By manipulating water flow rates, sediment transport and structural configurations, engineers can replicate various flood scenarios and observe how different factors interact. Physical scale models provide tangible insights into flow patterns, hydraulic gradients and structural responses, allowing engineers to assess the effectiveness of flood control measures and identify potential vulnerabilities. These models facilitate hands-on experimentation and validation of design concepts before implementation in the field. In general, these models are scaled on a very low scale since the scale effect is determined by the available space and discharge in the hydraulic laboratories. Photographs of physical model tests of flood investigations are shown in Fig. 11.7.

In addition to these models, structural flood models are used to investigate fluid–structure–soil interactions on a large or small scale. In general, it is recommended to scale structural models on a large scale to avoid scale effects. Typical investigations in coastal engineering research focus on phenomena such as overflow, wave overtopping, piping, and the stability of armor units, along with other related wave-structure interaction problems. Large-scale overflow experiments conducted in the Dike Flume at the Institute of Hydraulic Engineering and Water Resources Management, RWTH Aachen University are illustrated in Fig. 11.8. These experiments are designed to analyze the flow properties of wave overtopping on various sections of the dike, including the seaward slope, dike crest, and landward side. By conducting these experiments on a large scale, researchers can ensure that the results are not distorted by scale effects and hence, more accurate and reliable data can be provided. This setup allows for comprehensive studies of fluid-structure interactions, enabling the examination of how waves interact with different dike materials and configurations. Additionally, the experiments can include tests with soil or vegetation to further

Fig. 11.7. Physical model of flood investigations.

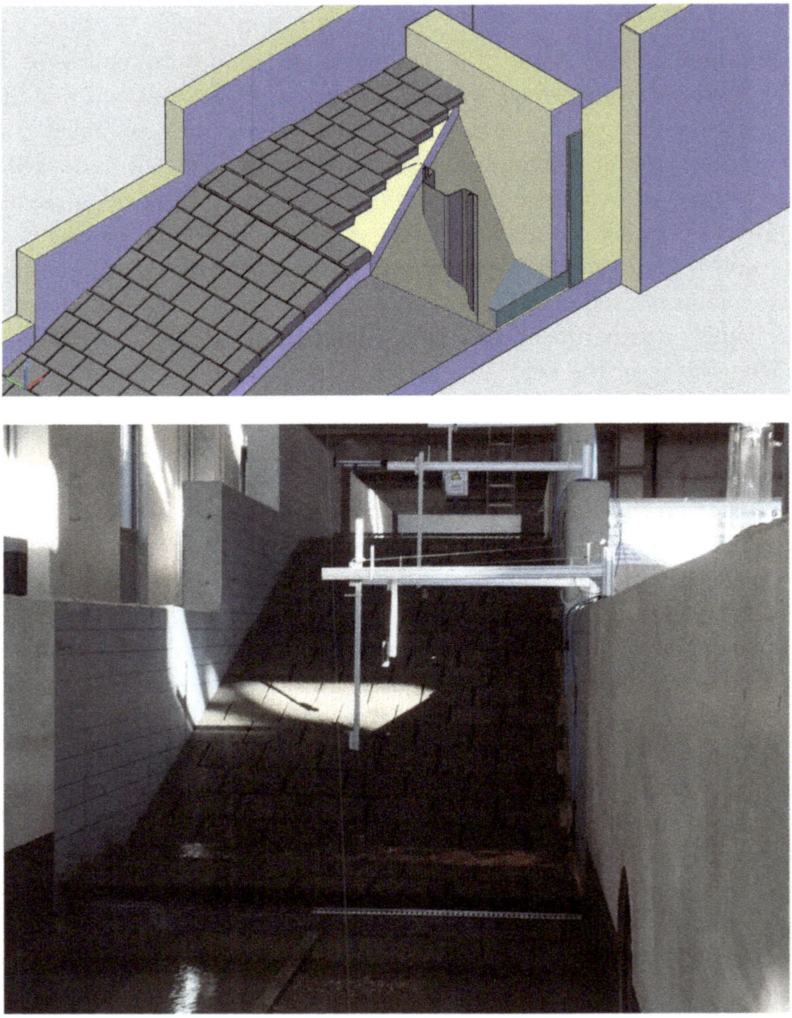

Fig. 11.8. Physical model of overflow investigations on a large scale.

enhance the understanding of how natural elements can influence and be integrated into coastal protection structures.

11.6.2 *Computational models*

Computational models utilize mathematical equations and computer algorithms to simulate flow dynamics, structural behaviour and flood inundation processes. Numerical simulations, such as finite element

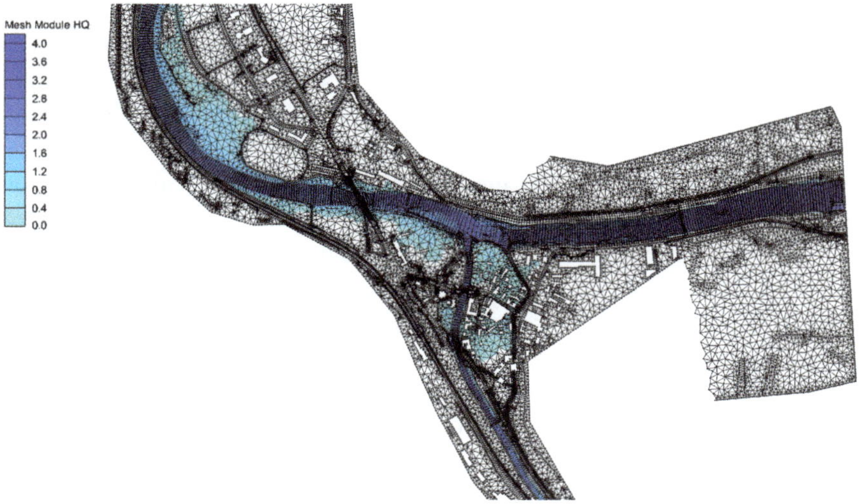

Fig. 11.9. Numerical model of flood investigations.

analysis and computational fluid dynamics, offer detailed insights into flow behaviour, turbulence and pressure distribution within flood control systems. Hydrodynamic modelling, which combines computational fluid dynamics with geographical data, enables engineers to predict flood extents, inundation depths and evacuation routes under different scenarios. Computational models provide a cost-effective and flexible tool of exploring a wide range of flood control strategies, optimizing design parameters and assessing the resilience of infrastructure to extreme events. Computational models are available to cover small- and large-scale domains. In general, restrictions are based on the available computer power, computation times and the possible spatial and temporal resolution of the model domain. A typical computational domain model is shown in Fig. 11.9.

11.6.3 *Integration of experimental and computational models*

The fusion of experimental and computational models harnesses the complementary advantages of each method to craft resilient flood management tactics tailored to particular geographic and hydrological contexts. Physical scale models furnish empirical data and validation for computational simulations, while computational models furnish intricate insights and predictive abilities surpassing the confines of laboratory tests. Through amalgamating these methodologies, engineers can fine-tune design parameters, verify theoretical frameworks and pinpoint potential design inadequacies or performance constraints prior to deploying flood management measures in practical scenarios. This integrated methodology amplifies the dependability and efficacy of flood management strategies, thereby enhancing communities' resilience against flood risks.

In summary, experimental models, including physical scale models and computational simulations, play a crucial role in the development and optimization of flood control strategies. By simulating the interaction between incoming flows, flood control structures and protected buildings, these models enable engineers to design resilient infrastructure, mitigate flood risk, and safeguard lives and property in flood-prone areas. In addition, the understanding of the different processes is improved and mistakes during the design process of flood protection structures and systems can be avoided.

11.7 Case Studies

The case studies provide insights into the causes of, impacts of and responses to severe floods worldwide, highlighting the importance of effective flood control measures and resilience planning in mitigating the impacts of such events. Although a number of case studies have been reported, only a few are presented in this chapter:

- China floods in 1931 Lillian M. Li. (2019),
- Mississippi River floods, USA, in 1993 (Parrett *et al.*, 1993),
- UK floods in 2007 (Pitt, 2008),
- Bihar floods, India, in 2008 (Kumar and Singh, 2010),
- Thailand floods in 2011 (Haraguchi and Lall, 2015),
- Central European floods in 2013 (Merz *et al.*, 2014),
- Chennai floods, India, in 2015 (Bandyopadhyay *et al.*, 2021),
- Venice floods, Italy, in 2019 (Cavaleri *et al.*, 2020),
- Yangtze river floods, China, in 2020 (China Daily, 2020),
- Western Europe floods in 2021(Lehmkuhl *et al.*, 2022),
- Christmas floods in Germany in 2023.

Only salient details of the most devastating floods are discussed in this chapter.

11.7.1 *Chennai floods, India, in 2015 (Bandyopadhyay et al., 2021)*

The 2015 floods in Chennai, India, stand as one of the most devastating natural disasters to strike the city, with heavy monsoon rains and inadequate urban planning converging to create a catastrophic flooding event. Torrential rainfall overwhelmed existing drainage systems, inundating large areas and causing extensive damage to property, infrastructure and livelihoods. The floods exposed the vulnerabilities inherent in Chennai's urban landscape, highlighting the urgent need for comprehensive flood management strategies and sustainable urban development practices. In response to the recurring threat of flooding, Chennai had implemented various flood control measures, including the construction of reservoirs and drainage channels to manage rainwater runoff and mitigate the impacts of monsoon season. However, these measures proved insufficient in the face of the unprecedented rainfall and rapid urbanization that characterized the 2015 floods. One of the key factors contributing to the severity of the flooding was poor urban planning and unchecked growth, which led to encroachment on water bodies, wetlands and natural drainage pathways. As urban areas expanded and land was developed for residential, commercial and industrial purposes, vital floodplains and water retention areas were lost, exacerbating the city's vulnerability to flooding. Moreover, inadequate maintenance of drainage infrastructure further compounded the problem, limiting the capacity of existing systems to cope with the deluge of water. In the aftermath of the disaster, there was a renewed focus on the need for holistic flood management strategies that address the root causes of flooding, including urbanization, land use changes and environmental degradation. Efforts were made to strengthen drainage infrastructure, restore natural water bodies and enforce regulations to prevent encroachment on floodplains and wetlands. The 2015 floods in Chennai serve as a sobering reminder of the urgent need to prioritize sustainable urban planning and resilience-building initiatives to mitigate the impacts of extreme weather events and ensure the long-term safety and well-being of cities and their inhabitants.

11.7.2 *Western Europe floods in 2021 (Lehmkuhl et al., 2022)*

The floods that struck Western Europe, notably affecting Germany, Belgium, and the Netherlands, in July 2021 were unprecedented in their scale and intensity, resulting from a convergence of extreme weather events and inadequate or partly missing flood control infrastructure. The event was estimated to have a return interval of more than 10.000 years on the basis of long-term statistical analysis. The floods of 2021 caused severe destruction in Germany in particular. The most affected parts were North Rhine-Westphalia and Rhineland-Palatinate. In the catchment areas of the

Fig. 11.10. Flood damage after the Western Europe floods in 2021 in Ahr Valley.
Source: Schüttrumpf (2021).

Ahr, Erft, Inde, Vicht and Wupper rivers, the water devastated some sections of the affected valleys. The greatest damage occurred along the Ahr river in Rhineland-Palatinate (Fig. 11.10). The flood event in July 2021 is the second most severe natural disaster in Germany after the storm surge on 16/17 February 1962 on the German North Sea coast in which more than 340 people lost their lives. In total, more than 180 people lost their lives during this flooding in July 2021 (CEDIM, 2021; Mühr *et al.*, 2021) and economic damage of more than 30 billion euros was caused.

Nevertheless, the floods experienced in July 2021 not only caused significant destruction and devastation but also offered a chance to glean valuable insights to prevent similar flash flood disasters in low-mountain regions in the future. Researchers, in particular, should seize this opportunity to reassess and refine their models, methodologies, strategies and concepts based on the wealth of available data and information.

The floods that struck Western Europe in July 2021 serve as a stark reminder of the urgent need to strengthen flood resilience and adapt to the growing risks posed by climate change. Investment in robust flood control infrastructure, improved forecasting and early warning systems, and sustainable land use planning is essential to mitigate the impacts of future flooding events and protect vulnerable communities in the region.

11.7.3 *Christmas floods in Germany in 2023*

The Christmas floods of 2023/24 were instigated by sustained precipitation across Germany during the final three months of 2023, significantly surpassing the long-term average. October saw 100 mm, November 126 mm and December 114 mm of rainfall.[1] This precipitation pattern triggered a prolonged flooding event from mid-December 2023 to mid-January 2024, with flood crests occurring between the Christmas holidays and New

Fig. 11.11. Large flooded areas in winter 2023/2024 in Germany.
Source: Burghardt (2024).

[1] https://de.statista.com/.

Year. Distinguishing itself from the flash flood of 2021 in Germany's mid-mountain regions, the 2023 floods persisted for several weeks. Despite the duration, flow velocities and water depths remained relatively low, resulting in inundated ground floors and basements but sparing structural damage to homes, thanks in part to the presence of river levees safeguarding settlements. Nevertheless, extensive rural areas experienced flooding, leading to the degradation of winter cereals (see Fig. 11.11). The December 2023 floods underscored the critical importance of flood protection measures. Initiatives such as water retention in upstream regions via large reservoirs, creating room for rivers in low-lying areas and the fortification of properties with river levees have proven effective in safeguarding settlements, villages and cities against flood hazards.

11.8 Summary

Flood control structures are indispensable elements of comprehensive flood risk management strategies, especially as extreme weather events become more frequent and severe due to factors like climate change. These structures play a pivotal role in mitigating flood impacts and protecting lives, infrastructure and ecosystems. By integrating probability aspects into flood event estimation and leveraging experimental models to comprehend the intricate interaction between water flows and structures, engineers can develop mitigation measures that effectively enhance resilience and safeguard vulnerable communities.

11.8.1 *Probability aspects in flood event estimation*

Incorporating probability aspects into flood event estimation involves assessing the likelihood of different flood scenarios based on historical data, hydrological models and climate projections. By quantifying the probability of various flood magnitudes and frequencies, engineers can prioritize resources and design flood control structures to withstand worst-case scenarios. This probabilistic approach enables decision-makers to allocate resources more efficiently and implement targeted mitigation measures in areas with the highest flood risk.

11.8.2 *Leveraging experimental models*

Experimental models, including physical scale models and computational simulations, are invaluable tools for understanding how flood control structures interact with incoming flows and mitigate flood impacts. By conducting controlled experiments and numerical simulations, engineers can evaluate the performance of different flood control strategies, optimize design parameters and identify potential vulnerabilities. These models enable engineers to test innovative solutions, validate design concepts and improve the effectiveness of flood control measures before implementation in real-world settings.

11.8.3 *Advancing flood control technologies*

Continued research, innovation and collaboration are essential for advancing flood control technologies and developing more resilient infrastructure. Engineers and scientists are constantly exploring new materials, construction techniques and monitoring systems to enhance the durability, efficiency and adaptability of flood control structures. Additionally, interdisciplinary collaboration between engineering disciplines, climate scientists, urban planners and policymakers is crucial for integrating flood control measures into broader climate adaptation and disaster resilience strategies.

11.8.4 *Safeguarding against future flooding*

As the frequency and intensity of extreme weather events continue to rise, proactive measures are essential for safeguarding against future flooding. These include investing in resilient infrastructure, enhancing early warning systems, promoting sustainable land use planning, and fostering community preparedness and awareness. By embracing a holistic approach to flood risk management that integrates structural and non-structural measures, stakeholders can reduce vulnerability, mitigate impacts and build adaptive capacity to withstand future flood events.

In summary, flood control structures are critical components of comprehensive flood risk management strategies. By incorporating probability aspects into flood event estimation, leveraging experimental models, advancing flood control technologies and fostering collaboration, engineers can develop effective mitigation measures that enhance resilience and protect vulnerable communities from the impacts of future flooding.

References

Bandyopadhyay, C., Bindal, M.K. and Manna, M. (2021). Chennai Floods 2015. National Institute of Disaster Management, Ministry of Home Affairs, Govt. of India.

Cavaleri, L., Bajo, M., Barbariol, F., Bastianini, M., Benetazzo, A., Bertotti, L., Chiggiato, J., Ferrarin, C., Trincardi, F. and Umgiesser, G. (2020). The 2019 flooding of Venice and its implications for future predictions. *Oceanography*, 33(1). https://doi.org/10.5670/oceanog.2020.105.

CEDIM. (2021). Hochwasser Mitteleuropa, Juli 2021 (Deutschland). Report No. 1. https://www.cedim.kit.edu/.

China Daily. (2020). Floods along Yangtze river persist as more rain forecast. *China Daily.*

EurOtop Manual. (2018). Van der Meer, J.W., Allsop, N.W.H., Bruce, T., De Rouck, J., Kortenhaus, Pullen, T., Schüttrumpf, H., Troch, P. and Zanuttigh, B. Manual on Wave Overtopping of Sea Defences and Related Structures an Overtopping Manual Largely Based on European Research, but for Worldwide Application.

Haraguchi, M. and Lall, U. (2015). Flood risks and impacts: A case study of Thailand's floods in 2011 and research questions for supply chain decision making. *International Journal of Disaster Risk Reduction*, 14(3), 256–272. DOI: 10.1016/j.ijdrr.2014.09.005.

Kumar, D. and Singh, R.B. (2010). The 2008 floods in Bihar: A study on preparedness, response and impact of the disaster. ResearchGate.

Lehmkuhl, F., Schüttrumpf, H., Schwarzbauer, J., Brüll, C., Dietze, M., Letmathe, P., Völker, C. and Hollert, H. (2022). Assessment of the 2021 summer flood in Central Europe. *Environmental Sciences Europe*, 34, 107. https://doi.org/10.1186/s12302-022-00685-1.

Lillian M. Li. (2019). Review Of "The Nature Of Disaster In China: The 1931 Yangzi River Flood By C. Courtney. *Journal of Interdisciplinary History.* 50(2), 310–311. https://works.swarthmore.edu/fac-history/531.

Merz, B., *et al.* (2014). Floods in Central Europe: A case study on vulnerability, impacts and adaptation. ScienceDirect.

Mühr, B., Mohr, S. and Kunz, M. (2021). CEDIM Forensic Disaster Analysis Group (FDA) Starkregen 2021 (USA, Kalifornien), Technical Report, October, https://www.researchgate.net/publication/355675469.

Parrett, C., Melcher, N.B. and James Jr., R.W. (1993). Flood discharges in the Upper Mississippi River Basin, 1993. U.S. Geological Survey Circular 1120–A, 14 p.

Pitt, M. (2008). The Pitt Review: Learning Lessons from the 2007 Floods. GOV.UK

Chapter 12

Field Monitoring

12.1 Introduction

Coasts are exposed to natural hazards (global warming-induced sea level rise, river mouth closure by siltation, changing climate with increased frequency of cyclonic storm surges, coastal flooding, tsunamis, etc.) and hazards due to anthropogenic activities (shoreline erosion, sewage treatment plants, desalination plants, etc.). Alongside combating the hazards, there is rapid progress in coastal development. These factors might have led to the realization of the necessity for coastal management information to be available to the scientific and engineering coastal working community.

Various types of disasters entail distinct hazards affecting different objects. For example, powerful winds have the potential to cause rural homes and school structures to collapse, leading to losses in both agriculture and rural communities. Likewise, excessive rainfall can halt or reduce agricultural and industrial production by flooding fields and facilities, leading to significant economic losses. This can also damage water conservation infrastructure, such as dams and reservoirs, causing long-term harm to water supply systems. Consequently, these disruptions can result in reduced income for farmers and businesses, increased repair costs, and a weakened local economy.

Storm surges often lead to breached dikes and the destruction of fishing boats, significantly affecting the aquaculture industry. Additionally, colossal waves may cause seawater intrusion and breach dikes, resulting in extensive damage to the coastal ecological environment. Consequently, a thorough analysis of the disaster-prone targets specific to each type of disaster is imperative.

Over the past two decades, a series of intensified tropical cyclones cited as Sidr in 2007, Aila in 2009, Giri in 2010, Phailin in 2013, Roanu in 2016, Titli in 2018, Fani in 2019, Bulbul in 2019, Amphan in 2020, and Yaas in (2021), have wreaked havoc on the lives and livelihoods of coastal communities in West Bengal. Particularly vulnerable are the residents of the coastal

Sundarbans, where high storm surges, breaches in embankments, salinization from inundation and erosion-related pressures pose significant threats before and after cyclonic landfall, especially during the pre-monsoon and post-monsoon periods (Yin *et al.*, 2013; Ghosh and Mistri, 2020; Mondal *et al.*, 2021).

A *natural hazard* as per the United Nations International Strategy for Disaster Reduction (UN/ISDR) (2009) is defined as "any natural process or phenomenon that may cause loss of life, injury or other health impacts, property damage, loss of livelihoods and services, social and economic disruption or environmental damage."

Exposure is the condition of encountering something or being impacted by it due to being in a specific situation or location. For instance, an asset might face a hazard but could have the resilience to endure it without sustaining damage, thereby averting a disaster.

Vulnerability denotes the potential for human casualties, physical harm or economic detriment to livelihoods and assets when confronted with hazardous occurrences.

A *disaster* is an event, or series of events, that causes significant disruption, destruction, harm or distress, often resulting in widespread adverse impacts on communities, economies, environments or infrastructure. Catastrophes can arise from natural phenomena such as earthquakes, floods and hurricanes, as well as from human activities such as industrial accidents, terrorist attacks or war. They typically overwhelm local capacities to cope and require external assistance for recovery and rebuilding efforts. A disaster is defined as

$$\textbf{Disaster} = (\textbf{Hazard} + \textbf{Vulnerability})/\textbf{Capacity}.$$

Herein, *capacity* refers to the amalgamation of strengths, attributes and resources existing within a community, society or organization, which can be harnessed to accomplish mutually agreed-upon objectives. It encompasses infrastructure, tangible assets, institutional frameworks, societal resilience, as well as the collective intelligence, expertise and interpersonal dynamics within the community. Capacity, often synonymous with capability, extends to the evaluation of a group's potential vis-à-vis its objectives, whereby any deficiencies in capacity are pinpointed for remedial measures through a process known as capacity assessment.

Risk refers to the probability or likelihood of an undesirable event or outcome occurring, along with the potential consequences or impacts associated with that event. It is often assessed by considering both the likelihood of an event happening and the severity of its consequences. Risk is defined as the product of hazard and vulnerability.

$$\textbf{Risk} = \textbf{Hazard} * \textbf{Vulnerability}.$$

Field monitoring stands as a crucial tool in overseeing fundamental water management and hydrographic data, facilitating comprehension across various physical, chemical and biological processes. It serves as a cornerstone for coastal engineering, hydraulic engineering and water resources management, enabling the strategic planning of measures and the detection of changes in water bodies prompted by construction activities. Effective field monitoring necessitates meticulous planning, adequate financial backing and a proficient workforce to ensure the acquisition of high-quality, actionable data. Key considerations in planning encompass the scope and duration of measurements, selection and deployment of appropriate monitoring devices, seamless data transmission to end users and a comprehensive approach to addressing myriad logistical and technical aspects.

Compared to laboratory measurements, field monitoring demands significantly more meticulous planning and preparation. In the field, measuring devices are subjected to diverse environmental forces and weather conditions, necessitating resilience against extreme pressures, forces and flow velocities while ensuring the reliability of data collection. Moreover, these devices must be safeguarded against potential impacts from external objects, as well as deliberate sabotage or vandalism. Operating in natural settings exposes the devices to harsh climatic conditions such as storms, ice and high temperatures, necessitating design and dimensioning tailored to withstand these challenging circumstances. Consequently, field monitoring requires the utilization of substantially more robust and consequently costlier measuring devices compared to laboratory settings.

Field monitoring frequently serves as a crucial foundation for designing structures and implementing measures. It also forms an essential prerequisite for hydraulic engineering models conducted in laboratories and for hydro-numerical modelling tools. However, despite its significance, field monitoring is often underutilized in research, primarily due to its inherent complexity and the multitude of uncertainties associated with it.

Given the multitude of processes and parameters pertinent to coastal, estuarine, riverine and lacustrine environments, it is impractical to delve into every measuring device and parameter in this chapter. Instead, we briefly outline the most common measurements: bathymetry and topography, water depth and levels, waves, current velocities, turbidity and sediment transport. Additional physical, chemical and biological parameters are omitted here but can be explored further in relevant literature.

This chapter aims to offer recommendations to interested readers regarding the selection and decision-making process for individual measuring devices based on their specific inquiries. We will not delve into specific types or brands of measuring devices, as the global market offers a vast array of options from various companies catering to most standard measuring tasks. Additionally, it is important to note that choosing a particular measuring device also hinges on factors such as compatibility with existing equipment.

Hence, this chapter is structured as follows: First, an examination of natural coastal hazards (Section 12.2), anthropogenic activities (Section 12.3) and coastal development (Section 12.4) within the coastal zone is presented. These sections furnish crucial context for understanding the need to monitor processes in the coastal zone. Subsequently, Section 12.5 addresses the monitoring of coastal structures, followed by an overview of coastal monitoring in Section 12.6, and a deeper exploration of shoreline monitoring in Section 12.7. Structural health monitoring is elaborated upon in Section 12.8, and lastly, Section 12.9 presents an overview of the methodology for measuring key parameters.

12.2 Natural Coastal Hazards

Natural hazards can be broadly classified as global warming-induced sea level rise, river mouth closure by siltation, changing climate with increased frequency of cyclonic storm surges, coastal flooding, tsunamis, etc.

Global warming-induced sea level rise: With the Earth's temperature on the rise, the expansion of seawater due to heat and the melting of glaciers and ice sheets are key factors contributing to the escalation of sea levels. This rise in sea levels has numerous implications for coastal communities, including increased coastal erosion, inundation of low-lying areas and saltwater intrusion into freshwater sources, impacting ecosystems and human settlements alike.

River mouth closure by siltation: The escalation in precipitation and the heightened occurrence of extreme weather events linked to climate change can result in soil erosion and increased runoff of sediment into rivers. This sedimentation can accumulate at river mouths, causing them to silt up. As a result, navigation channels may become shallower, hindering maritime transportation and impacting ecosystems that depend on these estuarine environments.

Changing climate with increased frequency of cyclonic storm surges: Climate change can alter atmospheric conditions, leading to changes in weather patterns and the frequency and intensity of cyclonic storms. Such storms, including hurricanes or typhoons, have the potential to unleash powerful storm surges, flooding coastal regions and resulting in extensive damage to infrastructure, loss of life and economic upheaval.

Coastal flooding: The combination of rising sea levels, storm surges and intensified weather phenomena heightens the likelihood of coastal flooding. Low-lying coastal areas, including densely populated urban centres, face increased vulnerability to flooding, resulting in community displacement, property and infrastructure damage, as well as disruptions to livelihoods and economies.

Tsunamis: While not directly caused by climate change, the risk of tsunami events can be exacerbated by changes in coastal dynamics resulting from rising sea levels and alterations in seabed topography due to human activities such as dredging or coastal development. Tsunamis, triggered by seismic activity such as earthquakes or volcanic eruptions under the ocean, can result in devastating coastal inundation and widespread destruction.

These phenomena underscore the interconnected nature of climate change impacts, affecting not only the physical environment but also socio-economic systems and human well-being. Tackling these challenges necessitates synchronized action at local, national and international scales to mitigate greenhouse gas emissions, adapt to shifting conditions and bolster resilience in communities that are particularly susceptible. Monitoring of these challenges is a prerequisite for understanding processes, impacts and changes. Management plans, strategies and structural design depend on detailed field monitoring.

12.3 Anthropogenic Activities

Shoreline erosion: Anthropogenic activities such as coastal development, including the construction of ports, harbours, and residential or commercial infrastructure, can disrupt natural coastal processes. For example, the alteration of natural shorelines through the construction of seawalls, jetties or groynes can impede the natural movement of sediment along the coast, leading to localized erosion or accretion. Additionally, the removal of coastal vegetation for development purposes can weaken natural coastal defences against erosion, leaving shorelines more vulnerable to wave action, currents and storm surges.

Sewage treatment plants: Improperly managed sewage treatment plants can contribute to coastal pollution, posing risks to marine ecosystems and human health. Untreated or poorly treated sewage discharged into coastal waters can introduce pathogens, nutrients and pollutants, leading to eutrophication, harmful algal blooms and contamination of shellfish beds and recreational waters. These pollution events can have detrimental effects on coastal biodiversity, fisheries, tourism and public health.

Desalination plants: While desalination plants provide a valuable source of freshwater in water-stressed coastal regions, they also pose environmental challenges. The intake and discharge of seawater for desalination processes can disrupt marine ecosystems and coastal hydrodynamics. For instance, the extraction of seawater for desalination can harm marine organisms entrained in intake structures, while the discharge of concentrated brine and other byproducts back into coastal waters can alter salinity levels, impact marine habitats and affect species composition.

Coastal development and land reclamation: Anthropogenic activities such as coastal land reclamation for urban expansion, industrial facilities or infrastructure projects can alter coastal landscapes and ecosystems. Land reclamation activities typically involve dredging and filling operations, which can disrupt sediment transport processes, degrade habitats and exacerbate shoreline erosion. Additionally, coastal development can increase impervious surfaces, leading to runoff pollution, habitat fragmentation and loss of natural coastal buffers such as wetlands and mangroves, which provide important ecosystem services and protection against coastal hazards.

Addressing coastal hazards associated with anthropogenic activities requires integrated coastal management approaches that prioritize ecosystem-based solutions, sustainable development practices and stakeholder engagement. This includes measures such as coastal zone planning, habitat restoration, green infrastructure implementation and the adoption of innovative technologies for wastewater treatment and desalination. By harmonizing human needs with the preservation and rehabilitation of coastal ecosystems, communities can fortify resilience and alleviate the repercussions of human-induced activities on coastal environments.

Continuous field monitoring of the effects of human-induced activities on coastal ecosystems is essential.

12.4 Coastal Development

Apart from combating the above-stated hazards, there is rapid progress in coastal development as discussed in the following:

Infrastructure development: Coastal regions are often hubs of economic activity due to their strategic locations for trade, transportation and tourism. Rapid progress in coastal development involves the construction of ports, harbours and shipping facilities to support maritime commerce and logistics. Additionally, the development of transportation infrastructure such as roads, bridges and railways connects coastal areas to inland regions, facilitating trade and regional development.

Urbanization and residential development: Coastal cities and towns attract population growth and urbanization due to their scenic landscapes, recreational opportunities and access to marine resources. Rapid progress in coastal development includes the expansion of urban areas through the construction of residential complexes, commercial centres and tourist resorts. This urban sprawl can lead to increased land use pressures, environmental degradation and challenges related to infrastructure provision and service delivery.

Tourism and recreation: Coastal areas are popular destinations for tourism and recreation, attracting millions of visitors each year for activities such as beach holidays, water sports and ecotourism. Rapid progress in coastal development involves the development of tourist infrastructure such as hotels, resorts, marinas and recreational facilities to cater to growing tourist demand. However, uncontrolled tourism development can lead to overcrowding, habitat destruction and cultural impacts on local communities.

Industrial and commercial expansion: Coastal regions are also sites for industrial and commercial activities such as manufacturing, fishing, aquaculture and energy production. Rapid progress in coastal development includes the establishment of industrial zones, fish-processing facilities, offshore oil and gas platforms, and renewable energy installations such as wind farms and tidal energy converters. These developments contribute to economic growth but can also pose risks to coastal ecosystems, water quality and marine biodiversity.

Climate adaptation and resilience: In response to the increasing risks posed by climate change and natural hazards, rapid progress in coastal development includes investments in climate adaptation and resilience measures. This may involve the construction of coastal defences such as seawalls, breakwaters, and dikes to protect against storm surges and sea level rise. Additionally, the implementation of nature-based solutions such as wetland restoration, dune stabilization and mangrove conservation can enhance coastal resilience and ecosystem services.

Ecosystems: Coastal regions harbour vital ecosystems that play a crucial role in supporting marine life and ocean health, while also serving as significant sources of sustenance for human populations. Consequently, fisheries represent a key economic driver in coastal areas. However, this reliance on fishing often leads to detrimental consequences such as overfishing and the decline of species and populations.

Overall, while rapid progress in coastal development offers opportunities for economic growth, job creation and improved living standards, it also raises concerns about environmental sustainability, social equity and resilience to climate change. Balancing the benefits and risks of coastal development requires integrated planning, stakeholder engagement and sustainable management practices to ensure the long-term health and resilience of coastal ecosystems and communities. All of these might have provoked the realization of the necessity for coastal management information to be available to the scientific and engineering coastal working community.

12.5 Monitoring of Coastal Structures

Early detection of structural issues: Coastal structures are exposed to harsh environmental conditions such as wave action, tidal forces, saltwater corrosion and weathering. Over time, these factors can lead to wear and tear, deterioration and potential structural weaknesses. Regular health monitoring allows for the early detection of signs of damage or deterioration, including cracks, corrosion, erosion, settlement and deformation. Identifying these issues promptly enables timely intervention and preventive maintenance measures to address them before they escalate into more severe problems.

Risk assessment and management: Health monitoring provides valuable data for assessing the condition and performance of coastal structures and evaluating potential risks and vulnerabilities. By monitoring factors such as structural deformation, stability and load-bearing capacity, engineers and stakeholders can identify areas of concern and prioritize maintenance and repair activities accordingly. This proactive approach to risk management helps minimize the likelihood of structural failures, accidents and disruptions to coastal infrastructure and operations.

Optimization of maintenance strategies: Regular health monitoring allows for the optimization of maintenance strategies and resource allocation by providing insights into the condition and deterioration rates of coastal structures. By monitoring changes in structural performance over time, decision-makers can make informed decisions about the timing and scope of maintenance interventions, such as repair, retrofitting or replacement. This proactive maintenance approach helps maximize the lifespan and cost-effectiveness of coastal structures while minimizing downtime and operational disruptions.

Compliance with regulatory standards: Many coastal structures are subject to regulatory requirements and standards for safety, environmental protection and asset management. Regular health monitoring helps to ensure compliance with these regulations by providing evidence of ongoing maintenance and structural integrity. By documenting monitoring data and inspection reports, asset owners and operators can demonstrate due diligence in managing risks and maintaining the safety and performance of coastal structures, thereby avoiding potential legal liabilities and penalties.

Sustainability and resilience: Effective health monitoring contributes to the sustainability and resilience of coastal infrastructure by promoting proactive maintenance practices, extending the service lifetime of structures and reducing the need for costly emergency repairs or replacements. By preserving coastal assets and protecting them from premature deterioration or failure, health monitoring supports the long-term viability of coastal communities, economies and ecosystems in the face of climate change, natural hazards and sea level rise.

In summary, regular health monitoring of coastal structures is essential for proactive maintenance, risk management, regulatory compliance, sustainability and resilience. By investing in monitoring programmes and technologies, coastal stakeholders can ensure the safety, reliability and longevity of critical infrastructure assets, thereby safeguarding coastal communities and supporting sustainable development along coastlines.

12.6 Monitoring of the Coast

12.6.1 *General*

Overall, field monitoring of coasts and rivers exposed to natural and man-made hazards is a multidisciplinary endeavour that integrates scientific research, technological innovation and community engagement to safeguard these vulnerable environments and the communities that depend on them. Field monitoring of the coasts and rivers exposed to natural and man-made hazards involves the systematic observation, measurement and assessment of environmental conditions along coastal areas and river systems that are susceptible to various threats. This process typically involves the deployment of monitoring equipment, data collection techniques and analysis methodologies to gather comprehensive information about these vulnerable areas. An elaboration on different aspects of field monitoring in this context in provided in the following:

Environmental monitoring equipment: Field monitoring often utilizes a variety of specialized equipment tailored to the specific needs of coastal and river environments. This may include water quality sensors, tide gauges, weather stations, current meters, sediment samplers, GPS units and remote sensing devices such as drones or satellites. These tools help in collecting data related to parameters like water level, salinity, temperature, flow velocity, turbidity, erosion rates and vegetation cover.

Data collection and analysis: Field monitoring involves the systematic collection of data at regular intervals over time. The data can be gathered through manual measurements, automated sensors or a combination of both. Once collected, the data undergo analysis to identify trends, patterns and anomalies that may indicate potential hazards or changes in the environment. Statistical analysis, geographic information systems (GIS) and remote sensing techniques are commonly utilized to analyze and interpret the gathered data.

Risk assessment and hazard identification: Field monitoring helps in assessing the risks posed by natural phenomena such as coastal erosion, storm surges, tsunamis, flooding and riverbank erosion, as well as human-induced hazards like pollution, industrial activities and infrastructure development. By continuously monitoring these areas, researchers and decision-makers can identify vulnerable hotspots, predict potential hazards and develop strategies to mitigate risks and enhance resilience.

Early warning systems: Real-time data collected through field monitoring play a crucial role in the development of early warning systems for coastal and riverine hazards. By monitoring parameters such as water levels, currents and weather conditions, warning systems can be activated to alert communities and authorities about impending threats, allowing for timely evacuation, preparedness measures and emergency response planning.

Ecological monitoring and management: Field monitoring not only focuses on physical parameters but also encompasses ecological monitoring to assess the health and biodiversity of coastal and riverine ecosystems. This includes monitoring the populations of flora and fauna, tracking habitat changes and evaluating the impact of human activities on these sensitive environments. By understanding the ecological dynamics, stakeholders can implement conservation measures and sustainable management practices to preserve these valuable ecosystems.

Community engagement and stakeholder collaboration: Successful field monitoring initiatives typically entail cooperation among scientists, governmental bodies, local communities and various stakeholders. Community involvement and engagement are crucial for gathering local insights, increasing awareness of hazards and cultivating a sense of ownership in environmental management endeavours. Stakeholder collaboration also facilitates the sharing of resources, expertise and data, leading to more comprehensive and impactful monitoring initiatives.

12.7 Monitoring the Shoreline

12.7.1 *General*

Monitoring the shoreline refers to the systematic observation, measurement and assessment of the area where the land meets the water along a coast or shoreline. This process involves collecting data on various aspects of the coastal environment, including changes in shoreline position, erosion rates, sediment transport, water quality, habitat health and the impacts of natural events such as storms and tides.

The primary objective of monitoring the shoreline is to gather information that can serve as a basis for decision-making related to coastal management, environmental protection, hazard mitigation, and sustainable development. By continuously monitoring the shoreline, scientists, policymakers, and coastal managers can better understand the dynamic processes shaping coastal areas, identify trends and patterns over time, assess risks associated with coastal hazards and climate change, and develop strategies to mitigate negative impacts and enhance resilience.

12.7.2 *Most common methods*

The most commonly used methods of shoreline monitoring are as follows:

Aerial photography and remote sensing: Aerial photography involves utilizing satellites, drones or aircraft to capture images of the shoreline to monitor changes over time (Fig. 12.1). Remote sensing can provide detailed data on coastal features, inundation areas, erosion, sediment movement and vegetation cover.

Fig. 12.1. Drone survey.
Source: Burghardt.

GPS (Global Positioning System) surveys: GPS technology is used to accurately measure and track the positions of shoreline features. GPS surveys can provide precise data on shoreline position, erosion rates and sediment transport.

LiDAR (Light Detection and Ranging): LiDAR technology uses laser pulses to measure distances to the Earth's surface, providing highly detailed elevation data. LiDAR surveys are valuable for assessing shoreline topography, erosion and coastal geomorphology.

Ground Surveys and Transects: Field surveys involve collecting data directly from the shoreline using instruments like total stations, GPS devices or tape measures. Transects are lines established perpendicular to the shoreline to measure changes in beach width, sediment characteristics and vegetation cover.

Beach Profiling: Beach profiling involves measuring the slope and elevation of the beach surface along predetermined transects. Profiling helps to assess changes in beach morphology, erosion, sediment deposition and beach volume.

Wave and Current Monitoring: Instruments such as wave buoys, current meters and tide gauges are deployed to measure wave characteristics, water levels and current velocities near the shoreline. Monitoring waves and currents provides insights into coastal processes and their effects on shoreline dynamics (Fig. 12.2).

Satellite Imagery Analysis: Satellite images are analyzed using Geographic Information Systems (GIS) software to monitor changes in shoreline position, land cover and coastal features. Satellite imagery analysis enables large-scale assessment of coastal areas and long-term trends.

Storm Surge Monitoring: During extreme weather events, monitoring stations and tide gauges are used to measure storm surge heights and inundation levels along the shoreline. The data help to assess the impacts of storms on coastal erosion and flooding.

Sonar Surveys: Sonar technology, including side-scan (Fig. 12.3) and multibeam sonar, is used to map the seafloor morphology and underwater features near the shoreline. Sonar surveys provide valuable information on coastal bathymetry, submerged structures and sediment distribution.

Fig. 12.2. Wave monitoring stations.
Source: Schüttrumpf.

Fig. 12.3. Side-scan sonar measurements.

Sometimes, multiple methods are combined to facilitate effective management and conservation of coastal environments.

12.7.3 *Case studies in brief*

Although, a comprehensive description of the methods is discussed in detail by Sundar *et al.* (2023), along with a few cases of the shoreline changes due to proposed coastal protection measures as well as river mouth training works, only two typical case studies are briefly presented herein.

12.7.3.1 *Shoreline changes in between the mouth of two estuaries*

Chennai, situated on the southeast coast of India, boasts a prominent metropolitan status. Its coastal landscape is characterized by the presence of the world's second-longest urban beach, which evolved over the course of a century as a result of breakwater construction for harbour development. Positioned between two estuaries, this beach serves as a testament to the city's maritime history and expansion as a pivotal port. The coastal dynamics of Chennai primarily revolve around two distinct zones: Zone I, centred on the Cooum estuary (13° 4′N and 80° 17′20″E), and Zone II, adjacent to the Adyar estuary (13° 0′49″N and 80° 16′38″E). Through the utilization of Geographic Information Systems (GIS) and Digital Shoreline Analysis System (DSAS), the period spanning from 2000 to 2019 witnessed a meticulous examination of shoreline alterations, categorizing them into erosion and accretion phenomena.

The findings of Gracy Margret Mary *et al.* (2021) reveal a remarkably dynamic coastal environment, with Chennai experiencing an average erosion rate of −1.67 m/year. Conversely, accretion, attributed to littoral drift, demonstrates an average increase of +1.31 metres per year. This accretion phenomenon is predominantly influenced by the net alongshore current directing sediment northward, particularly evident in the vicinity of the Adyar River. Spatial quantification of shoreline changes from 2000 to 2019 was conducted across various transects, utilizing statistical outputs such as EPR (End Point Rate), LRR (Linear Regression Rate) and LMS (Least Median of Squares). The EPR and LRR values depicted in Fig. 12.4 represent accretion and erosion through positive and negative measurements

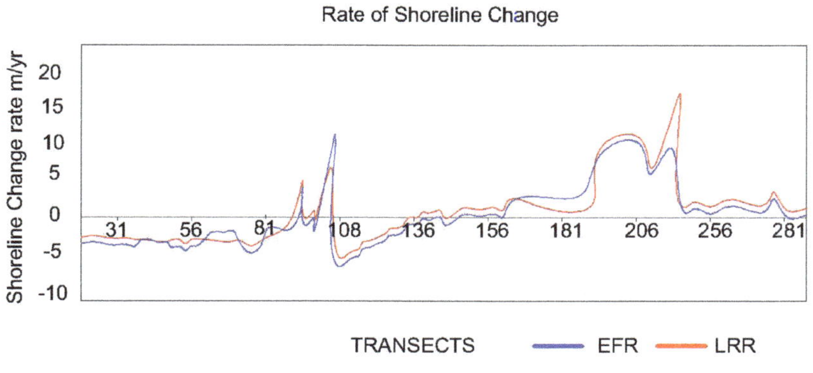

Fig. 12.4. Rate of shoreline transformation.

relative to the zero line. Positive values denote accretion above the zero line, while negative values signify erosion below it.

12.7.3.2 *Response of shoreline in between groynes to the 2004 tsunami*

The Chennai port resides on the eastern shore of the Indian peninsula (13.0844° N, 80.2899° E). Since the establishment of Chennai port's harbour with breakwaters, the northern coastline has experienced erosion at an approximate rate of 8 metres per year, primarily due to the prevailing northerly drift.

The implementation of a groyne field in May 2004 led to an immediate advancement of the shoreline to the south of the constructed groyne, providing compelling evidence of the efficacy of the chosen design as a remedial measure. The construction of adjacent groynes resulted in consistent shoreline advancement, reflected in the beach widths. However, following the 2004 tsunami, a recession occurred, leading to the formation of an offshore bar. Subsequently, this bar migrated towards the shoreline, contributing to shoreline advancement once again. This underscores the necessity for continuous monitoring of shoreline dynamics. For additional insights, consult Sundar (2005) and Sundar and Sannasiraj (2019).

12.8 Field Monitoring

12.8.1 *Topography*

Topography refers to the terrain features above the water level, playing a crucial role in various hydraulic engineering projects and initiatives. In addition to bathymetry, which focuses on the sea bed, accurately determining topography is essential. Seamless integration of land surface surveys (topography) with sea bed surveys (bathymetry) is imperative to prevent data gaps, especially along the shoreline. When selecting survey methods for both topography and bathymetry, it is vital to prioritize those capable of operating effectively in shallow-water depths. The ultimate aim of topographic measurements is to generate detailed topographic maps, as depicted in Fig. 12.5, aiding in project development and planning.

Topographic measurements can be conducted using traditional surveying/geodetic techniques or modern remote sensing methods. Traditional methods encompass levelling instruments and total stations, while innovative approaches include laser-based, airborne or satellite remote

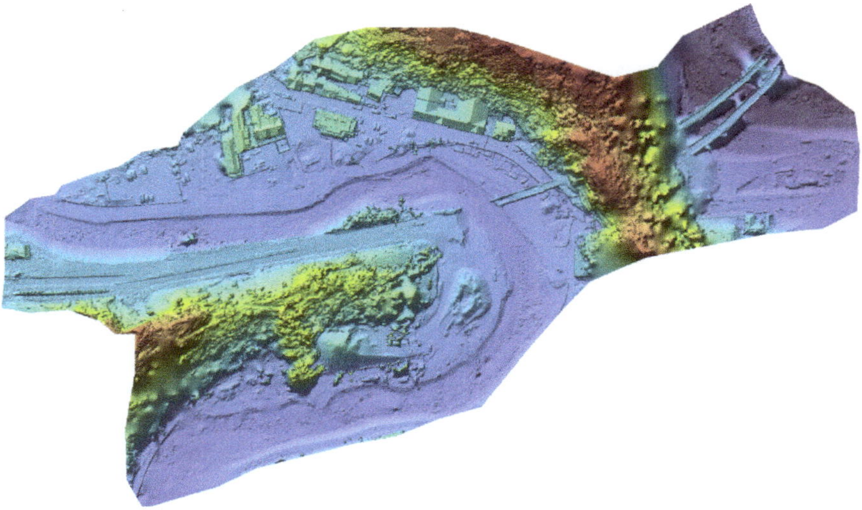

Fig. 12.5. Digital elevation model.
Source: Burghardt.

sensing technologies. The choice of method hinges upon the specific requirements of the project, such as measurement quality, accuracy and resolution. However, practical considerations like equipment availability, trained personnel and budget constraints also shape the selection of surveying methods and influence the attainable data quality. The overarching goal is to strive for the highest quality and accuracy in the collected data.

12.8.2 *Bathymetry and water depth*

The assessment of water body bathymetry and depths stands as a pivotal task within hydraulic engineering. Frequent bathymetric surveys are essential, especially due to the dynamic nature of river and sea beds, subject to continual alterations. These changes may stem from rapid current velocities during extreme events, coastal structures or sedimentation during periods of low current. Consequently, consistent monitoring of sea bed fluctuations, particularly in coastal regions, estuaries and navigable rivers, is imperative to facilitate effective planning and implementation of maintenance measures.

Accurate knowledge of water depths is crucial for maritime activities. Ensuring navigable water depths is essential to guarantee adequate keel clearance for ships. Depending on the waterway system, regular assessments of the waterway bed are necessary to uphold safe navigation conditions.

Constructing hydraulic structures on or within water bodies demands meticulous water data to effectively plan, design and build structures, considering factors such as statics, soil mechanics and functionality. The data necessary for this process should be current, of high quality and possess adequate resolution to prevent costly over- or under-sizing.

Building hydraulic structures in coastal areas can alter flow velocities and subsequently impact coastal morphodynamics. Monitoring these changes is essential to implement the necessary measures effectively.

Moreover, it is crucial to emphasize the necessity of extensive foundational data for simulating coastal hydraulics and addressing hydrological concerns. Calculating parameters like flow velocities, water depths, water levels, shear stresses and sediment transport relies on accurate bathymetric data.

This underscores the significance of prioritizing water depth determination in the planning and execution of coastal projects. Consequently, a variety of measurement devices are employed for this purpose.

Measurement techniques can be categorized into point and spatial measurements, as well as direct and indirect methods for assessing the sea bed. Point measurements involve tools like plumb bobs, which ascertain water depth from a boat or fixed points (e.g. bridges or quay walls) by suspending a weight from a rope. Alternatively, echosounders utilize reflected radar pulses to gauge water depth, offering an alternative to traditional sounding devices. Multibeam echosounders have emerged as a prominent choice in hydraulic engineering due to their ability to sweep signals over larger areas of the water bed, facilitating precise and high-resolution depth determination down to the centimetre scale. In recent years, multibeam echosounders have gained widespread acceptance across various facets of hydraulic engineering.

12.8.3 *Water levels*

In addition to water depths, water levels must also be monitored. In principle, a large number of gauges along coasts and rivers measure the changes in water levels in relation to a reference level. Determining the water level is relevant for many issues. The water level itself is particularly important for shipping in order to determine the nautical depth with knowledge of the bed topography. Knowledge of the water depths is also relevant for the realization of construction measures in maritime and inland waterways. In tidal areas, construction works can sometimes only be carried out at low or high tide. The water levels must therefore be recorded continuously.

However, water levels change as a result of many processes such as runoff, floods, tides and storms. This leads to daily changes in water levels, which initially only provide an indication of the water level at a specific point in time. These water levels must be analyzed statistically in order to determine design water levels. In many cases, this requires sufficiently long data time series of many years in order to determine extreme design water levels, which, for example, are decisive for the structural design of a coastal structure.

There are currently two basic methods for measuring water levels. The classic method uses the so-called staff gauges (Fig. 12.6), which can be used to visually read the water level at a point in time. The advantage of this method is that even an inexperienced observer can read the water level on the staff gauge at any time. The disadvantage of the method is that continuous measurement is not possible and therefore the water level is only documented at long intervals. This is why there are automatic water gauges along many rivers, lakes and the coast, which record the water level continuously using a float principle. Automatic water gauges therefore allow continuous observation of the water level data and thus also permit the transient processes to be recorded.

12.8.4 *Waves*

Waves represent transient phenomena characterized by their high temporal resolution. Various types of waves exist, such as tidal waves, wind waves, swell waves and tsunami waves, each differing in wave height, period and direction. Particularly in seas and oceans, waves often overlap, forming short- or long-crested waves, where waves of varying heights, periods and directions converge at a single point. It is essential to recognize that measuring a single wave is insufficient; continuous recording of waves over extended periods is necessary, considering different water levels, wind speeds and current conditions, followed by statistical analysis.

These considerations impose specific requirements for planning wave measurements. The choice of measurement equipment hinges on expected wave heights (measuring range), wave periods (recording rate) and wave

Fig. 12.6. Water level gauge.
Source: Schüttrumpf.

directions. In the simplest scenario, a wave gauge (wave wire) positioned at one point can determine wave characteristics (refer to Fig. 12.2). However, such a setup lacks the ability to discern individual wave directions or analyze wave superposition. In a case where a multidirectional wave field should be recorded, more sophisticated wave recorders (bottom seated or floating) are adopted. Sundar *et al.* (2023) have discussed in detail the instruments that are used for monitoring the flow fields and other parameters in the coastal and ocean environments.

12.8.5 *Flow velocities*

Flow measurements are essential for assessing flow velocities. Prior to planning such measurements, it is crucial to estimate the magnitude, direction and temporal variations of flow velocities. These parameters are contingent upon local conditions and the prevailing processes at the measurement site. Conducting flow measurements on rivers is comparatively straightforward, given the dominant downstream flow direction. However, in estuarine or coastal areas, the complexity increases due to various factors. Additionally, when measuring currents, variations in velocity across water depth must be considered. Thus, both the measurement location and water depth are pivotal and require thorough documentation.

Various measuring devices are available for current measurements. Traditional instruments like flow propellers are effective for measuring flow velocities in a single direction at a specific water depth. However, they lack the capability to capture changes in flow direction or variations across water depth. Consequently, numerous other measuring devices have been developed to address these limitations, allowing for the determination of two- or even three-dimensional flow patterns.

Furthermore, a distinction must be made between the Eulerian and Lagrangian approaches to flow measurements. Flow propellers, ADV and ADCP devices (Figs. 12.7 and 12.8) can be assigned to the Eulerian approach, i.e. the measurement takes place at a fixed location and changes can be recorded over time. Float measurements, on the other hand, can be assigned to the Lagrangian approach. Here, a measurement takes place depending on a variable location and variable time.

12.8.6 *Other measurements*

In addition to the aforementioned processes, numerous other phenomena and parameters are measured along coastlines, estuaries and rivers. While it is impractical to detail every possible measurement within the scope of this book, it is important to acknowledge the significance of measuring sediment transport, heat transport and biological/chemical processes, depending on the specific circumstances. These measurements necessitate specialized devices capable of capturing the required data with adequate quantity and quality.

Fig. 12.7. ADCP measurements.
Source: Meier.

12.9 Structural Health Monitoring (SHM) of Inland Navigation and Port Structures

12.9.1 *General*

Developing a structural health monitoring (SHM) system generally comprises three main components:

Planning a comprehensive approach of the system, which includes determining data acquisition and management methods, and configuring workflow.
Implementing the plan by configuring instruments and establishing models.
Analyzing data, extracting features, interpreting results and presenting findings.

Depending on factors such as structural complexities, costs and the significance of the structure or its components, an SHM system may be designed to achieve one of four different objectives:

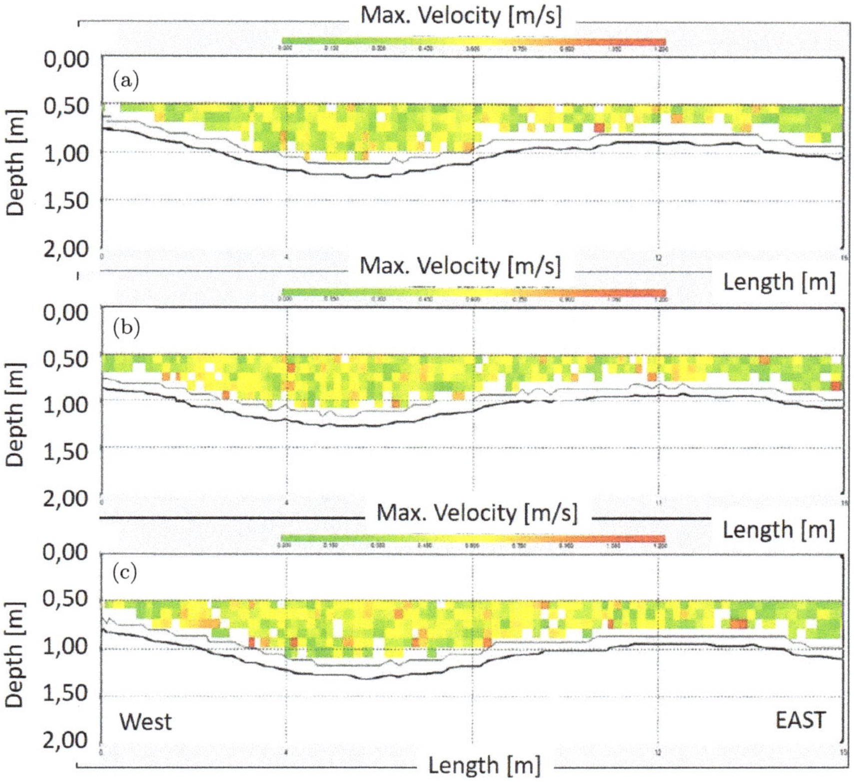

Fig. 12.8. ADCP measurements.

(a) Assessing the presence of damage.
(b) Locating the damage.
(c) Evaluating the extent of damage.
(d) Estimating the remaining lifespan of the structure (Jiang *et al.*, 2022; Sajedi and Liang, 2022).

Moreover, the demand for SHM tools extends to quality control of high-profile mechanical components to ensure safe service performance (Qarib and Adeli, 2014). For additional details, the readers may refer to Pezeshki *et al.* (2023).

Transporting goods via waterways is inherently more energy-efficient than using roads and railways. Like other civil structures, inland navigation structures (INSs) are subject to ageing and necessitate regular condition assessments and maintenance. Timely maintenance and early detection of alterations in material or geometric properties, which determine the extent of damage, can be facilitated through SHM. This process involves analyzing monitored data, including loads, structural responses and environmental actions, to support such efforts. Integral to structural health monitoring are sensors, which measure physical (e.g. displacements, accelerations, strains), environmental (e.g. temperature, humidity) and chemical (e.g. chloride contamination) parameters. Continuously monitored, these parameters generate time histories or signals.

After collection, signals undergo preprocessing and analysis using suitable algorithms and machine learning techniques to identify and characterize changes within the monitored structure. Sensors and sensing technologies are classified based on measured parameters, physical phenomena and their proximity to the host structure. In this section, they are categorized according to proximity as contact and non-contact sensors, alongside remote sensing. Notably, within the literature, non-contact methods are often considered within the realm of remote sensing. However, their classification may differ depending on the necessity for physical access to the site or areas of observation/monitoring.

Inland navigation structures and harbours vary significantly in size, spanning from a few metres to hundreds of kilometres. Consequently, non-contact sensors are delineated from satellite-based remote sensing methodologies to accommodate these differences.

12.9.2 *Typical case studies*

12.9.2.1 *Case study 1*

12.9.2.1.1 Study location and background

Kamarajar Port, positioned at coordinates 13.0975° N and 80.2935° E, is India's 12th major port, inaugurated in 2001 on the southeastern coast of the country. Illustrated in Fig. 12.9, its layout reveals a strategic design. The north breakwater stretches across a length of 3.080 m, with a water depth at the head section measuring (−) 11.0 m CD. Conversely, the south breakwater spans 1.070 m in length. At the head section of the north breakwater, the primary armour comprises Accropodes of 6 m^3.

The harbour structures endured formidable sea conditions such as those brought about by cyclones Fanoos, Nisha, Thane and Nilam. Following the

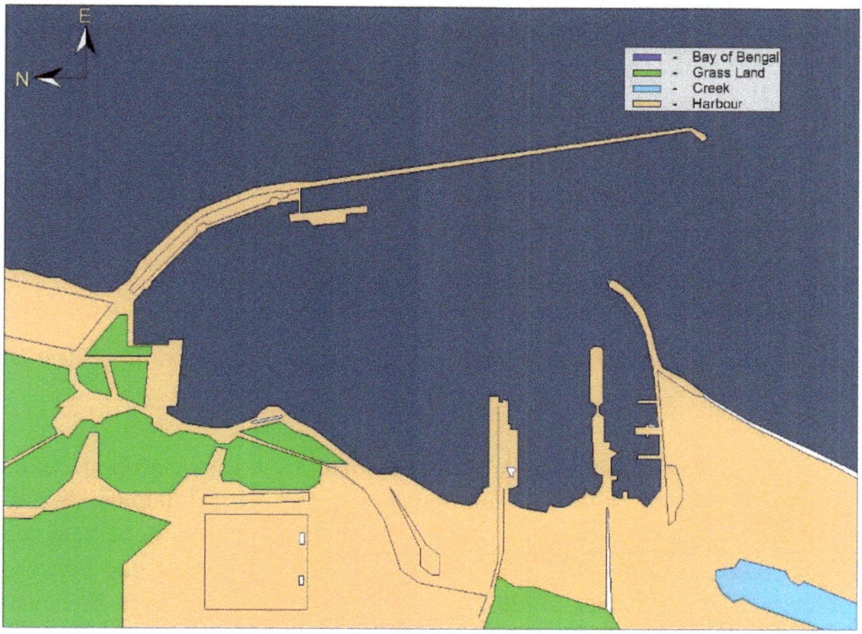

Fig. 12.9. Layout of Kamarajar Port, India.

impacts of the Thane and Nilam cyclones, the port authorities noted damage to the armour layers of the Accropodes, prompting a comprehensive evaluation of the northern breakwater's condition. This study encompassed various methodologies including field visits, examination of existing reports and literature, analysis of construction techniques, physical modelling, evaluation of extreme wave scenarios and a thorough survey of the north breakwater extending approximately 200 m. Underwater surveys were conducted, profiling the side slopes and employing cameras for underwater photo and video inspections.

12.9.2.1.2 Damage assessment

The assessment aimed to determine the extent of damage and propose suitable mitigation measures accordingly. Subsequently, the study was conducted, and its findings, including recommended mitigation measures and the rehabilitation plan for the structure, were outlined in detail.

Fig. 12.10. (a) Structural damage to Accropodes and (b) displacement of Accropodes.

Upon physical examination, it was evident that the Accropode units had been disrupted, with some units exhibiting structural damage. Significantly, there was noticeable movement of the armour units in comparison to the completion drawings of the breakwater. A substantial gap between the wave wall and armour units, spanning 15 m from the north breakwater head section, was observed, exposing the underlying units. Stability is predominantly governed by interlocking and gravity.

The displacement and structural damage observed in the Accropode units exceeded acceptable limits. The affected areas are delineated in Figs. 12.10(a) and 12.10(b).

12.9.2.1.3 Topographical survey

A comprehensive topographical survey was conducted on and around the northern breakwater (KPL), covering an area of approximately 190 m × 235 m. Post-cyclone cross-sectional data of the north breakwater were juxtaposed with the initially designed cross-sectional data. It was observed that the shape of the breakwater's top level had shifted due to previous cyclones, spanning from chainage 3070 to the head of the breakwater. The topographic survey, accompanied by bathymetric data, is depicted in Fig. 12.11.

Based on the test results, it was determined that the design life has been exceeded by approximately 6 years (refer to Fig. 12.15), representing a shortfall of approximately 12% from its designated service life of 50 years. Such findings highlight the urgent need for proactive interventions to mitigate further deterioration and uphold the structural integrity of the

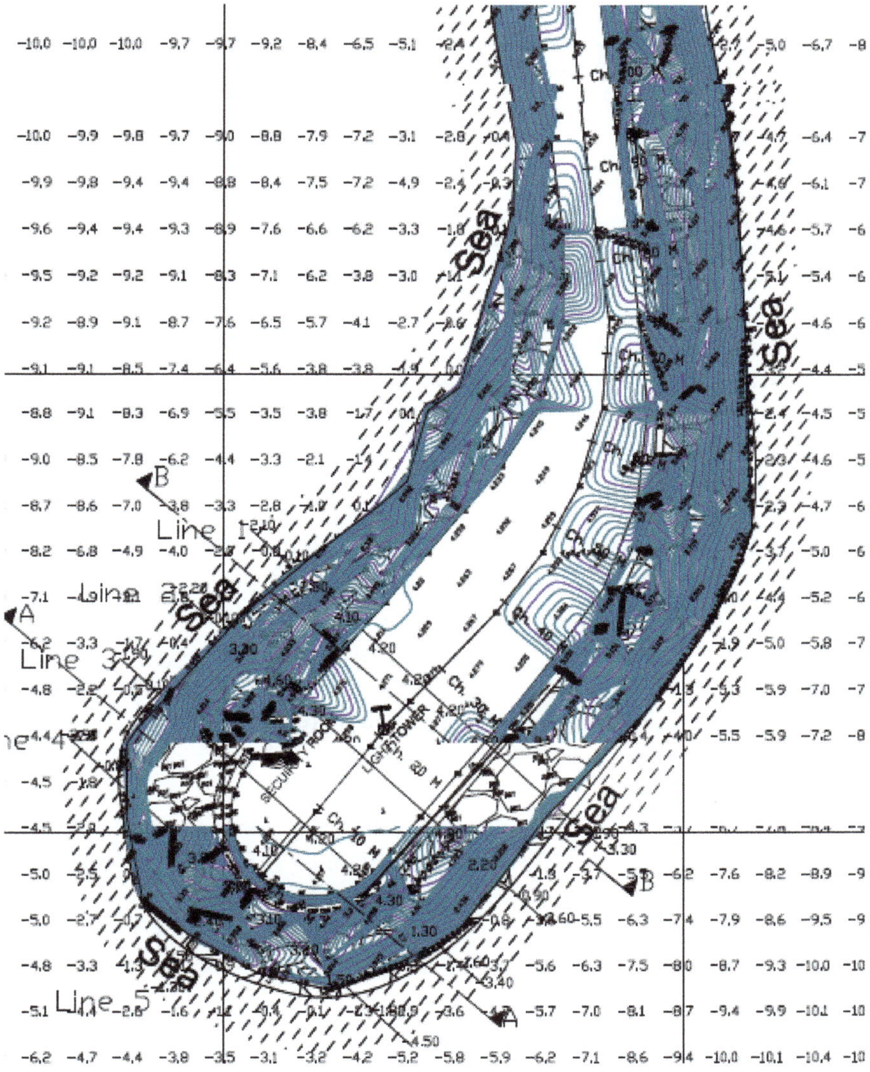

Fig. 12.11. Topography superposed on the bathymetry along the breakwater.

berthing facility. This case study underscores the critical importance of diligent maintenance practices and regular structural assessments to ensure the longevity and functionality of coastal infrastructure in the face of persistent environmental challenges. Implementing targeted remedial measures based on comprehensive assessment findings is essential for safeguarding coastal

Table 12.1. Topographical survey data.

Cross-section	Slope as per built section	Slope of damaged section
AT chainage 3030	1:0.75	1:0.65
AT chainage 3050	1:0.75	1:0.32
AT chainage 3079	1:0.75	1:0.88
AT chainage 3090	1:0.75	1:10
AT head section	1:0.75	1:0.75

structures and maintaining their operational efficiency amidst evolving climatic dynamics and ageing infrastructure.

Topographical survey details revealed the status of the breakwater section as provided in Table 12.1.

A cost-effective approach was utilized to conduct impact analysis and shoreline inventory of submerged aquatic resources involving an underwater survey using echosounder technology. This method employed geo-located photo documentation to identify, map and record the x & y coordinates of submerged aquatic resources in low-visibility waters, achieving minimal error. Subsequently, commercial software was employed to correlate GPS information with digital photos, generating geo-located digital images that can be seamlessly integrated into a map or Geographic Information System (GIS) for comprehensive analysis of the seafloor environment.

The echosounder was utilized to plot the profile of the northern breakwater within the designated survey zone, aligning it with the original profile developed during the project. Upon scrutinizing both profiles, it was deduced that underground damage has been present in the profile since its initial construction. Analysis revealed top-level changes ranging between 10 cm and 20 cm. Figure 12.12 illustrates the displacement of armour units as detected by the echosounder.

A noticeable gap exists between the waves and all armour units, exposing the underlying layer. This situation indicates a lack of stability between the underlocking and gravitational forces, rendering the units unstable even for the design wave height.

The failure of the north breakwater at Kamarajar Port Limited primarily stemmed from inadequate crest elevation and hydraulic instability of the Accropodes during severe cyclones such as Thane and Nilam. Furthermore,

(a)

(b)

(c)

Fig. 12.12. (a) A view of armour stones displaced one above the other in a triangular manner, (b) view of armour stones displaced in the harbour side of the NBW at chainage 150 m and (c) a view of armour stones displaced in the sea side of the NBW at chainage 200 m.

the failure can be attributed to prolonged wave action, cyclic loading from a series of waves with significant height, the occurrence of freak waves or a combination of these factors. Detailed field surveys and investigations revealed a flattening of the breakwater side slope, settlement of sections and deterioration of interlocking due to displacements and structural damage of the primary armour, specifically the Accropodes, with instances of arms being missing.

12.9.2.2 *Case study 2*

12.9.2.2.1 Background

Coastal structures are conventionally engineered to endure for 50 years or more. Yet, despite meticulous planning, signs of deterioration often manifest within their anticipated design lifespan, owing to the relentless forces of climate extremes. The imperative to utilize exceptionally resilient materials for construction stems from the continuous exposure of these structures to the harsh marine environment. Moreover, the escalating effects of global climate change exacerbate these challenges, demanding frequent repair and retrofitting endeavours.

Climate-induced changes in seawater chemistry play a substantial role in the deterioration of coastal structures. As the oceans absorb roughly 30% of atmospheric carbon dioxide, exacerbated by human activities, the subsequent rise in seawater acidity presents a significant threat. The acidic conditions of the marine environment alter the chemical composition of concrete within the structure, diminishing its alkalinity and leading to the corrosion of embedded rebars. This reduced alkalinity intensifies the corrosion of embedded reinforcement bars, setting off a cyclical process of deterioration. The corrosion process involves the expansion of reinforcement bars, resulting in the formation of cracks, spalling of concrete cover, reduced bond strength and eventual compromise of structural integrity and functionality. Amplified by elevated levels of salinity, chlorides and sulphates in the atmospheric components of the marine environment, along with significant temperature fluctuations, these challenges collectively hasten the degradation of susceptible areas within coastal structures. Essentially, the convergence of environmental factors, such as increased carbon dioxide levels, heightened salinity and temperature variations, accelerates the deterioration of coastal structures.

Fig. 12.13. Exposed rebars in slab, column and RC fender.

In 1975, a berthing structure was constructed in the backwaters of Kochi city, located along the southwest coast of India, to support maritime operations. Nevertheless, recent preliminary inspections have revealed troubling indications of structural deterioration, such as extensive corrosion of reinforcement bars, cracks, delamination and spalling of concrete cover (refer to Fig. 12.13).

12.9.2.2.2 Details of the study

To determine the extent of deterioration and assess structural integrity, a thorough damage assessment was conducted using Non-Destructive Testing (NDT) techniques. The assessment included Ultrasonic Pulse Velocity

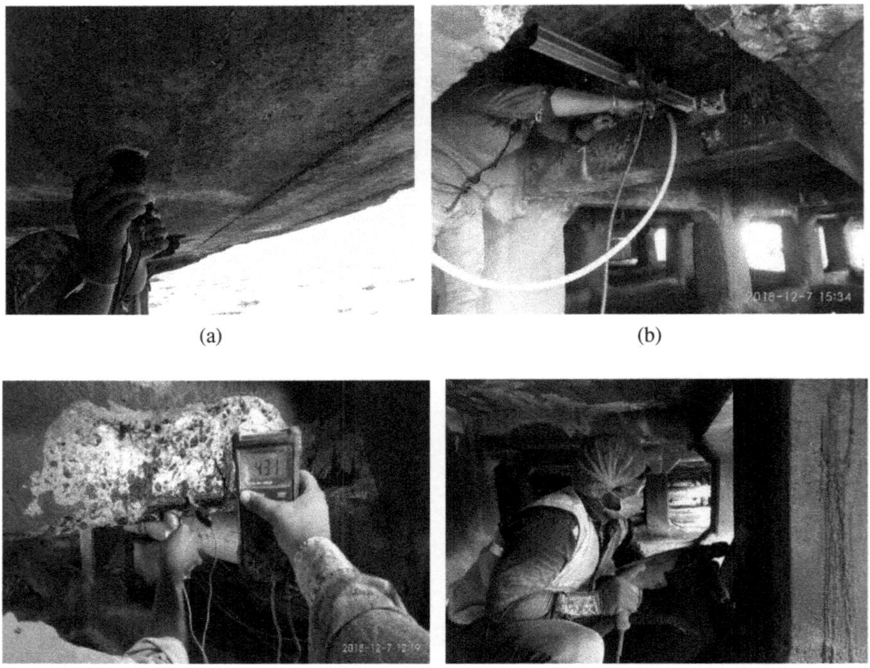

(a) (b) (c) (d)

Fig. 12.14. (a)–(d) show the NDT methods of UPV, core sampling, half-cell potential and powder sampling, respectively.

(UPV), half-cell potential analysis, powder sampling and core sampling methods (refer to Fig. 12.14). The UPV test played a crucial role in assessing the uniformity and integrity of the concrete, while the half-cell potential analysis aided in identifying corrosion within structural elements. Core and powder samples were collected for chemical analyses.

12.10 Summary

This chapter emphasizes the need for comprehensive analyses of disaster-prone areas, tailored to each specific type of disaster, to effectively mitigate their impact. It also presents a concise formula showing that the occurrence of a disaster is influenced by the interaction between hazard, vulnerability, and the capacity to cope. This chapter not only delineates various hazards stemming from natural and anthropogenic activities but also underscores the significance of addressing rapid coastal development. Furthermore, it

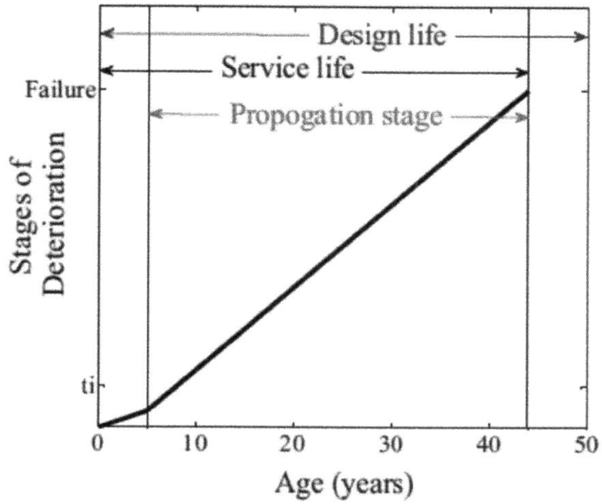

Fig. 12.15. Variation of stages of deterioration over years.

emphasizes the importance of regular monitoring of coastal areas, especially following the implementation of structures, as evidenced by the case studies.

References

Ghosh, S. and Mistri, B. (2020). Geo-historical appraisal of embankment breaching and its management on active tidal land of Sundarban: A case study in Gosaba Island, South 24 Parganas, West Bengal. *Space and Culture, India*, 7(4), 166–180. https://doi.org/10.20896/saci.v7i4.587.

Gracy Margret Mary, R., Sundar, V. and Sannasiraj, S.A. (2021). Analysis of shoreline change between inlets along the coast of Chennai, India. *Marine Georesources & Geotechnology Journal*. https://doi.org/10.1080/1064119X.2020.1856241.

Jiang, K., Han, Q. and Du, X. (2022). Lost data neural semantic recovery framework for structural health monitoring based on deep learning. *Computer-Aided Civil and Infrastructure Engineering*, 37(9), 1160–1187. https://doi.org/10.1111/mice.12850.

Mondal, M., Haldar, S., Biswas, A., Mandal, S., Bhattacharya, S. and Paul, S. (2021). Modeling cyclone-induced multi-hazard risk assessment using analytical hierarchical processing and GIS for coastal West Bengal, India. *Regional Studies in Marine Science*, 44, 101779.

Pezeshki, H., Adeli, H., Pavlou, D. and Siriwardane, S.C. (2023). State of the art in structural health monitoring of offshore and marine structures. *Proceedings of the Institution of Civil Engineers – Maritime Engineering*, 176(2), 89–108. https://doi.org/10.1680/jmaen.2022.027.

Qarib, H. and Adeli, H. (2014). Recent advances in health monitoring of civil structures. *Scientia Iranica*, 21(6), 1733–1742.

Sajedi, S.M. and Liang, X. (2022). Deep generative Bayesian optimization for sensor placement in structural health monitoring. *Computer-Aided Civil and Infrastructure Engineering*, 37(9), 1109–1127. https://doi.org/10.1111/mice.12799.

Sundar, V. (2005). Behaviour of shoreline between groin field and its effect on the tsunami propagation. In *Fifth International Symposium on Wave Measurement and Analysis, WAVES 2005*, Madrid, Spain, 3–7 July, Paper No: 323.

Sundar, V. and Sannasiraj, S.A. (2019). *Coastal Engineering- Theory and Practice.* Advanced Series on Ocean Engineering, Vol. 47. https://www.worldscientific.com/worldscibooks/10.1142/11148.

Sundar, V. Sannasiraj, S.A. and Murali, K. (2023). *Coasts and Estuaries-Management and Engineering.* Advanced Series on Ocean Engineering, Vol. 57. World Scientific. https://doi.org/10.1142/13011.

UN/International Strategy for Disaster Reduction (UN/ISDR). (2009). Terminology on Disaster Risk Reduction. United Nations, Geneva, Switzerland.

Yin, J., Yin, Z. and Xu, S. (2013). Composite risk assessment of typhoon-induced disaster for China's coastal area. *Natural Hazards*, 69(3), 1423–1434.

Index

Q

R

S

www.ingramcontent.com/pod-product-compliance
Ingram Content Group UK Ltd.
Pitfield, Milton Keynes, MK11 3LW, UK
UKHW020353150926
8624IPUK00004B/147

9 789811 294167